1991. 1. 11.

임 종성.

Molecular Cloning

Associate Author | **Nina Irwin**

Managing Editor | **Nancy Ford**

Editor | **Chris Nolan**

Associate Editor | **Michele Ferguson**

Illustrator | **Michael Ockler**

3

Molecular Cloning

A LABORATORY MANUAL

SECOND EDITION

J. Sambrook
UNIVERSITY OF TEXAS SOUTHWESTERN MEDICAL CENTER

E.F. Fritsch
GENETICS INSTITUTE

T. Maniatis
HARVARD UNIVERSITY

Cold Spring Harbor Laboratory Press
1989

Molecular Cloning

A LABORATORY MANUAL

SECOND EDITION

All rights reserved
© 1989 by Cold Spring Harbor Laboratory Press
Printed in the United States of America

9 8 7 6 5 4 3 2 1

Book and cover design by Emily Harste

Cover: The electron micrograph of bacteriophage λ particles
stained with uranyl acetate was digitized and assigned false color
by computer. (Thomas R. Broker, Louise T. Chow, and James I.
Garrels)

Cataloging in Publications data

Sambrook, Joseph
    Molecular cloning : a laboratory manual / E.F.
Fritsch, T. Maniatis—2nd ed.
       p.   cm.
    Bibliography: p.
    Includes index.
    ISBN 0-87969-309-6
    1. Molecular cloning—Laboratory manuals. 2. Eukaryotic cells-
-Laboratory manuals. I. Fritsch, Edward F. II. Maniatis, Thomas
III. Title.
QH442.2.M26 1987
574.87'3224—dc19   87-35464

All Cold Spring Harbor Laboratory Press publications may be ordered directly from Cold
Spring Harbor Laboratory, Box 100, Cold Spring Harbor, New York 11724. Phone: 1-800-843-
4388. In New York (516)367-8423.

Contents

BOOK 1

1

Plasmid Vectors

2

Bacteriophage λ Vectors

3

Cosmid Vectors

4

Single-stranded, Filamentous Bacteriophage Vectors

Cloning into Bacteriophage M13 Vectors and Transfection of Competent Bacteria   4.33

Identification and Analysis of Recombinants   4.39

Cloning in Phagemids   4.44

References   4.51

5

Enzymes Used in Molecular Cloning

Restriction and DNA Methylation Enzymes   5.3

6

Gel Electrophoresis of DNA

7

Extraction, Purification, and Analysis of Messenger RNA from Eukaryotic Cells

BOOK 2

8

Construction and Analysis of cDNA Libraries

9

Analysis and Cloning of Eukaryotic Genomic DNA

10

Preparation of Radiolabeled DNA and RNA Probes

11

Synthetic Oligonucleotide Probes

12

Screening Expression Libraries
with Antibodies and Oligonucleotides

13

DNA Sequencing

14

In Vitro Amplification of DNA by the Polymerase Chain Reaction

15

Site-directed Mutagenesis of Cloned DNA

BOOK 3

16

Expression of Cloned Genes in Cultured Mammalian Cells

17

Expression of Cloned Genes in Escherichia coli

18

Detection and Analysis of Proteins Expressed from Cloned Genes

Appendixes

Index

Molecular Cloning

16
Expression of Cloned Genes in Cultured Mammalian Cells

The development of methods for the introduction of DNA into cultured mammalian cells has made it possible to express cloned genes in a broad range of cell types from different species. These methods have been used to overproduce proteins for structural and biochemical studies and to identify elements involved in the control of gene expression. In both types of studies, the cloned sequence of interest is inserted into the appropriate expression vector, cloned in bacteria, amplified by replication, and then used to transfect mammalian cells.

In this chapter, we describe a number of commonly used mammalian expression vectors, and we provide protocols for introducing cloned genes into mammalian cells. We begin by discussing expression of proteins and then go on to describe methods used to study gene regulation. In general, different vectors are required for the two types of studies, but many of the basic components used in the construction of these vectors are the same.

Expression of Proteins

EXPRESSION OF PROTEINS FROM CLONED GENES

A few eukaryotic proteins have been expressed efficiently and inexpensively in prokaryotic hosts (see Chapter 17). However, many eukaryotic proteins synthesized in bacteria fold incorrectly or inefficiently and, consequently, exhibit low specific activities. In addition, production of authentic, biologically active eukaryotic proteins from cloned DNA frequently requires posttranslational modifications such as accurate disulfide bond formation, glycosylation, phosphorylation, oligomerization, or specific proteolytic cleavage—processes that are not performed by bacterial cells. This problem is particularly severe when expression of functional membrane or secretory proteins such as cell surface receptors and extracellular hormones or enzymes is required.

Because of these problems, considerable effort has been made to develop systems to express mammalian proteins in mammalian cells. These systems can be divided into two types: those that involve transient or stable expression of transfected DNA and those that involve the use of viral expression vectors derived from simian virus 40 (SV40) (Elder et al. 1981; Gething and Sambrook 1981; Rigby 1982, 1983; Doyle et al. 1985; Sambrook et al. 1986), vaccinia virus (Mackett et al. 1985; Moss 1985; Fuerst et al. 1986, 1987), adenovirus (Solnick 1981; Thummel et al. 1981, 1982, 1983; Mansour et al. 1985; Karlsson et al. 1986; Berkner 1988), retroviruses (Dick et al. 1986; Gilboa et al. 1986; Eglitis and Anderson 1988), and baculoviruses (Luckow and Summers 1988). The diversity of these animal viruses is so great that an account of their molecular biology is beyond the scope of this chapter. In addition, effective utilization of viral vectors requires some prior experience in the methods used to grow, quantitate, and plaque-purify different viruses. An entire chapter would be required to provide all of the information necessary to use each of these vectors. For these reasons, we have chosen to focus entirely on expression methods that involve DNA transfection. (*Note*: An excellent manual describing methods for the use of baculovirus vectors and procedures for culture of insect cells has been published by Summers and Smith [1987].)

Expression of proteins from cloned eukaryotic genes in mammalian cells has been used for a number of different purposes:

- To confirm the identity of a cloned gene by using immunological or functional assays to detect the encoded protein

- To express genes encoding proteins that require posttranslational modifications such as glycosylation or proteolytic processing

- To produce large amounts of proteins of biological interest that are normally available in only limited quantities from natural sources

- To study the biosynthesis and intracellular transport of proteins following their expression in various cell types

- To elucidate structure-function relationships by analyzing the properties of normal and mutant proteins

- To express intron-containing genomic sequences that cannot be transcribed correctly into mRNA in prokaryotes or yeasts

- To identify DNA sequence elements involved in control of gene expression

When choosing a mammalian expression vector, the following parameters should be taken into account:

- *The species and types of host cells that are available.* Not all types of mammalian cells can be transfected efficiently and not all of them will necessarily carry out exactly the same set of posttranslational modifications. For example, Chinese hamster ovary (CHO) cells typically add more terminal sialic acid residues to secretory and transmembrane proteins than do simian CV-1 cells or mouse NIH-3T3 cells. Whether or not this difference is significant may vary according to the nature of the protein that is expressed. In addition, some lines of cultured cells may endogenously synthesize high levels of the protein that is expressed. This can complicate functional and immunological assays.

- *Whether the experiment can be carried out with cells that transiently express the foreign protein or whether it will be necessary to isolate cell lines that permanently express the protein.* For example, immunofluorescent localization of a protein can be carried out just as well with transiently transfected cells as with stable cell lines. However, production of more than a few micrograms of foreign protein can usually be achieved only by the development of suitable stable cell lines.

- *The size of the gene that is to be transfected and expressed.* Some mammalian viral vectors have strict packaging requirements and will not accept large pieces of foreign DNA.

- *The presence of controlling elements in the transfected DNA.* Cloned cDNAs can only be expressed if they are correctly placed in a vector that supplies a promoter and other elements such as enhancers, splice acceptor and/or donor sequences, and polyadenylation signals. Genomic DNA sequences may already carry these controlling elements, but there is no guarantee that they will work normally in the lines of cultured cells that are available. This is a problem particularly when dealing with genes that are expressed in a tissue-specific fashion.

Plasmid vectors that have been used to introduce and express cloned genes in mammalian cells can be divided into three major classes:

1. Simple plasmid-based vectors that contain no eukaryotic replicon.

2. More complex plasmid vectors that incorporate elements from the genomes of eukaryotic viruses to increase the copy number of the transfected DNA and the efficiency with which foreign proteins are expressed.

3. Vectors designed to facilitate amplification of transfected sequences that become integrated into the host genome.

FUNCTIONAL COMPONENTS OF MAMMALIAN EXPRESSION VECTORS

Mammalian expression vectors contain both prokaryotic sequences that facilitate the propagation of the vector in bacteria and one or more eukaryotic transcription units that are expressed only in eukaryotic cells. The eukaryotic transcription unit consists of noncoding sequences and sequences coding for selectable markers. It is frequently assembled as a composite of elements derived from different, well-characterized viral or mammalian genes. The components that are used in various expression vectors are described briefly below.

Prokaryotic Plasmid Sequences That Facilitate the Construction, Propagation, and Amplification of Recombinant Vector Sequences in Bacteria

The essential prokaryotic elements include a replicon that functions in *Escherichia coli*, a gene encoding antibiotic resistance to permit selection of bacteria that harbor the recombinant plasmids, and a limited number of unique restriction sites in nonessential regions of the plasmid to allow insertion of eukaryotic sequences. Most of the mammalian vectors in current use contain prokaryotic sequences from derivatives of the plasmid pBR322 (e.g., pXf3, pBRd, and pML) that lack sequences that seem to interfere with the replication of the transfected DNA in eukaryotic cells (Lusky and Botchan 1981). Deletion of unnecessary segments of plasmid DNA also reduces the size of the vector and facilitates the placing of unique restriction sites that can be utilized for the insertion and manipulation of eukaryotic sequences.

A Eukaryotic Expression Module That Contains All of the Elements Required for the Expression of Foreign DNA Sequences in Eukaryotic Cells

The most basic eukaryotic expression module contains a promoter element to mediate transcription of foreign DNA sequences and signals required for efficient polyadenylation of the transcript. Additional elements of the module may include enhancers and introns with functional splice donor and acceptor sites.

PROMOTER AND ENHANCER ELEMENTS

Unlike the signals required for RNA processing, which function efficiently in all types of mammalian cells, the activities of elements that control transcription—promoters and enhancers—vary considerably among different cell types. Promoters and enhancers consist of short arrays of DNA sequences that interact specifically with cellular proteins involved in transcription (for review, see Dynan and Tjian 1985; Serfling et al. 1985; McKnight and Tjian 1986; Sassone-Corsi and Borrelli 1986; Maniatis et al. 1987). The combination of different recognition sequences and the amounts of the cognate transcription factors determine the efficiency with which a given gene is transcribed in a particular cell type.

Many eukaryotic promoters contain two types of recognition sequences: the

TATA box and the *upstream promoter elements*. The TATA box, located 25–30 bp upstream of the transcription initiation site, is thought to be involved in directing RNA polymerase II to begin RNA synthesis at the correct site. In contrast, the upstream promoter elements determine the rate at which transcription is initiated. These elements can act regardless of their orientation, but they must be located within 100 to 200 bp upstream of the TATA box. *Enhancer elements* can stimulate transcription up to 1000-fold from linked homologous or heterologous promoters. However, unlike upstream promoter elements, enhancers are active when placed downstream from the transcription initiation site or at considerable distances from the promoter. Many enhancers of cellular genes work exclusively in a particular tissue or cell type (for review, see Voss et al. 1986; Maniatis et al. 1987). In addition, some enhancers become active only under specific conditions that are generated by the presence of an inducer, such as a hormone or metal ion (for review, see Sassone-Corsi and Borrelli 1986; Maniatis et al. 1987). Because of these differences in the specificities of cellular enhancers, the choice of promoter and enhancer elements to be incorporated into a eukaryotic expression vector will be determined by the cell type(s) in which the recombinant gene is to be expressed. Conversely, the use of a prefabricated vector containing a specific promoter and cellular enhancer may severely limit the cell types in which expression can be obtained.

Many enhancer elements derived from viruses have a broader host range and are active in a variety of tissues, although significant quantitative differences are observed among different cell types. For example, the SV40 early gene enhancer is promiscuously active in many cell types derived from a variety of mammalian species, and vectors incorporating this enhancer have consequently been widely used (Dijkema et al. 1985). Two other enhancer/promoter combinations that are active in a broad range of cells are derived from the long terminal repeat (LTR) of the Rous sarcoma virus genome (Gorman et al. 1982b) and from human cytomegalovirus (Boshart et al. 1985).

TERMINATION AND POLYADENYLATION SIGNALS

During the expression of eukaryotic genes, RNA polymerase II transcribes through the site where polyadenylation will occur. Consequently, the 3′ terminus of the mature mRNA is formed by site-specific posttranscriptional cleavage and polyadenylation (for review, see Birnstiel et al. 1985; Proudfoot and Whitelaw 1988; Proudfoot 1989). Although discrete sites for the termination of the primary transcript have not yet been identified, general regions of DNA a few hundred nucleotides in length and downstream from the polyadenylation site have been identified where transcription randomly terminates.

Two distinct sequence elements are required for accurate and efficient polyadenylation: (1) GU- or U-rich sequences located downstream from the polyadenylation site and (2) a highly conserved sequence of six nucleotides, AAUAAA, located 11–30 nucleotides upstream, which is necessary but not sufficient for posttranscriptional cleavage and polyadenylation (for review, see Mason et al. 1986; Proudfoot and Whitelaw 1988). The practical implication of these observations is that sequences downstream from the polyadenylation site must be included in eukaryotic expression vectors to ensure efficient polyadenylation of the mRNA of interest. Although a full-length cDNA clone may encode the conserved AAUAAA sequence and a tract of poly(A), these endogenous elements are not by themselves sufficient to guarantee polyadenylation. The downstream GU- or U-rich sequences necessary for cleavage and polyadenylation must therefore be incorporated into the vector. The most frequently utilized signals are those derived from SV40; a 237-bp *Bam*HI-*Bcl*I restriction fragment contains the cleavage/polyadenylation signals from both the early and the late transcription units. These signals are positioned in opposite orientations, one on each DNA strand, and both sets of signals have been shown to be extremely efficient for the processing of hybrid mRNAs. Less frequently, polyadenylation signals have been provided by fusing a full-length cloned cDNA onto a partial genomic copy of a gene already resident in an expression vector (O'Hare et al. 1981; Kaufman et al. 1986b).

Sequences within the 3' noncoding regions of eukaryotic genes may play a role in mRNA stability. For example, the presence of an AU-rich sequence, derived originally from the 3' noncoding region of granulocyte-macrophage colony-stimulating factor (GM-CSF), has been shown to destabilize mRNAs transcribed from mammalian expression vectors (Shaw and Kamen 1986). Although similar motifs have been found in analogous locations within mRNAs encoding a variety of growth factors and oncogenes, relatively little is known about the way they function. To obtain maximal expression of a cloned gene, it may therefore be necessary to remove the nucleotide sequences 3' of the termination codon.

SPLICING SIGNALS

The DNA sequences coding for a eukaryotic protein are rarely contiguous; usually, they are separated in the genome by intervening noncoding sequences that may vary in size from tens to many thousands of nucleotides. Following polyadenylation of the primary transcript, the introns are removed by splicing to generate the mature mRNA, which is then transported from the nucleus to the cytoplasm (for review, see Nevins 1983; Green 1986; Padgett et al. 1986; Krainer and Maniatis 1988).

The minimal sequences required for splicing of mRNA are located at the 5' and 3' boundaries of the intron. Comparison of a large number of these sequences has led to the identification of consensus sequences in which the first two and the last two nucleotides of the intron are essentially invariant:

A G : **G U** (A) A G U . . . intron . . . (U/C) N_{11} **C A G** : G
   5' splice site   3' splice site

The development of in vitro splicing systems has led to the elucidation of much of the biochemistry of the splicing reaction, but the processes that guarantee correct matching of 5' and 3' splice sites are not yet understood. The fact that hybrid pre-mRNAs containing 5' and 3' splice sites derived from different introns can be accurately spliced (Chu and Sharp 1981) indicates the importance of the conserved consensus sequences in this process. However, these sequences cannot be the sole determinants of splice-site selection, since identical, but ordinarily inactive, consensus sequences can be found within both exons and introns of many eukaryotic genes. Such "cryptic" splice sites can be efficiently utilized when the normal splice sites are inactivated by mutation (Treisman et al. 1983; Wieringa et al. 1983).

Both the distance between splice sites and the DNA sequences surrounding them may influence the pathway of splicing in pre-mRNAs that contain multiple introns (Reed and Maniatis 1986). Alterations to the exon sequences flanking 5' or 3' splice sites can dramatically affect the efficiency with which the adjacent splice site is utilized. These findings are relevant to the design of eukaryotic expression vectors: Substitution of exon sequences or juxtaposition of normally noninteracting splice sites in a hybrid transcription unit might lead to the appearance of inappropriately spliced transcripts that cannot be translated.

Early studies of the expression of β-globin cDNA clones in cultured mammalian cells suggested that splicing is required for the production of cytoplasmic β-globin mRNA (Hamer and Leder 1979a,b,c). Furthermore, the expression of a gene with a mutation at a natural splice site could be rescued by insertion of a heterologous intron into the transcription unit (Gruss et al. 1979; Gruss and Khoury 1980). It is now known that this requirement for splicing signals is not absolute: Many cDNAs have been efficiently expressed from vectors that lack splicing signals (see, e.g., Gething and Sambrook 1981; Treisman et al. 1981). However, because the presence of an intron has proven to be deleterious in only a few cases and because some genes appear to be expressed more efficiently when introns are present, we recommend the use of vectors that contain a splice donor and acceptor site within the mammalian transcription unit.

ELEMENTS FOR REPLICATION AND SELECTION

In addition to the elements already described, eukaryotic vectors may contain other specialized elements intended to increase the level of expression of cloned genes or to facilitate the identification of cells that carry the transfected DNA.

Viral replicons

A number of animal viruses contain DNA sequences that promote the extrachromosomal replication of the viral genome in permissive cell types. Plasmids bearing these viral replicons are replicated episomally as long as the appropriate *trans*-acting factors are provided by genes either carried on the plasmid or within the genome of the host cell. Different viral replicons work with different efficiencies. Plasmid vectors containing the replicons of papovaviruses such as SV40 or polyomavirus replicate to extremely high copy number in cells that express the appropriate viral T antigen. Because the transfected cells die after 3 or 4 days, when the number of plasmid molecules exceeds 10^4 copies/cell, these systems are used for the transient, but abundant, expression of the transfected genes (see pages 16.17–16.22). Plasmid vectors containing replicons from viruses such as bovine papillomavirus (see pages 16.23–16.26) and Epstein-Barr virus (see pages 16.26–27) are propagated episomally at lower copy numbers (usually < 100 copies/cell) and do not generally cause cell death. These vectors can be used to isolate stable lines of cells that permanently express more modest levels of the transfected genes.

Genes encoding selectable markers

DNA, which enters only a small proportion of mammalian cells in a given culture, becomes stably maintained in an even smaller fraction. In a very few cases—for example, when the cells are transformed by an oncogene—stably transfected cells can be identified because they express an altered phenotype such as morphological transformation, loss of contact inhibition, or increased growth rate. However, in the great majority of cases, isolation of cell lines that express the transfected gene is achieved by introduction into the same cells of a second gene that encodes a selectable marker, i.e., an enzymatic activity that confers resistance to an antibiotic or other drug. Some of the markers described below are dominant and can be used with any type of mammalian cell; others must be used with particular cell lines that lack the relevant enzyme activity.

In early experiments, the genes encoding the protein of interest and the selectable marker were included on a single vector. However, Wigler et al. (1979) found that mammalian cells capable of taking up DNA do so efficiently, so that two unlinked plasmids can be cotransfected with high frequency (>90%). Cotransfection, which obviates the need to construct complex recombinants, has become the standard method of introducing a selectable marker (on one plasmid) and the gene of interest (on another plasmid) into mammalian cells. The selectable markers that are currently used include:

- *Thymidine kinase.* The thymidine kinase gene (*tk*), which is expressed in most mammalian cells, codes for an enzyme that is involved in the salvage pathway for synthesis of thymidine nucleotides. A number of *tk*⁻ cell lines have been isolated from different mammalian species, including mouse (Ltk⁻ cells) (Kit et al. 1963; Wigler 1977), human (143tk⁻ cells) (Bacchetti and Graham 1977), and rat (Rat-2 fibroblast cells) (Topp 1981). These mutant cell lines, in contrast to their wild-type parents, will grow in medium that contains the thymidine analog 5-bromodeoxyuridine. Szybalska and Szybalski (1962) and Littlefield (1964, 1966) developed a selective medium containing hypoxanthine, aminopterin, and thymidine (HAT medium; see Appendix A) in which only cells expressing the *tk* gene will grow. By the appropriate use of this medium, it is therefore possible to select for or against cells that express the *tk* gene.

  Early cotransfection experiments utilized purified fragments of herpes simplex virus (HSV) DNA that contained the viral *tk* gene (Wigler et al. 1977). Subsequent cloning of the *tk* gene both from HSV (Colbère-Garapin et al. 1979) and from chicken cells (Perucho et al. 1980) made it possible to construct plasmids such as that shown in Figure 16.1A for use in cotransfection experiments. The primary limitation of these vectors is that they can be used only in *tk*⁻ cell lines.

- *Dihydrofolate reductase.* Mutants of CHO cells that lack the enzyme dihydrofolate reductase (Urlaub and Chasin 1980) cannot synthesize tetrahydrofolate and therefore can grow only in media supplemented with thymidine, glycine, and purines. Transfection of these cells with vectors that express a cloned copy of the dihydrofolate reductase gene (*dhfr*) gives rise to clones that can grow in the absence of these supplements (Subramani et al. 1981; Kaufman and Sharp 1982a,b; Kaufman et al. 1985; see Figures 16.1B and 16.3C).

DHFR can be inhibited by methotrexate, a folate analog. Progressive selection of cells that are resistant to increasing concentrations of methotrexate leads to amplification of the *dhfr* gene, with concomitant amplification of extensive regions of the DNA that flank the *dhfr* sequences (Schimke 1982). DNAs that are cotransfected with the *dhfr* gene tend to become integrated into the same region of the cellular chromosome and therefore can frequently be coamplified with *dhfr*. Alternatively, cells lacking DHFR activity can be transfected with a recombinant construct containing the gene of interest linked to the *dhfr* gene. The linked gene is then amplified by selecting with successively higher concentrations of methotrexate. The resulting cell lines express very high levels of the desired recombinant protein product (Kaufman and Sharp 1982a,b; Kaufman et al. 1985). This approach is described in more detail on page 16.28.

The coamplification method has also been adapted for use with cells that synthesize wild-type levels of DHFR. In one approach, the *dhfr* gene was placed under the control of a strong promoter, thereby conferring on transfected cells the ability to grow in concentrations of methotrexate that would be lethal to cells expressing normal, wild-type levels of the enzyme (Murray et al. 1983). Alternatively, cells transfected with a plasmid that carries a dominant selectable marker (e.g., resistance to geneticin [G418]), the *dhfr* gene, and the gene of interest are selected first for their ability to grow in G418 and then for their ability to grow in progressively higher concentrations of methotrexate (Kim and Wold 1985). Finally, an altered form of the *dhfr* gene encoding an enzyme that is more resistant to methotrexate has been utilized as a dominant selectable marker for cotransformation experiments in a broad range of cell types (Spandidos and Siminovitch 1977; O'Hare et al. 1981; Simonsen and Levinson 1983).

Note: G418 is now commercially available. Because cultured lines of mammalian cells differ widely in their sensitivity to this antibiotic, the concentration appropriate for the selection of stably transfected cells must be determined empirically.

• *Aminoglycoside phosphotransferase.* The mostly widely used dominant selection system utilizes the bacterial gene encoding aminoglycoside 3′ phosphotransferase (APH). Two distinct APH enzymes, encoded by the bacterial transposons Tn5 and Tn601, confer resistance to aminoglycoside antibiotics such as kanamycin, neomycin, and geneticin, which inhibit protein synthesis in both prokaryotic and eukaryotic cells. Eukaryotic cells do not normally express an endogenous APH activity, but they are capable of expressing the enzymes encoded by the bacterial transposons. When fused to eukaryotic transcriptional regulatory elements, the genes encoding APH can be used as dominant markers to select cells that take up exogenous DNA (Jimenez and Davies 1980; Colbère-Garapin et al. 1981). The first APH (*neo*[r]) vectors designed for mammalian cells expressed the Tn5 *neo*[r] gene under the control of the HSV *tk* promoter and polyadenylation sequences (Colbère-Garapin et al. 1981). Subsequently, vectors were

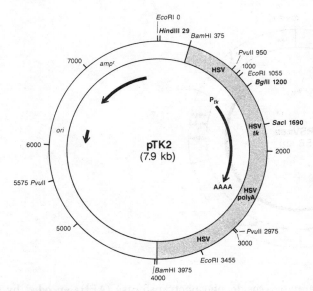

FIGURE 16.1A

pTK2 is a derivative of pBR322 that carries a 3.6-kb *Bam*HI fragment of herpes simplex virus (HSV) encoding thymidine kinase (*tk*). The positions of the *tk* promoter (P*tk*) and the polyadenylation site (polyA; AAAA) are indicated.

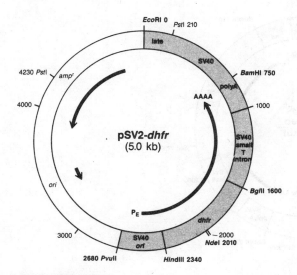

FIGURE 16.1B

pSV2-*dhfr* carries the SV40 origin (SV40 *ori*) and expresses dihydrofolate reductase (*dhfr*) from the SV40 early promoter (P_E). The SV40 small T intron and polyadenylation site (polyA; AAAA) are shown.

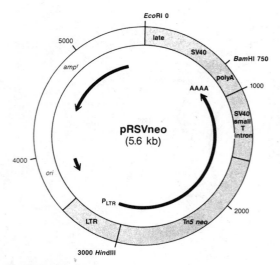

FIGURE 16.1C

pRSVneo expresses aminoglycoside phosphotransferase (APH) encoded by the bacterial transposon gene Tn*5 neo*^r from the Rous sarcoma virus (RSV) LTR promoter (P$_{LTR}$). The SV40 small T intron and polyadenylation site (polyA; AAAA) are located downstream from Tn*5 neo*.

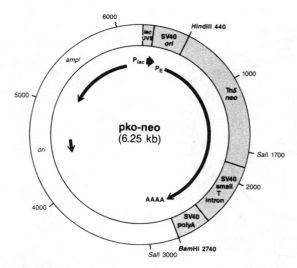

FIGURE 16.1D

pko-neo expresses aminoglycoside phosphotransferase encoded by the bacterial transposon gene Tn*5 neo*^r from the eukaryotic SV40 early promoter (P$_E$) or the prokaryotic *E. coli lac*UV5 promoter (P$_{lac}$). The SV40 origin (SV40 *ori*), SV40 small T intron, and SV40 polyadenylation sites (polyA; AAAA) are present.

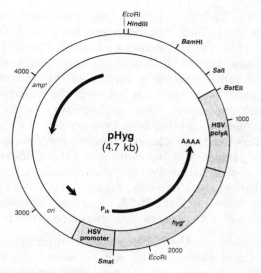

FIGURE 16.1E

pHyg directs the expression of the *E. coli* gene encoding hygromycin B phosphotransferase (*hyg*^{r}) using the herpes simplex virus promoter (P_{tk}) and polyadenylation site (HSV polyA; AAAA).

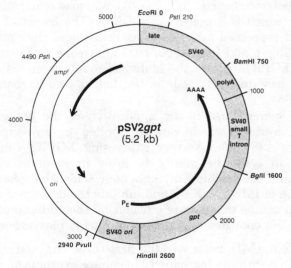

FIGURE 16.1F

In pSV2*gpt*, the *E. coli* xanthine-guanine phosphoribosyl transferase gene (*gpt*) is expressed using the SV40 early promoter (P_E) located in the SV40 origin (SV40 *ori*), the SV40 small T intron, and the SV40 polyadenylation site (polyA; AAAA).

developed that express the Tn5 *neo*[r] gene under the control of SV40 regulatory elements (Chia et al. 1982; Southern and Berg 1982; Okayama and Berg 1983; Van Doren et al. 1984). Vectors such as pSV2-*neo* (Southern and Berg 1982) and pRSVneo (Figure 16.1C), which have been widely used in cotransformation experiments, contain a version of the Tn5 *neo*[r] gene that retains prokaryotic promoter sequences between the eukaryotic promoter and the APH coding sequences. This configuration yields a vector that can confer antibiotic resistance upon both prokaryotic and eukaryotic cells. However, perhaps because the bacterial promoter contributes several upstream AUG codons, the efficiency of translation of APH mRNAs synthesized from these vectors is comparatively low in mammalian cells (Chen and Okayama 1987). Vectors such as pko-neo (Figure 16.1D) (Van Doren et al. 1984) and pcDneo (Okayama and Berg 1983; Chen and Okayama 1987), which lack prokaryotic promoter sequences, are therefore preferred.

- *Hygromycin B phosphotransferase.* The *E. coli* gene encoding hygromycin B phosphotransferase (Gritz and Davies 1983) can be used as a dominant selectable marker in much the same way as the APH gene. When the hygromycin B phosphotransferase gene (*hyg*) is introduced into mammalian cells on an appropriate expression vector (e.g., pHyg, Figure 16.1E) (Sugden et al. 1985), the transfected cells become resistant to the antibiotic hygromycin. Resistance to neomycin and to hygromycin can be selected for independently and simultaneously in cell lines that have been transfected with both genes. Thus, two different vectors can be introduced into one cell line, either simultaneously or sequentially.

- *Xanthine-guanine phosphoribosyl transferase.* The *gpt* gene of *E. coli* encodes the enzyme xanthine-guanine phosphoribosyl transferase (XGPRT), which is the bacterial analog of the mammalian enzyme hypoxanthine-guanine phosphoribosyl transferase (HGPRT). Whereas only hypoxanthine and guanine are substrates for HGPRT, XGPRT will also efficiently convert xanthine into XMP, which is a precursor of GMP. The bacterial *gpt* gene has been cloned and expressed in mammalian cells under the control of an SV40 promoter (Mulligan and Berg 1980, 1981a,b) (see, e.g., Figure 16.1F). Vectors expressing XGPRT restore the ability of mammalian cells lacking HGPRT activity to grow in HAT medium (Szybalska and Szybalski 1962; Littlefield 1964, 1966).

  Of much greater general use is the application of the *gpt* gene as a dominant selection system, which can be applied to any type of cell (Mulligan and Berg 1981a,b). Vectors expressing XGPRT confer upon wild-type mammalian cells the ability to grow in medium containing adenine, xanthine, and the inhibitor mycophenolic acid. Mycophenolic acid blocks the conversion of IMP into XMP and inhibits the de novo synthesis of GMP. The selection can be made more efficient by the addition of aminopterin, which blocks the endogenous pathway of purine biosynthesis.

- *CAD.* A single protein, CAD, possesses the first three enzymatic activities of de novo uridine biosynthesis (carbamyl phosphate synthetase, aspartate transcarbamylase, and dihydroorotase). Transfection of vectors expressing the CAD protein from Syrian hamsters into CAD-deficient (UrdA) mutants of CHO cells allows selection of CAD[+] transfectants that are able to grow in the absence of uridine (Robert de Saint Vincent et al. 1981).

L-Phosphonacetyl-L-aspartate (PALA) is a specific inhibitor of the aspartate transcarbamylase activity of CAD. Growth of wild-type or transfected mammalian cells in the presence of increasing concentrations of PALA leads to the amplification of the CAD gene and DNA sequences linked to it (Kempe et al. 1976; Robert de Saint Vincent et al. 1981; Wahl et al. 1984). The *E. coli* gene encoding aspartate transcarbamylase (*pyrB*), when expressed in CHO cells deficient in aspartate transcarbamylase, is also amplified by PALA selection (Ruiz and Wahl 1986).

- *Adenosine deaminase.* Adenosine deaminase (ADA) is present in virtually all animal cells, but it is normally synthesized in minute quantities and is not essential for cell growth. However, because ADA catalyzes the irreversible conversion of cytotoxic adenine nucleosides to their respective nontoxic inosine analogs, cells propagated in the presence of toxic concentrations of adenosine or its analog 9-β-D-xylofuranosyl adenine (Xyl-A) require ADA for survival (for references and review, see Kaufman 1987). Under conditions where ADA is required for cell growth, amplification of the gene can be achieved in the presence of increasing concentrations of 2′-deoxycoformycin (dCF), a transition-state analog of adenine nucleotides that strongly inhibits the enzyme. In cells selected for their ability to resist high concentrations of 2′-deoxycoformycin, it has been shown that ADA was overproduced 11,400-fold and represented 75% of the soluble protein synthesized by the cells (Ingolia et al. 1985).

- *Asparagine synthetase.* The *E. coli* gene coding for asparagine synthetase (AS) is a potentially useful, dominant, amplifiable marker for mammalian cells. Because the bacterial enzyme uses ammonia as an amide donor—in contrast to the mammalian enzyme, which uses glutamine—cells that express the bacterial AS gene will grow in asparagine-free medium containing the glutamine analog albizziin. Subsequently, the transfected AS gene can be amplified by selection in medium containing increasing concentrations of β-aspartyl hydroxamate, an analog of aspartic acid.

Foreign DNA Sequences

DNAs encoding the foreign protein of interest are usually cloned as cDNAs that lack all of the controlling elements required for expression in mammalian cells but may contain ancillary sequences introduced during the construction of the cDNA library (e.g., homopolymeric stretches of guanine or cytosine residues, synthetic linkers, etc.). No consensus exists as to whether or not these ancillary sequences need to be removed before the cDNA can be expressed in mammalian cells. However, since such sequences never enhance, and in some circumstances may suppress, the level of expression of foreign DNAs in mammalian cells (Simonsen et al. 1982), most workers prefer to remove as many extraneous sequences as is conveniently possible. Less frequently, DNAs encoding the foreign protein of interest are obtained as a genomic copy in which the coding sequences may be interrupted by one or more introns. A complete genomic copy will have all the controlling sequences necessary for the expression of the protein in some, but not necessarily all, cell types. Because the specificity of these sequences determines the range of cell types in which the gene will be active, replacement of the enhancer and/or promoter sequences may be necessary to allow efficient expression of the gene in lines of cultured cells.

Cloned sequences that are expressed in mammalian cells almost invariably include the ribosome-binding site and initiation codon found in the natural gene. In contrast, in prokaryotic expression systems, the ribosome-binding site and initiation codon of the natural gene are almost invariably replaced by a Shine-Dalgarno ribosome-binding sequence and an ATG codon that have been optimally spaced for efficient expression. With mammalian mRNAs, the number of nucleotides between the starting sites of transcription and translation is not critical, but the initiation codon of the cloned gene should be the first AUG in the transcript. The presence of AUG codons upstream of the intended site of initiation may significantly decrease the efficiency of translation of the desired product (Hughes et al. 1984; Liu et al. 1984; Perez et al. 1987; Kozak 1989). In a transcription unit coding for two polypeptides, the efficiency of translation of the downstream reading frame is decreased several hundredfold, despite the presence of an in-frame termination codon between the two open reading frames (Kaufman et al. 1987). The consensus sequence for initiation of translation by eukaryotic ribosomes is the sequence

$$G\,C\,C\,G\,C\,C\,A^{-3}\,/\,G\,C\,C\,A^{1}\,U\,G\,G^{+4}$$

(Kozak 1989). However, Kozak points out that, for practical purposes, an initiaton codon can usually be designated "strong" or "weak" by considering only positions -3 and $+4$. As long as there is a purine in position -3, deviations from the rest of the consensus sequence only marginally impair initiation. In the absence of a purine in position -3, however, G^{+4} is essential for efficient translation and the contributions of other nearby nucleotides can be detected (Kozak 1989).

VECTOR SYSTEMS

A good way to establish rapidly the feasibility of expressing a cloned gene in mammalian cells is to use a transient expression system such as that provided by the simian COS cell line (Gluzman 1981). In addition to functional SV40 large T antigen, COS cells produce the permissivity factors required for replication of DNAs that contain the SV40 origin of DNA replication. In many cases, transient expression of cloned genes in COS cells will provide all of the experimental data required to solve the biological question at hand. In other situations, it may be necessary to choose another vector–host system—for example, when sustained high levels of production of a protein encoded by a cloned gene are required or when there is a need to obtain expression of a cloned gene in specialized cell types. Until recently, most mammalian expression vectors were designed to express a gene or cDNA of interest in a particular cell line. They were therefore tailored to contain controlling elements of the desired specificity and restriction sites to match those at the 5' and 3' termini of the sequences to be expressed. Consequently, most vectors in current use are idiosyncratic in their design and frequently do not contain either restriction sites or controlling elements that are universally useful. Construction of new recombinant vectors often requires extensive engineering (e.g., deletion or addition of restriction sites or substitution of controlling elements) of either the vector or the new gene.

In the following sections, various vector–host systems that are currently available are described in detail.

Plasmid-based Vectors That Do Not Carry a Eukaryotic Replicon

In these simple systems, a complete mammalian transcription unit (see pages 16.5–16.9) and a selectable marker (see pages 16.9–16.15) are inserted into a prokaryotic plasmid. The resulting vector is then amplified in bacteria before being transfected into cultured mammalian cells. Because these vectors do not contain a eukaryotic replicon, no episomal amplification of the transfected DNA occurs. Instead, the transfected DNA integrates into the genomes of a minority of the transfected cells, where it may direct the expression of low levels of the protein of interest. The selectable marker facilitates the isolation of the very small numbers of transfected cells that take up and express the foreign gene. Examples of simple vectors of this type that contain a range of different selectable markers and are suitable vehicles for the transfection of complete mammalian transcription units are shown in Figure 16.1 (e.g., pTK2, pHyg, and pRSVneo). In addition, all of the plasmids shown in Figure 16.1 can be used in cotransformation experiments to allow selection of those cells that have taken up the foreign DNA.

Plasmid DNA Expression Vectors Containing Regulatory Elements from Eukaryotic Viruses

SIMIAN VIRUS 40 VECTORS

SV40 is a member of the papova group of small, nonenveloped DNA viruses and causes lytic infection of permissive monkey cells. The virus has been studied extensively, and most aspects of its molecular biology are understood in great detail (for review, see Tooze 1980). In cultured African green monkey kidney cells, SV40 undergoes a conventional replication cycle, in which the expression of the viral early genes is followed by replication of viral DNA, synthesis of the late viral proteins, and, finally, assembly of progeny virus particles. In rodent cells, infection is blocked at the level of DNA replication.

The viral genome is a covalently closed circular double-stranded DNA molecule of 5243 bp. The genome is functionally divided into early and late regions, which are transcribed from the two DNA strands in opposite directions. The early region is transcribed throughout the lytic cycle, and differential splicing generates two mRNAs that encode the large T and small T antigens. The late region is transcribed only after the onset of DNA replication and encodes the viral capsid proteins VP1, VP2, and VP3 from spliced mRNAs.

A number of plasmid-based expression vectors carry individual regulatory regions derived from SV40 but lack most of the coding region of the viral genome. After transfection into mammalian cells, foreign DNAs cloned into these vectors are transiently expressed, but no virus particles are produced. The regulatory region most commonly used is a 300-bp segment of SV40 DNA that lies between the viral early and late transcription units and contains a number of different controlling *cis* elements (see Figure 16.2) (for review, see McKnight and Tjian 1986). These elements include (1) the origin of DNA replication, (2) the promoters and sites of initiation of transcription of early and late mRNAs, (3) T-antigen binding sites involved in activation of the origin of replication and in autoregulation of early transcription, (4) a sequence of three G/C-rich, 21-bp direct repeats that are recognized by cellular transcription factors, and (5) two 72-bp direct repeats that comprise the SV40 early enhancer. In addition to these regulatory sequences, some SV40-based plasmid vectors contain a complete transcription unit that encodes the SV40 large T antigen, whose expression is required to activate the SV40 origin of replication in simian cells.

The basic properties of these vectors are:

- They have low to moderate levels of expression in a wide variety of transfected mammalian cells and high levels of expression in transfected COS cells

- Both genomic DNA and cDNA sequences can be expressed, and there are no constraints on the size of the foreign DNA that can be inserted

- They are generally used as a transient system, although cell lines that express low levels of the gene of interest can be isolated if a selectable marker is present in the expression vector or is cotransfected into the cells

- Typically, these vectors contain the SV40 origin of replication, a promoter with a broad host range (e.g., the SV40 early promoter), and a polyadenylation signal (almost always from SV40)

- Many vectors of this class also carry splice donor and acceptor signals (usually the intron from the SV40 small T antigen gene)

The early promoter and/or enhancer sequences work at low to moderate levels in a wide range of mammalian cells in the absence of DNA replication.

However, much higher levels of expression can be obtained in transfected simian cells, where the viral origin of DNA replication (minimum size 85 bp) drives replication of the plasmid when viral T antigen is supplied in *trans* (Gluzman 1981). The resulting increase in copy number of the plasmid DNA provides many more templates that become transcriptionally active over the course of the transfection. The use of origin-containing transient expression vectors was greatly facilitated by the development of lines of cells (COS cells) that were derived by transformation of simian CV-1 cells with an origin-defective SV40 genome (Gluzman 1981). These cells constitutively express wild-type SV40 T antigen and contain all of the cellular factors required to drive the replication of SV40 origin-containing plasmids. The efficiency of replication seems to be increased if the transfected plasmids lack certain ill-defined sequences that appear to interfere with replication in eukaryotic cells (Lusky and Botchan 1981). Most currently available vectors are free of these "poison" sequences. Over the course of a transfection experiment, COS cells accumulate $>10^5$ copies per cell of recombinant expression plasmids containing the SV40 origin of replication (Mellon et al. 1981) and express high levels of foreign DNA sequences. Expression in this system is transient because replication of the transfected plasmids continues unchecked until the cells die (at ~70–90 hours posttransfection), presumably because they cannot tolerate the high levels of extrachromosomally replicating DNA.

Transient expression in COS cells is the most widely used of all eukaryotic expression systems. The recombinant vectors are easy to construct and to use, and there are no constraints on the amount of DNA that can be inserted or on the use of genomic DNA sequences. This system has provided convenient positive verification of cDNA clones by expression (Toole et al. 1984; Wood et al. 1984), has facilitated rapid analysis of mutations introduced into cloned cDNAs (Mishina et al. 1984), and, following the pioneering efforts of Okayama and Berg (1982, 1983, 1985), has made possible the screening of cDNA libraries constructed in expression vectors to isolate cDNAs via expression of a desired activity (Lee et al. 1985; Wong et al. 1985; Aruffo and Seed 1987; Seed and Aruffo 1987). Expression screening is discussed on page 16.69.

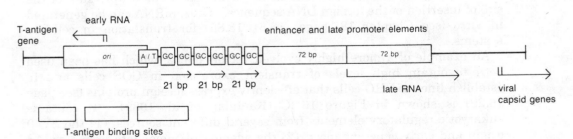

FIGURE 16.2
Regulatory elements in SV40-based expression vectors (see text for details). (Adapted from McKnight and Tjian 1986).

Although COS cells remain the most widely used host for transient expression of foreign DNAs, two more recently developed cell systems offer the possibility of controlling the replication of transfected plasmids. A line of temperature-sensitive COS cells produces high levels of thermolabile large T antigen under the transcriptional control of the Rous sarcoma virus LTR (Rio et al. 1985). These cells, which support replication of SV40-origin-containing vectors at 33°C but not at 40°C, can therefore be used to regulate the copy number of transfected plasmids. In addition, vectors that carry a selectable marker can be maintained either as integrated DNA or as autonomously replicating episomes. Similarly, Gerard and Gluzman (1985) have isolated transformed monkey cell lines (CMT and BMT) that express T antigen under the control of the murine metallothionein promoter. A direct correlation was found between the level of T antigen synthesis (which could be induced five- to tenfold by addition of heavy metals) and the extent of episomal replication of transfected plasmids containing the SV40 origin.

A number of vectors have been tailored to simplify expression of cDNAs in mammalian cells (Okayama and Berg 1982, 1983, 1985). In addition to the minimal sequences required for expression of cloned genes, many vectors used for transient expression in COS cells now carry ancillary sequences that simplify the cloning process or serve a variety of specialized functions. Included among these specialized sequences are: (1) polycloning sites; (2) selectable markers that can be used to establish permanent cell lines carrying the sequences of the recombinant plasmid; (3) bacteriophage promoters that can be used to generate transcripts of the foreign gene, which can then be translated in vitro; and (4) additional transcription units that encode the SV40 large T antigen, expression of which is required to activate the SV40 origin of replication in normal monkey cells. However, expression of an exogenous T antigen gene is unnecessary if COS cells are used as the hosts for transfection.

Figure 16.3 shows maps of three currently used transient expression vectors. pMSG (Figure 16.3A; available from Pharmacia) is suitable for expression of cDNA sequences inserted into a polycloning site downstream from the mouse mammary tumor virus LTR promoter, which can be induced with glucocorticoid in hormone-responsive cells. The plasmid contains the *E. coli gpt* gene, which is expressed from the SV40 early promoter. pSVT7 (Figure 16.3B) contains a polycloning site for insertion of cDNAs downstream from the SV40 early promoter. The vector also contains a promoter for bacteriophage T7 DNA-dependent RNA polymerase located upstream of the site of insertion of the foreign DNA sequence. Thus, mRNA can be generated in vitro (see Chapter 18, pages 18.81–18.85) for translation in cell-free systems.

An example of a more highly evolved vector, pMT2, which has been used both to obtain high levels of transient expression in COS cells and to establish lines of CHO cells that efficiently express foreign proteins (see page 16.29) is shown in Figure 16.3C (Kaufman et al. 1989). It contains eukaryotic regulatory elements from several different sources: (1) the SV40 origin and early gene enhancer, (2) the adenovirus major late promoter (Ad MLP) coupled to a cDNA copy of the adenovirus tripartite leader, (3) a hybrid intron consisting of a 5′ splice site from the first exon of the tripartite leader

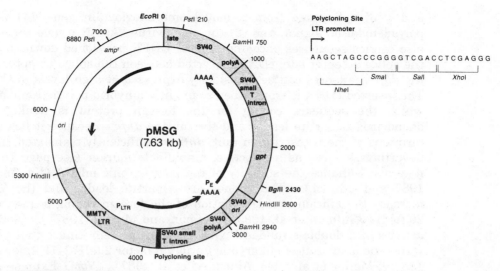

FIGURE 16.3A

pMSG is a transient expression vector that uses the mouse mammary tumor virus (MMTV) LTR promoter (P_{LTR}) to express cDNA sequences cloned into the polycloning site. The *E. coli* xanthine-guanine phosphoribosyl transferase gene (*gpt*) is expressed from the SV40 promoter (P_E) located in the SV40 origin (SV40 *ori*). The SV40 small T intron and polyadenylation site (SV40 polyA; AAAA) are found downstream from both the polycloning site and the *gpt* gene.

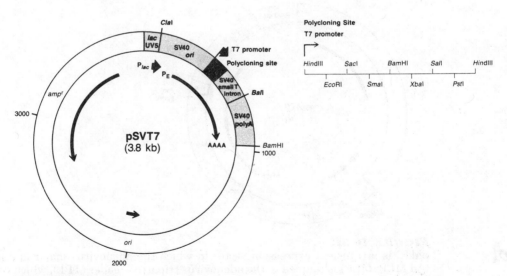

FIGURE 16.3B

pSVT7 is a transient expression vector that uses SV40 signals, the SV40 early promoter (P_E), the SV40 small T intron, and the SV40 polyadenylation site (polyA; AAAA) to express cDNA cloned into the polycloning site. The vector also carries the *lac*UV5 promoter (P_{lac}) for expression in *E. coli* and a bacteriophage T7 promoter for generation of mRNA in vitro.

and a 3′ splice site from a mouse immunoglobulin gene, (4) the SV40 polyadenylation signal, and (5) the adenovirus VA$_I$ RNA gene region. pMT2 also carries sequences encoding murine DHFR positioned downstream from the splice acceptor site. cDNAs inserted between the splice acceptor site and the *dhfr* sequences can be transiently expressed at high levels in COS cells. The inserted cDNA is transcribed to produce a hybrid polycistronic mRNA in which the sequence coding for the foreign protein is flanked by the adenovirus tripartite leader and the murine *dhfr*. Because it lies at the 3′ terminus of the transcription unit, *dhfr* is inefficiently translated, but it can nevertheless serve as a selective, amplifiable marker (see page 16.28) and may also enhance the stability of the polycistronic mRNA (Kaufman et al. 1987 and unpubl.). The adenovirus tripartite leader and the VA$_I$ RNA increase the efficiency of translation of the foreign coding sequences (3- to 20-fold) (Kaufman et al. 1985; Kaufman and Murtha 1987) by blocking the activity of a double-stranded RNA-dependent protein kinase that phosphorylates the α subunit of eukaryotic initiation factor 2 (eIF-2) (Kitajewski et al. 1986; O'Malley et al. 1986; Akusjärvi et al. 1987). (*Note:* Expression of the adenovirus VA RNAs is restricted to cells [e.g., COS cells] that contain a specific factor required for transcription of these RNA-polymerase-III-dependent genes.)

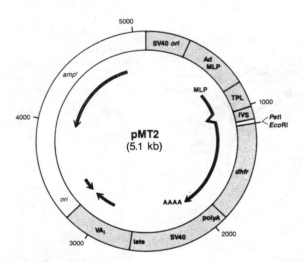

FIGURE 16.3C
pMT2 is a transient expression vector in which the adenovirus major late promoter (Ad MLP; MLP) is coupled to the adenovirus tripartite leader (TPL), which contains a 5′ splice site and part of a mouse immunoglobulin gene that contains a 3′ splice site. The intervening sequence (IVS) is followed by two cloning sites (*Pst*I and *Eco*RI), the dihydrofolate reductase gene (*dhfr*), the SV40 polyadenylation site (polyA; AAAA), and the adenovirus VA$_I$ gene. The SV40 origin (SV40 *ori*) and enhancer are also present on the vector.

BOVINE PAPILLOMAVIRUS VECTORS

Papillomaviruses are small, double-stranded DNA viruses that commonly infect higher vertebrates, including man. These viruses usually cause benign, self-limiting, proliferative lesions commonly known as warts (for review, see Orth and Favre 1985), but they are oncogenic under certain circumstances (for review, see Broker and Botchan 1986). Many different papillomaviruses have been isolated, but most studies at the molecular level have been carried out with a bovine papillomavirus, BPV-1. This virus is capable of transforming rodent cells in vitro but, like all other papillomaviruses, will not grow in tissue culture. Molecular characterization of the BPV-1 genome was therefore achieved only after the viral DNA was cloned in prokaryotic vectors (Howley et al. 1980). The sequence of BPV-1 DNA was then rapidly determined (Chen et al. 1982), and the genomic organization of the virus was elucidated (Figure 16.4) (for review, see Broker and Botchan 1986).

When rodent cells in culture are morphologically transformed by BPV DNA, 20 to 100 copies of the viral DNA persist as extrachromosomal DNA. This ability of BPV DNA to replicate as an episome has led to its use both as an expression vector and as a vehicle for studying gene regulation. Not all of the viral genome is required for transformation or for establishment of the episomal state. The observation that both of these functions can be carried out by a fragment of the viral DNA (the 69% transforming fragment) led directly to the use of BPV DNA as a vector to express rat preproinsulin in murine C127 cells (Sarver et al. 1981). Since then, a number of BPV shuttle vectors have been developed that replicate episomally both in transformed mammalian cells and in bacteria (DiMaio et al. 1982; Sarver et al. 1982; for review, see Campo 1985).

The basic properties of these vectors are:

- They have low to moderate levels of expression in a wide variety of mammalian cells.

- Both genomic DNA and cDNA sequences can be expressed, and there are no constraints on the size of the foreign DNAs that can be inserted.

- They are never used as a transient expression system; instead, these vectors are used to establish cell lines that contain multiple copies of the foreign gene.

- Typically, these vectors contain a segment of BPV DNA (either the entire viral genome or the 69% transforming fragment), a promoter with a broad host range (e.g., the SV40 early promoter or the murine metallothionein promoter), a polyadenylation signal (almost always from SV40), splice donor and/or acceptor signals (usually the intron from the SV40 small T antigen gene), and "poisonless" plasmid sequences that allow the vector to be propagated in *E. coli*.

- BPV vectors may also carry a selectable marker (e.g., neo^r), whose expression is controlled by a separate promoter, and sequences derived from mammalian genomes which may help to maintain episomal replication of the recombinant vector in mammalian cells.

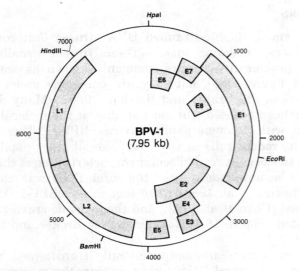

FIGURE 16.4
Bovine papillomavirus (BPV-1) encodes eight early gene products (E1–E8) and two late gene products (L1 and L2).

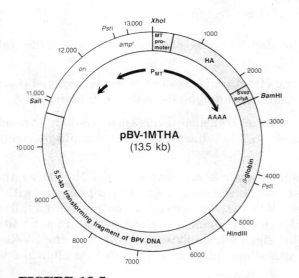

FIGURE 16.5
pBV-1MTHA carries the 69% transforming fragment (5.5-kb transforming fragment of BPV DNA), which allows episomal replication of this vector in mammalian cells. In addition, it encodes hemagglutinin (HA) expressed from the metallothionein (MT) promoter (P_{MT}) and human β-globin. It also has an SV40 polyadenylation site (SV40 polyA; AAAA).

Certain vectors (those of the BV-1 series [Figure 16.5]) carry an additional fragment of DNA derived initially from the human β-globin gene cluster. In some systems (DiMaio et al. 1982; Zinn et al. 1983; Sambrook et al. 1985), but not in all (DiMaio et al. 1984), this fragment appears to carry an activity that enhances the ability of BPV vectors containing only the 69% transforming fragment to replicate episomally in mammalian cells. Because of the large size of BPV-based vectors (8–14 kb) and the dearth of unique cloning sites, the insertion of a foreign gene into BPV vectors is not always straightforward. Most vectors contain only one or two restriction sites at which the coding sequences of interest may be inserted, attached either to their natural controlling elements or to heterologous controlling elements carried by the vector. To date, no BPV vectors have been produced that contain polycloning sites.

Following construction and amplification in bacteria, recombinant BPV DNAs are transfected into cultured mammalian cells, usually by the calcium phosphate coprecipitation technique (see pages 16.32–16.40). If a selectable marker is not used, cells that have been successfully transfected can sometimes be identified by their characteristic transformed morphology. Individual transformed clones are then isolated and tested for expression of the protein of interest. Typically, different transformants isolated from the same transfection vary up to tenfold in their level of expression of the protein of interest, and it is therefore worthwhile to screen a number of clones to identify those that express the protein at the desired level. By assaying a number of individual cell clones, it is usually possible to establish cell lines that yield 50- to 100-fold more protein than cell lines transformed with simple plasmid-based vectors (for review, see Campo 1985; DiMaio 1987).

In the absence of a selectable marker, the host range of BPV vectors is limited to those cells, such as the murine C127 line, that undergo identifiable changes in their morphology and growth after transfection by BPV DNA. Expansion of the host and tissue range to cells that are capable of supporting the autonomous replication of BPV-1-derived plasmids but do not manifest a transformed phenotype (e.g., mouse NIH-3T3, hamster CHO, canine MDCK, and pig PK-1 cell lines) can be achieved by the use of a dominant selectable marker. Several BPV vectors carry a copy of the bacterial neomycin resistance gene (neo^r) (see pages 16.10 and 16.14) under the control of efficient eukaryotic regulatory sequences (see, e.g., Law et al. 1983); alternatively, the neo^r gene may be cotransfected on a separate plasmid (e.g., pko-neo [Figure 16.1D]) (Sambrook et al. 1985). Resistance to high concentrations of toxic heavy metals such as cadmium is conferred by murine or human metallothionein genes and may also be used as a dominant selectable marker (Karin et al. 1983; Pavlakis and Hamer 1983; Krystal et al. 1986) (see Figure 16.5). Other potential markers, such as tk, $dhfr$, or E. coli gpt, have proven to be unsatisfactory because the BPV-based vector undergoes rearrangement and/ or integrates into the cellular chromosome (for references, see DiMaio 1987).

The copy number and physical state of the recombinant genomes are affected both by the composition of the vector and by the host cell (Campo 1985; Sambrook et al. 1985) in ways that are not understood. Recombinant vectors containing the entire BPV genome or the 69% transforming fragment are usually propagated as stable, multicopy (20–300 copies/cell) extrachromosomal elements. Less frequently, the recombinant sequences are maintained as oligomeric plasmids or as head-to-tail tandem arrays integrated into the cellular chromosome. Whatever the state of the DNA, the cell lines invariably carry many copies of the gene of interest. It is this high copy number that is responsible, at least in part, for the efficient expression of the foreign protein.

EPSTEIN-BARR VIRUS VECTORS

The human herpesvirus, Epstein-Barr virus (EBV), transforms resting human B lymphocytes into dividing blast cells that can proliferate indefinitely in culture (for review, see Sugden 1982). EBV-transformed lymphoblasts secrete immunoglobulin, carry multiple episomal copies of the 172-kb viral DNA, and are usually diploid. The discovery that a *cis*-acting element of EBV, *oriP*, permits maintenance of episomal DNA molecules in adherent cells that carry EBV DNA (Sugden et al. 1985) and express the *trans*-acting EBNA-1 antigen (Lupton and Levine 1985; Yates et al. 1985) led to the development of vectors for the expression of foreign genes in a broad range of mammalian cells. The basic properties of these vectors are:

- They have low to moderate levels of expression in a wide variety of mammalian cells.

- Both genomic DNA and cDNA sequences can be expressed, and there are no constraints on the size of the foreign DNAs that can be inserted.

- They are never used as a transient expression system; instead, these vectors are used to establish cell lines that contain multiple episomal copies of the foreign gene.

- Typically, these vectors contain a segment of EBV DNA that carries the *oriP* region (a *cis*-acting replication element), plasmid sequences that allow the vector to be propagated in *E. coli*, restriction sites at which a foreign transcription unit may be inserted, and a selectable marker (e.g., hygromycin resistance) whose expression is controlled by a broad-host-range promoter.

One example of an EBV-based vector, pHEBo, is shown in Figure 16.6A. This vector has been used to express cloned human class I major histocompatibility complex genes in human lymphoblastoid cells (Shimizu et al. 1986). More complex vectors have been designed to extend the host range beyond EBV-transformed human cells (Yates et al. 1985). For example, p205 (Figure 16.6B) contains both *oriP* and a variant EBNA-1 gene that has sustained a spontaneous deletion of a repetitive sequence of glycine and alanine codons and supports plasmid replication more efficiently than the wild-type EBNA-1 gene. p205 replicates episomally in a variety of animal cells at a copy number of 1–90 molecules per cell, depending on the cell type used. In some rodent cells, however, the plasmid appears to integrate into the host-cell chromosome. The vector contains unique *Nar*I and *Nru*I sites that can be used for the insertion of a foreign transcription unit. A vector similar to p205 has been used in a shuttle cosmid vector to express and rescue the gene coding for human tumor necrosis factor (Kioussis et al. 1987).

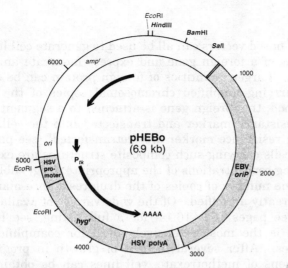

FIGURE 16.6A

pHEBo contains the Epstein-Barr virus origin *P* (EBV *oriP*) and directs the expression of the *E. coli* gene encoding hygromycin B phosphotransferase (*hyg*^r) using the herpes simplex virus promoter (P_{tk}) and polyadenylation site (HSV polyA; AAAA).

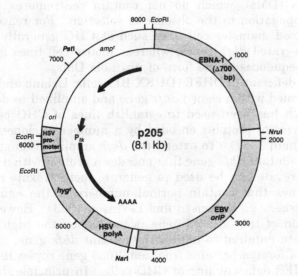

FIGURE 16.6B

p205 carries the Epstein-Barr virus origin *P* (EBV *oriP*) and expresses a *trans*-acting Epstein-Barr nuclear antigen (EBNA-1) from DNA containing a 700-bp deletion (Δ700 bp). It also directs the expression of the *E. coli* gene encoding hygromycin B phosphotransferase (*hyg*^r) using the herpes simplex virus promoter (P_{tk}) and polyadenylation site (HSV polyA; AAAA).

SV40-, BPV-, and EBV-based vectors can all be used to generate cell lines that contain multiple copies of a foreign gene and express moderate amounts of the protein of interest. Larger quantities of foreign protein can be obtained from lines of cells carrying amplified chromosomal copies of the gene of interest. In this method, the foreign gene is attached to a segment of DNA that carries a drug resistance marker and transfected into the cells, or the foreign gene and drug resistance marker are "cotransfected" (see page 16.9) into the cells. When cells carrying such composite structures are exposed to progressively increasing concentrations of the appropriate drug, sublines can be selected in which the number of copies of the drug resistance marker and the foreign gene are greatly amplified. Of the wide variety of available drug resistance markers (see pages 16.9–16.14; for a full listing, see Kaufman 1987), the *dhfr* gene is the most extensively used for coamplification of foreign DNA sequences. After several months of growth in progressively increasing concentrations of methotrexate, cell lines can be obtained that carry up to 1000 copies of the *dhfr* gene (for review, see Schimke 1984, 1988; Stark and Wahl 1984). The DNA unit that is amplified under selective conditions varies from cell line to cell line but is always much larger than the *dhfr* gene itself and may include up to 1000 kb of flanking DNA sequences. The amplified genes can be either stable or unstable when the cells are subsequently grown under nonselective conditions. In stable lines, the amplified genes are integrated into the chromosome and are associated with expanded chromosomal regions termed homogeneously staining regions (HSRs). In contrast, in unstably amplified lines, the *dhfr* genes are present on extrachromosomal, autonomously replicating elements called double-minute chromosomes (DMs), which do not contain centromeres and are rapidly lost upon propagation in the absence of selection. For reasons that are not well understood, hamster cell lines such as CHO generally contain stably amplified, integrated *dhfr* genes, whereas mouse cell lines generally carry the amplified sequences in the form of unstable DMs.

CHO cells that are deficient in DHFR (DUKX-B11 cells; Urlaub and Chasin 1980) can be transformed with a cloned *dhfr* gene and amplified as described above. This approach has been used to establish lines of CHO cells that express very high levels of proteins encoded by a number of cloned cDNAs (for review, see Kaufman 1987). To extend the *dhfr* amplification system to different cell types, a mutant *dhfr* gene that encodes a protein with a reduced sensitivity to methotrexate can be used to generate methotrexate-resistant derivatives of cell lines that contain normal numbers of the endogenous wild-type *dhfr* gene (see, e.g., Simonsen and Levinson 1983). However, the extent of amplification of this altered gene is limited by the high concentration of methotrexate required to inhibit the mutant *dhfr* gene. Thus, in practice, expression of foreign proteins from amplified gene copies has relied heavily upon the DHFR-deficient line of CHO cells. In principle, however, this method can be extended to any cell line by cotransfecting the *dhfr* gene and the foreign DNA together with a dominant selectable marker, for example, *neo*[r] (Kim and Wold 1985). Cells that become resistant to neomycin are then exposed to progressively increasing concentrations of methotrexate as described above.

Foreign DNA sequences can be introduced into cells (1) as part of a simple *dhfr* vector such as pSV2-*dhfr* (see Figure 16.1B), (2) as part of a separate plasmid that is cotransfected with the *dhfr* vector, or (3) as part of a complex vector (e.g., pMT2) that can also be used to obtain transient expression in transfected COS cells. pMT2 (see Figure 16.3C) (Kaufman et al. 1989), which is described in detail on pages 16.20 and 16.22, contains the sequences coding for murine DHFR immediately downstream from the site at which foreign cDNAs are inserted. This arrangement results in the production of a dicistronic mRNA that carries the *dhfr* sequences towards its 3' terminus. Originally, the *dhfr* sequences were placed in this position because they appeared to enhance the stability of the hybrid mRNA (Kaufman et al. 1985, 1986b; R. Kaufman, unpubl.). In these experiments, the separate plasmid pAdD26SV(A)-3 (Kaufman and Sharp 1982a,b), which contains a monocistronic *dhfr* transcription unit lacking an enhancer element, was cotransfected with a derivative of pMT2 [p91023(B)] containing the cDNA of interest. Under these conditions, it was believed that efficient expression of DHFR was dependent on continued physical association of the *dhfr* sequences with the enhancer provided by the p91023(B). Resistance to high concentrations of methotrexate would therefore require coamplification of the two separate transcription units. More recently, it has been shown that the *dhfr* gene can be translated from the distal region of the dicistronic transcript expressed from p91023(B), obviating the need for the second *dhfr* vector (Kaufman et al. 1987).

Two general procedures have been used to obtain cell lines carrying amplified copies of the foreign DNA sequences. First, individual clones of DHFR$^+$ transformants can be screened for expression of the heterologous gene and then amplified separately by growth in increasing concentrations of methotrexate. Alternatively, DHFR$^+$ transformants can be pooled, screened for expression of the gene of interest, and then grown en masse in the presence of progressively higher concentrations of methotrexate. In either case, the cells are then cloned, and the level of expression of the foreign protein in individual cell lines is measured. Although both of these procedures yield cell lines that produce large quantities of the protein of interest, Kaufman et al. (1985) found the second procedure to be better, perhaps because the mass selection favors clones that readily amplify the newly acquired DNA rather than clones that merely receive large amounts of DNA during transfection.

Although it has not yet been used frequently, an alternative amplification system based on selection for adenosine deaminase (ADA; see page 16.15) shows great promise (see, e.g., Yeung et al. 1985; Kaufman et al. 1986a). In contrast to *dhfr,* the gene encoding ADA is a dominant selectable marker and its use is therefore not limited to particular types of mammalian cells. An ADA coamplification vector has been constructed in which a transcription unit encoding ADA (Orkin et al. 1985) expressed from the SV40 early promoter has been introduced into a pMT2-based expression plasmid (Bonthron et al. 1986). The resultant expression vector is pMT3SV2.

Introduction of Recombinant Vectors into Mammalian Cells

Many methods have been developed for introducing cloned eukaryotic DNAs into cultured mammalian cells, several of which are discussed below.

- *Calcium phosphate- or DEAE-dextran-mediated transfection.* The most widely used method is transfection mediated by either calcium phosphate or DEAE-dextran. Although the mechanism remains obscure, it is believed that the transfected DNA enters the cytoplasm of the cell by endocytosis and is transferred to the nucleus. Depending on the cell type, up to 20% of a population of cultured cells can be transfected at any one time. Because of its high efficiency, transfection mediated by calcium phosphate or DEAE-dextran is the method of choice for experiments that require transient expression of the foreign DNA in large numbers of cells. Calcium phosphate-mediated transfection is also used to establish cell lines that carry integrated copies of the foreign DNA, which are usually arranged in head-to-tail tandem arrays.

- *Polybrene.* The polycation Polybrene allows the efficient and stable introduction of low-molecular-weight DNAs (e.g., plasmid DNAs) into cell lines (e.g., CHO cells) that are relatively resistant to transfection by other methods (Kawai and Nishizawa 1984; Chaney et al. 1986).

- *Protoplast fusion* (Schaffner 1980; Rassoulzadegan et al. 1982). In this method, protoplasts derived from bacteria carrying high numbers of copies of a plasmid of interest are mixed directly with cultured mammalian cells. After fusion of the cell membranes (usually with polyethylene glycol), the contents of the bacteria are delivered into the cytoplasm of the mammalian cells and the plasmid DNA is transferred to the nucleus. Protoplast fusion is not as efficient as transfection for many of the cell lines that are commonly used for transient expression assays, but it is useful for cell lines in which endocytosis of DNA occurs inefficiently. Protoplast fusion frequently yields multiple copies of the plasmid DNA tandemly integrated into the host chromosome (Robert de Saint Vincent et al. 1981).

- *Electroporation.* The application of brief, high-voltage electric pulses to a variety of mammalian and plant cells leads to the formation of nanometer-sized pores in the plasma membrane (Neumann et al. 1982; Zimmermann 1982). DNA is taken directly into the cell cytoplasm either through these pores or as a consequence of the redistribution of membrane components that accompanies closure of the pores. Electroporation can be extremely efficient and can be used both for transient expression of cloned genes and for establishment of cell lines that carry integrated copies of the gene of interest. Electroporation, in contrast to calcium phosphate-mediated transfection and protoplast fusion, frequently gives rise to cell lines that carry one, or at most a few, integrated copies of the foreign DNA (Boggs et al. 1986).

- *Liposomes.* Artificial membrane vesicles (liposomes) are being intensively studied for their usefulness as delivery vehicles in vitro and in vivo. For a review of the current procedures for liposome preparation, targeting, and delivery of contents, see Mannino and Gould-Fogerite (1988). Most of these procedures involve encapsulation of DNA or RNA within liposomes, followed by fusion of the liposomes with the cell membrane. However, Felgner et al. (1987) have reported that DNA that is coated with a synthetic cationic lipid can be introduced into cells by fusion. Although this method is simple and appears to be efficient, it is comparatively new and untested (but see Felgner and Holm 1989; Maurer 1989).

- *Direct microinjection into nuclei* (Capecchi 1980). Although this method has the advantage of not exposing DNA to cellular compartments such as low-pH endosomes, it cannot be used to introduce DNA on a scale large enough for biochemical analysis. Microinjection is therefore used primarily as a method to establish lines of cells that carry integrated copies of the DNA of interest.

Irrespective of the method used to introduce DNA into cells, the efficiency of transient or stable transfection is determined largely by the cell type that is used. Different lines of cultured cells vary by several orders of magnitude in their ability to take up and express exogenously added DNA. Furthermore, a method that works well for one type of cultured cell may be useless for another. Many of the techniques described below have been optimized for the standard lines of cultured cells. When using more exotic lines of cells, it is important to compare the efficiencies of several different methods. Below, we present detailed protocols for the most commonly used transfection techniques (calcium phosphate coprecipitation and DEAE-dextran transfection) and we outline alternative methods (transfection using Polybrene [page 16.47], protoplast fusion [pages 16.48–16.53], and electroporation [16.54–16.55]) that can be used with cell lines that are resistant to transfection by more standard techniques.

TRANSFECTION OF COPRECIPITATES OF CALCIUM PHOSPHATE AND DNA

The uptake of DNA by cells in culture is markedly enhanced when the nucleic acid is presented as a calcium phosphate–DNA coprecipitate. Graham and van der Eb (1973), who developed the procedure for the introduction of adenovirus and SV40 DNA into adherent cells, described the concentrations of calcium (125 mM) and DNA (5–30 μg/ml) that were optimal for formation of calcium phosphate–DNA coprecipitates at neutral pH (7.05). In addition, they established the optimal times for the precipitation reaction (20–30 minutes) and for the subsequent exposure of cells to the precipitate (5–24 hours). Their work laid the foundation for the introduction of cloned DNA into many different kinds of mammalian cells and led directly to reliable methods for stable transformation of cells and for transient expression of cloned DNAs. Many minor modifications of the procedure have been described, mostly involving permutations of the order and manner of mixing of ingredients in the precipitation reaction. Increases in the efficiency of the procedure have been achieved by incorporating additional steps such as glycerol shock (Parker and Stark 1979) and/or chloroquine treatment (Luthman and Magnusson 1983) following the transfection protocol. Treatment with sodium butyrate has also been shown to enhance the expression in simian and human cells of plasmids that contain the SV40 enhancer (Gorman et al. 1983a,b).

We describe below variants of calcium phosphate-mediated transfection that are used to introduce DNA into (1) adherent cells, (2) adherent cells that have been released from the substratum with trypsin, and (3) nonadherent cells. The second procedure, although applicable for all types of adherent cells, appears to be of particular advantage for polarized epithelial cells, which do not efficiently take up material by endocytosis through the apical plasma membrane, which is normally exposed to the culture medium. The final protocol differs from the standard protocol in that the calcium phosphate–DNA coprecipitate is formed slowly in the tissue culture medium during prolonged incubation (15–24 hours) under controlled conditions of pH (6.96) and CO_2 tension (2–4%). This is a highly efficient method (Chen and Okayama 1987, 1988) to achieve stable transformation of mammalian cells with supercoiled plasmid DNAs.

Standard Protocol for Calcium Phosphate-mediated Transfection of Adherent Cells

1. Prepare the following solutions:

> *2 × HEPES-buffered saline (HBS)*
>
> 280 mM NaCl
> 10 mM KCl
> 1.5 mM $Na_2HPO_4 \cdot 2H_2O$
> 12 mM dextrose
> 50 mM HEPES
>
> Dissolve 1.6 g of NaCl, 0.074 g of KCl, 0.027 g of $Na_2HPO_4 \cdot 2H_2O$, 0.2 g of dextrose, and 1 g of HEPES in a total volume of 90 ml of distilled H_2O. Adjust the pH to 7.05 with 0.5 N NaOH, and then adjust the volume to 100 ml with distilled H_2O. Sterilize the solution by passage through a 0.22-micron filter. Store in 5-ml aliquots at −20°C.
>
> *2 M $CaCl_2$*
>
> Dissolve 10.8 g of $CaCl_2 \cdot 6H_2O$ in 20 ml of distilled H_2O. Sterilize the solution by passage through a 0.22-micron filter. Store in 1-ml aliquots at −20°C.
>
> *0.1 × TE (pH 8.0)*
>
> 1 mM Tris·Cl (pH 8.0)
> 0.1 mM EDTA (pH 8.0)
>
> Sterilize the solution by passage through a 0.22-micron filter. Store in aliquots at 4°C.
>
> *DNA*
>
> Dissolve the DNA (~ 20 $\mu g/10^6$ cells) in 0.1 × TE (pH 8.0) at a concentration of 40 $\mu g/ml$.
>
> To obtain the highest transformation efficiencies, plasmid DNAs should be purified by equilibrium centrifugation in CsCl–ethidium bromide density gradients. If smaller amounts of DNA are used, carrier DNA should be added to adjust the concentration to 40 $\mu g/ml$. Eukaryotic carrier DNA prepared in the laboratory (see Chapter 9) usually gives higher transfection efficiencies than commercially available DNA such as calf thymus or salmon sperm DNA. Carrier DNA should be sterilized before use by ethanol precipitation or extraction with chloroform.

2. Twenty-four hours before transfection, harvest exponentially growing cells by trypsinization and replate them at a density of 1×10^5–2×10^5 cells/cm^2 in 60-mm tissue culture dishes in the appropriate serum-containing medium. Incubate the cultures for 20–24 hours at 37°C in a humidified incubator in an atmosphere of 5–7% CO_2.

3. For each 60-mm monolayer of cells to be transfected, prepare the calcium phosphate–DNA coprecipitate as follows: Place 220 μl of the DNA prepared in step 1 (40 μg/ml in 0.1 $\times$ TE [pH 8.0]) and 250 μl of 2 $\times$ HBS in a disposable, sterile 5-ml plastic tube (e.g., Falcon 2509). Slowly add 31 μl of 2 M $CaCl_2$ (with gentle mixing over a period of ~30 seconds).

   Incubate the mixture for 20–30 minutes at room temperature, during which time a fine precipitate should form. At the end of the incubation, pipette the mixture up and down once to resuspend the precipitate.

   A commonly used alternative procedure is to add a dilute mixture of $CaCl_2$ and DNA to the 2 $\times$ HBS solution. For example, to 250 μl of 2 $\times$ HBS, add dropwise with gentle mixing 250 μl of the DNA prepared in step 1 and $CaCl_2$ (250 mM). Incubate the mixture as described above.

   Both types of reaction mixtures can be doubled or quadrupled in volume if a larger number of cells are to be transfected. When the volumes are quadrupled, a larger tube (e.g., a 15-ml Falcon 2059 tube) should be used. Normally, 0.5 ml of the calcium phosphate–DNA coprecipitate is added to 5 ml of medium in a 25-cm² flask or 60-mm dish; 1 ml of the precipitate is normally added to 10 ml of medium in a 90-mm dish.

   Published procedures differ widely in the manner and rate of mixing of ingredients. Some advise against anything but the gentlest agitation and suggest that air, bubbled from an electric pipetting device, should be used to mix the solution. Others advocate continuous slow mixing during addition of the DNA solution, followed by gentle vortexing. The object is to avoid the rapid formation of coarse precipitates that results in a decreased efficiency of transformation. In practice, several factors other than the speed of mixing affect the size of the precipitate, including the concentration and size of the DNA (high-molecular-weight DNA can be sheared by passage through a fine syringe needle) and the exact pH of the buffer (some workers make up several batches of HBS buffer over the pH range 6.95–7.1 and test each batch for the quality of precipitates and for the efficiency of transfection). If it is crucial to achieve the highest transfection efficiencies, time should be spent to optimize these factors for your particular system. Once an efficacious batch of reagents has been prepared and stored as indicated above, reproducible results can be obtained over long periods of time.

4. Transfer the calcium phosphate-DNA suspension into the medium above the cell monolayer. (Use 0.5 ml of suspension for 5 ml of medium in a 60-mm dish [see note 2 to step 3].) Rock the dish gently to mix the medium, which will become yellow-orange and turbid. An alternative method is to remove the medium and add the precipitate directly to the cells. Incubate the cells for 15 minutes at room temperature, and then add medium to the dish. Following either procedure, incubate the transfected cells for up to 24 hours (depending on the choice of subsequent treatments; see step 5) at 37°C in a humidified incubator in an atmosphere of 5–7% CO_2.

5. The transfected cells can then be treated in one of the following ways:

   a. If no additional treatments (with reagents such as chloroquine, glycerol, or sodium butyrate; see below) are to be employed, incubate the cells for 16–24 hours; then remove the medium and precipitate by aspiration. Wash the monolayer once with phosphate-buffered saline (PBS; see Appendix B), and add 5 ml of prewarmed complete growth medium. Return the cells to the incubator for 24–60 hours before assaying for transient expression of the transfected DNA or replating the cells in the appropriate selective medium for the isolation of stable transformants.

b. In many instances, uptake of DNA is increased by concurrent treatment of the cells with chloroquine, which may act by inhibiting the degradation of the DNA by lysosomal hydrolases (Luthman and Magnusson 1983). The concentration of chloroquine to be added to the growth medium and the time of treatment are limited by the sensitivity of the cells to the toxic effect of the drug. The optimal concentration of chloroquine for the particular cell type that is used should therefore be determined in preliminary experiments. However, most cell types respond well to treatment with chloroquine at a final concentration of 100 μM for 3–5 hours. A stock solution (100 mM) of chloroquine diphosphate is diluted (1:1000) directly into the medium either before or after the addition of the calcium phosphate–DNA coprecipitate to the cells (see alternative procedures in step 4). During treatment with chloroquine, it is normal for cells to develop a vesicularized appearance. After the 3–5-hour treatment with DNA and chloroquine, remove the medium and the precipitate, wash with PBS, and add 5 ml of prewarmed complete growth medium. Return the cells to the incubator for 24–60 hours before assaying for transient expression of the transfected DNA or replating the cells in appropriate selective medium for the isolation of stable transformants.

The 100 mM stock solution of chloroquine diphosphate (60 mg/ml in water) should be sterilized by filtration and stored in foil-wrapped tubes at −20°C).

c. Brief treatment of transfected cells with glycerol has also been shown to increase the efficiency of transformation or transient expression of the introduced DNA. This procedure may be used following treatment with chloroquine. Because cells vary widely in their sensitivity to the toxic effects of glycerol, each cell type must be tested in advance to determine the optimum time (30 seconds to 3 minutes) of treatment. Tolerant cells may be given a glycerol shock after 3–5 hours of exposure of the cells to the calcium phosphate–DNA coprecipitate in growth medium (+/− chloroquine).

   i. Remove the growth medium by aspiration and wash the monolayer once with PBS.
   ii. Add 1.5 ml of 15% glycerol in 1× HBS to the monolayer, and incubate the cells for 30 seconds to 3 minutes at 37°C.
   iii. Remove the glycerol by aspiration, and wash the monolayers once with PBS.
   iv. Add 5 ml of complete growth medium, and incubate the cells for 24–60 hours before assaying for transient expression of the transfected DNA or replating the cells in the appropriate selective medium for the isolation of stable transformants.

d. Sodium butyrate has also been shown to enhance the expression of recombinant plasmids carrying the SV40 early promoter/enhancer in simian and human cells (Gorman et al. 1983a). Cells are exposed to sodium butyrate following glycerol shock by adding the agent directly to the growth medium (step 5c, iv). Different concentrations of sodium butyrate (prepared as a 500 mM stock solution by neutralizing butyric acid in a chemical hood with NaOH) are used, depending on the cell type, for example:

CV-1	10 mM
NIH-3T3	7 mM
HeLa S3	5 mM
CHO	2 mM

Incubate the cells for 24–60 hours before assaying for transient expression or replating the cells in the appropriate selective medium for the isolation of stable transformants.

6. a. *Transient expression:* Harvest the cells 48–60 hours after transfection for analysis of RNA or DNA by hybridization. Newly synthesized protein may be analyzed by radioimmunoassay, by western blotting, by immunoprecipitation following in vivo metabolic labeling, or by assays of enzymatic activity in cell extracts. For assays that involve replicate samples or treatment of transfected cells under multiple conditions or over a time course, it is desirable to avoid dish-to-dish variation in transfection efficiency. In these cases, it is best to transfect large monolayers of cells (90-mm dishes) and then to trypsinize the cells after 24 hours of incubation and distribute them among several smaller dishes.

   b. *Stable transformation:* Following 18–24 hours of incubation in non-selective medium to allow expression of the transferred gene(s) to occur, the cells are trypsinized and replated in the appropriate selective medium. This medium should be changed every 2–4 days for 2–3 weeks to remove the debris of dead cells and to allow colonies of resistant cells to grow.

   Individual colonies may be cloned and propagated for assay (for methods, see Jakoby and Pastan 1979). A permanent record of the numbers of colonies may be obtained by fixing the remaining cells with ice-cold methanol for 15 minutes and then staining them with 10% Giemsa for 15 minutes at room temperature before rinsing in tap water. The Giemsa stain should be freshly prepared in PBS or water and filtered through Whatman No. 1 filter paper.

The dilution at which the cells should be replated to yield well-separated colonies will be determined by the efficiency of stable transformation, which can vary over several orders of magnitude (see, e.g., Spandidos and Wilkie 1984). The efficiency is dependent on (1) the recipient cell type (significant differences have been observed even between different clones or different passage numbers of the same cell line [Corsaro and Pearson 1981; Van Pel et al. 1985]), (2) the nature of the introduced gene and the efficacy of the transcriptional control signals associated with it, and (3) the amount of donor DNA used in the transfection.

Calcium Phosphate-mediated Transfection of Adherent Cells in Suspension

1. Prepare the solutions and form the calcium phosphate–DNA coprecipitate as described on pages 16.33–16.34, steps 1 and 3.

2. While the precipitate is forming, harvest exponentially growing adherent cells by trypsinization. Resuspend the cells in medium containing serum, and centrifuge aliquots that contain 10^6 cells at 800g for 5 minutes at 4°C. Discard the supernatant.

3. Resuspend each aliquot of 10^6 cells in 0.5 ml of the calcium phosphate–DNA suspension, and incubate for 15 minutes at room temperature.

4. To each aliquot, add 4.5 ml of prewarmed complete medium (with or without chloroquine; see page 16.35), and plate the entire suspension (~5 ml) in a single 90-mm tissue culture dish. Incubate the cells for up to 24 hours (depending on the choice of subsequent treatments; see step 5, pages 16.34–16.35) at 37°C in a humidified incubator in an atmosphere of 5–7% CO_2.

Note

This technique can easily be modified to accommodate greater numbers of cells. For example, Chu and Sharp (1981) used 10^8 cells in 2 ml of calcium phosphate–DNA suspension containing 25 μg of DNA. After 15 minutes, the mixture was diluted with 40 ml of complete medium supplemented with 0.05 × HBS and 6.25 mM $CaCl_2$, and the cells were plated at 5×10^7 cells per 150-mm dish.

Calcium Phosphate-mediated Transfection of Cells Growing in Suspension

1. Prepare the solutions and form the calcium phosphate–DNA coprecipitate as described on pages 16.33–16.34, steps 1 and 3.

2. While the precipitate is forming, collect the exponentially growing cells by centrifugation at 800g for 5 minutes at 4°C. Discard the supernatant, and resuspend the cell pellet in 20 volumes of ice-cold phosphate-buffered saline (PBS; see Appendix B). Divide the suspension into aliquots that contain 1×10^7 cells. Recover the washed cells by centrifugation, and again discard the supernatant.

3. Gently resuspend 1×10^7 cells in 1 ml of calcium phosphate–DNA suspension (containing ~20 μg of DNA), and let the suspension stand for 20 minutes at room temperature.

4. Add 10 ml of serum-containing medium (with or without chloroquine; see page 16.35), and plate the entire suspension in a single 90-mm tissue culture dish. Incubate the cells for 6–24 hours at 37°C in a humidified incubator in an atmosphere of 5–7% CO_2.

5. Recover the cells by centrifugation at 800g for 5 minutes at room temperature, and wash them once with PBS. Resuspend the cells in 10 ml of prewarmed complete growth medium. Return the cells to the incubator for 48 hours before assaying for transient expression of transfected genes (step 6a, page 16.36) or replating the cells in selective medium for isolation of stable transformants (step 6b, page 16.36).

Note

If the cells have been proved to survive treatment with glycerol, a glycerol shock (detailed below) may be used 4–6 hours after exposure of the cells to the calcium phosphate–DNA coprecipitate begins in order to improve the frequency of transfection:

1. Collect the cells by centrifugation as described in step 5, and wash them once with PBS.

2. Resuspend the washed cells in 1 ml of 15% glycerol in 1× HBS, and incubate for 30 seconds to 3 minutes at 37°C.

3. Dilute the suspension with 10 ml of PBS, and recover the cells by centrifugation as described in step 5. Wash the cells in PBS.

4. Resuspend the cells in 10 ml of serum-containing medium, and plate them in a 90-mm tissue culture dish. Incubate the culture for 48 hours at 37°C in a humidified incubator in an atmosphere of 5–7% CO_2 before assaying for transient expression of transfected genes or replating the cells in selective medium.

Modified Calcium Phosphate-mediated Transfection Procedure

This is a highly efficient method to obtain stable transformation of mammalian cells with supercoiled plasmid DNAs. The calcium phosphate–DNA coprecipitate is allowed to form in the tissue culture medium during prolonged incubation (15–24 hours) under controlled conditions of pH (6.96) and CO_2 tension (2–4%) (Chen and Okayama 1987, 1988). It should be noted that linear DNAs yield very low transformation frequencies using this protocol, perhaps because the slow formation of the calcium phosphate–DNA coprecipitate delays protection of the DNA from nucleases. The best results have been obtained using supercoiled plasmid DNAs purified by two rounds of equilibrium centrifugation in CsCl–ethidium bromide density gradients.

1. Prepare the following solutions:

> ### 2 × BES-buffered saline
>
> 50 mM BES (N,N-bis[2-hydroxyethyl]-2-aminoethanesulfonic acid)
> 280 mM NaCl
> 1.5 mM $Na_2HPO_4 \cdot 2H_2O$
>
> Dissolve 1.07 g of BES, 1.6 g of NaCl, and 0.027 g of Na_2HPO_4 in a total volume of 90 ml of distilled H_2O. Adjust the pH of the solution to 6.96 with HCl at room temperature, and then adjust the volume to 100 ml with distilled H_2O. Sterilize the solution by passage through a 0.22-micron filter, and store in aliquots at −20°C.
>
> ### 2.5 M $CaCl_2$
>
> Dissolve 13.5 g of $CaCl_2 \cdot 6H_2O$ in 20 ml of distilled H_2O. Sterilize the solution by passage through a 0.22-micron filter. Store in 1-ml aliquots at −20°C.

2. Twenty-four hours before transfection, harvest exponentially growing cells by trypsinization and replate aliquots of 5×10^5 cells onto 90-mm tissue culture dishes. Add 10 ml of complete growth medium, and incubate the cultures overnight at 37°C in a humidified incubator in an atmosphere of 5–7% CO_2.

3. Mix 20–30 μg of superhelical plasmid DNA with 0.5 ml of 0.25 M $CaCl_2$, add 0.5 ml of 2 × BES-buffered saline, and incubate the mixture for 10–20 minutes at room temperature. Do not expect a visible precipitate to form during this time.

4. Add the $CaCl_2$/DNA/BES-buffered saline solution dropwise to the dishes of cells, swirling gently to mix well. Incubate the cultures for 15–24 hours at 37°C in a humidified incubator in an atmosphere of 2–4% CO_2. The calcium phosphate–DNA complex forms slowly in the medium under conditions of low pH and precipitates gradually onto the cells during the incubation in an atmosphere containing low concentrations of CO_2. The nature of the precipitate is affected by the amount of DNA used. A

transition from a coarse precipitate to a fine precipitate occurs at the optimal DNA concentration (2–3 μg/ml in the growth medium).

5. Remove the medium by aspiration, and rinse the cells twice with medium. Add 10 ml of fresh medium, and incubate the cultures for 24 hours at 37°C in a humidified incubator in an atmosphere of 5% CO_2.

6. Following 18–24 hours of incubation in nonselective medium to allow expression of the transferred genè(s) to occur, the cells are trypsinized and replated in the appropriate selective medium. This medium should be changed every 2–4 days for 2–3 weeks to remove the debris of dead cells and to allow colonies of resistant cells to grow.

The dilution at which the cells should be replated to yield well-separated colonies will be determined by the efficiency of stable transformation, which can vary over several orders of magnitude (see, e.g., Spandidos and Wilkie 1984). The efficiency is dependent on (1) the recipient cell type (significant differences have been observed even between different clones or different passage numbers of the same cell line [Corsaro and Pearson 1981; Van Pel et al. 1985]), (2) the nature of the introduced gene and the efficacy of the transcriptional control signals associated with it, and (3) the amount of donor DNA used in the transfection.

7. Individual colonies may be cloned and propagated for assay (for methods, see Jakoby and Pastan 1979). A permanent record of the numbers of colonies may be obtained by fixing the remaining cells with ice-cold methanol for 15 minutes and then staining them with 10% Giemsa for 15 minutes at room temperature before rinsing in tap water. The Giemsa stain should be freshly prepared in PBS or water and filtered through Whatman No. 1 filter paper.

TRANSFECTION MEDIATED BY DEAE-DEXTRAN

DEAE-dextran was originally used as a facilitator to introduce poliovirus RNA (Vaheri and Pagano 1965) and SV40 and polyomavirus DNAs (McCutchan and Pagano 1968; Warden and Thorne 1968) into cells. The procedure, with slight modifications (see below), continues to be widely used for transfection of viral genomes and plasmids carrying viral sequences. Although the mechanism of action of DEAE-dextran is not known, it is thought that the polymer might bind to DNA and inhibit the action of nucleases and/or bind to cells and promote endocytosis of the DNA.

Transfection mediated by DEAE-dextran differs from calcium phosphate coprecipitation in three important respects. First, it is generally used only for transient expression of cloned genes and not for stable transformation of cells. Second, it works very efficiently with lines of cells such as BSC-1, CV-1, and COS but is unsatisfactory with many other types of cells, perhaps because the polymer is toxic. Third, smaller amounts of DNA are used for transfection with DEAE-dextran than with calcium phosphate coprecipitation. Maximal transfection efficiency of 10^5 simian cells is achieved with 100–200 ng of supercoiled plasmid DNA; larger amounts of DNA (>2–3 μg) can be inhibitory. In contrast to transfection mediated by calcium phosphate, where high concentrations of DNA are required to promote the formation of a precipitate, carrier DNA is never used with the DEAE-dextran transfection method.

Since the method was introduced over 20 years ago, many variants of DEAE-dextran transfection have been described. In most cases, the differences are minor and reflect more the preferences of individual laboratories than fundamental improvements to the technique. However, there are two important variables that greatly affect the efficiency of the method: the concentration of DEAE-dextran that is used and the length of time that the cells are exposed to the DNA/DEAE-dextran mixture. It is possible to use either a relatively high concentration of DEAE-dextran (1 mg/ml) for short periods (30 minutes to 1.5 hours) or a lower concentration (250 μg/ml) for longer periods of time (up to 8 hours). The first of these transfection procedures is the more efficient, but it involves monitoring the cells for early signs of distress when they are exposed to the facilitator. This requires practice. The second technique is less stringent and is therefore more reliable.

Transfection Using DEAE-Dextran: Protocol I

1. Prepare the following solutions:

50 mg/ml DEAE-dextran

Dissolve 100 mg of DEAE-dextran ($M_r = 500,000$; Pharmacia) in 2 ml of distilled H_2O. Sterilize the solution by autoclaving for 20 minutes at 15 lb/sq. in. on liquid cycle. (*Note:* Autoclaving also assists dissolution of the polymer.)

The molecular weight of the DEAE-dextran originally used for transfection was $> 2 \times 10^6$ (McCutchan and Pagano 1968). Although this material is no longer available commercially, it is still occasionally found in chemical storerooms. The older batches of higher-molecular-weight DEAE-dextran are more efficient facilitators of transfection than those currently available.

Tris-buffered saline–dextrose (TBS-D)

See Appendix B for preparation of TBS. Immediately before use, add 1 ml of a 20% (w/v) solution of dextrose in water (autoclaved) to each aliquot of TBS. The final concentration of dextrose should be 0.1%.

Phosphate-buffered saline (PBS)

See Appendix B for preparation of PBS.

2. Twenty-four hours before transfection, harvest exponentially growing cells by trypsinization and replate 60-mm tissue culture dishes with 10^5 cells/dish (or 35-mm dishes with 5×10^4 cells/dish). Add 5 ml (or 3 ml for 35-mm dish) of complete growth medium, and incubate the cultures for 20–24 hours at 37°C in a humidified incubator in an atmosphere of 5–7% CO_2.

The cells should be approximately 75% confluent at the time of transfection. If the cells are grown for less than 12 hours prior to transfection, they will be less well anchored to the substratum and more likely to detach during exposure to DEAE-dextran.

3. Prepare the DNA/DEAE-dextran/TBS-D solution (250 μl per 60-mm dish; 150 μl per 35-mm dish) by mixing 0.1–4 μg/ml supercoiled or circular DNA and 1.0 mg/ml DEAE-dextran in TBS-D.

The amount of DNA required to achieve maximal levels of transient expression depends on the exact nature of the construct and should be determined in preliminary experiments. If the construct carries a replicon that will function in the transfected cells, 100–200 ng of DNA per 10^5 cells should be sufficient; if no replicon is present, larger amounts of DNA may be required (up to 2 μg per 10^5 cells).

4. Remove the medium by aspiration, and wash the monolayers twice with prewarmed (37°C) PBS and once with prewarmed TBS-D.

5. Add the DNA/DEAE-dextran/TBS-D solution (250 μl per 60-mm dish;

150 μl per 35-mm dish). Rock gently to spread the solution evenly across the monolayer. Return the cultures to the incubator for 30–90 minutes (the time will depend on the sensitivity of each batch of cells to the DNA/DEAE-dextran/TBS-D solution). At 15–20-minute intervals, remove the dishes from the incubator, swirl them gently, and check the appearance of the cells under the microscope. If the cells are still firmly attached to the substratum, continue the incubation. The incubation should be halted when the cells begin to shrink and round up.

6. Remove the DNA/DEAE-dextran/TBS-D solution by aspiration. Gently wash the monolayers once with prewarmed (37°C) TBS-D and then once with prewarmed PBS, taking care not to dislodge the transfected cells.

7. Add 5 ml (per 60-mm dish) or 3 ml (per 35-mm dish) of prewarmed (37°C) medium supplemented with serum and chloroquine diphosphate (100 μM final concentration), and incubate the cultures for 3–5 hours at 37°C in a humidified incubator in an atmosphere of 5–7% CO_2.

The efficiency of transfection is increased severalfold by treatment with chloroquine, which may act by inhibiting the degradation of the DNA by lysosomal hydrolases (Luthman and Magnusson 1983). Chloroquine diphosphate is stored as a sterile stock solution (100 mM; 60 mg/ml in water) in foil-wrapped tubes at −20°C. Note, however, that the cytotoxic effects of a combination of DEAE-dextran and chloroquine can be severe. It is therefore important to carry out preliminary experiments to determine the maximum permissible length of exposure to chloroquine after treatment of cells with DEAE-dextran.

8. Remove the medium by aspiration, and wash the monolayers three times with prewarmed (37°C) serum-free medium. Add to the cells 5 ml (per 60-mm dish) or 3 ml (per 35-mm dish) of medium supplemented with serum, and incubate the cultures for 36–60 hours at 37°C in a humidified incubator in an atmosphere of 5–7% CO_2 before assaying for expression of the transfected DNA. Depending on the experiment, continue with step 9a (transient expression) or 9b (stable transformation).

9. a. *Transient expression:* Harvest the cells 48–60 hours after transfection for analysis of RNA or DNA by hybridization. Newly synthesized protein may be analyzed by radioimmunoassay, by western blotting, by immunoprecipitation following in vivo metabolic labeling, or by assays of enzymatic activity in cell extracts. For assays that involve replicate samples or treatment of transfected cells under multiple conditions or over a time course, it is desirable to avoid dish-to-dish variation in transfection efficiency. In these cases, it is best to transfect large monolayers of cells (90-mm dishes) and then to trypsinize the cells after 24 hours of incubation and distribute them among several smaller dishes.

   b. *Stable transformation:* Following 18–24 hours of incubation in non-selective medium to allow expression of the transferred gene(s) to occur, the cells are trypsinized and replated in the appropriate selective medium. This medium should be changed every 2–4 days for 2–3 weeks to remove the debris of dead cells and to allow colonies of resistant cells to grow.

Individual colonies may be cloned and propagated for assay (for methods, see Jakoby and Pastan 1979). A permanent record of the numbers of colonies may be obtained by fixing the remaining cells with ice-cold methanol for 15 minutes and then staining them with 10% Giemsa for 15 minutes at room temperature before rinsing in tap water. The Giemsa stain should be freshly prepared in PBS or water and filtered through Whatman No. 1 filter paper.

The dilution at which the cells should be replated to yield well-separated colonies will be determined by the efficiency of stable transformation, which can vary over several orders of magnitude (see, e.g., Spandidos and Wilkie 1984). The efficiency is dependent on (1) the recipient cell type (significant differences have been observed even between different clones or different passage numbers of the same cell line [Corsaro and Pearson 1981; Van Pel et al. 1985]), (2) the nature of the introduced gene and the efficacy of the transcriptional control signals associated with it, and (3) the amount of donor DNA used in the transfection.

Transfection Using DEAE-Dextran: Protocol II

1. Prepare the following solutions:

   ### 50 mg/ml DEAE-dextran

   Dissolve 100 mg of DEAE-dextran ($M_r = 500,000$; Pharmacia) in 2 ml of distilled H_2O. Sterilize the solution by autoclaving for 20 minutes at 15 lb/sq. in. on liquid cycle. (*Note:* Autoclaving also assists dissolution of the polymer.)

   The molecular weight of the DEAE-dextran originally used for transfection was $> 2 \times 10^6$ (McCutchan and Pagano 1965). Although this material is no longer available commercially, it is still occasionally found in chemical storerooms. The older batches of higher-molecular-weight DEAE-dextran are more efficient facilitators of transfection than those currently available.

   ### HEPES-buffered Dulbecco's modified Eagle's medium (HEPES-buffered DMEM)

   DMEM lacking $NaHCO_3$ but containing 10 mM HEPES (pH 7.15).

   No serum should be added to this reagent.

2. Twenty-four hours before transfection, harvest exponentially growing cells by trypsinization and replate 60-mm tissue culture dishes with 10^5 cells/dish (or 35-mm dishes with 5×10^4 cells/dish). Add 5 ml (or 3 ml for 35-mm dish) of complete growth medium, and incubate the cultures for 20–24 hours at 37°C in a humidified incubator in an atmosphere of 5–7% CO_2.

   The cells should be approximately 75% confluent at the time of transfection. If the cells are grown for less than 12 hours prior to transfection, they will be less well anchored to the substratum and more likely to detach during exposure to DEAE-dextran.

3. Prepare the DNA/DEAE-dextran solution (500 µl per 60-mm dish; 250 µl per 35-mm dish) by mixing 0.1–4 µg/ml supercoiled or circular DNA and 250 µg/ml DEAE-dextran in HEPES-buffered DMEM.

   The amount of DNA required to achieve maximal levels of transient expression depends on the exact nature of the construct and should be determined in preliminary experiments. If the construct carries a replicon that will function in the transfected cells, 100–200 ng of DNA per 10^5 cells should be sufficient; if no replicon is present, larger amounts of DNA may be required (up to 2 µg per 10^5 cells).

4. Remove the medium from the cells by aspiration, and wash the monolayers twice with prewarmed (37°C) HEPES-buffered DMEM. This step is important for maximal efficiency of the procedure.

5. Add the DNA/DEAE-dextran/DMEM solution to the cells (0.5 ml per 60-mm dish; 0.25 ml per 35-mm dish), and return the cells to the incubator for up to 8 hours. Gently rock the dishes every 2 hours to ensure even exposure to the DNA/DEAE-dextran/DMEM solution.

The efficiency of transfection is increased severalfold by concurrent treatment of the cells with chloroquine diphosphate, which may act by inhibiting the degradation of the DNA by lysosomal hydrolases (Luthman and Magnusson 1983). The drug (final concentration 100 μM) is added to the DNA/DEAE-dextran solution just before it is applied to the cells. Because chloroquine is toxic, the time of incubation must then be limited to 3–5 hours. Chloroquine diphosphate is stored as a sterile stock solution (100 mM; 60 mg/ml in water) in foil-wrapped tubes at $-20°C$.

6. Remove the DNA/DEAE-dextran/DMEM solution from the cells by aspiration, and gently wash the cell monolayers twice with prewarmed (37°C) HEPES-buffered DMEM. Take care not to dislodge the transfected cells. Wash the cells once with prewarmed DMEM (buffered with $NaHCO_3$ and not HEPES) supplemented with serum. Add to the cells 5 ml (per 60-mm dish) or 3 ml (per 35-mm dish) of complete growth medium, and incubate the cultures for 36–60 hours at 37°C in a humidified incubator in an atmosphere of 5–7% CO_2 before assaying for expression of the transfected DNA. Depending on the experiment, continue with step 7a (transient expression) or step 7b (stable transformation).

7. a. *Transient expression:* Harvest the cells 48–60 hours after transfection for analysis of RNA or DNA by hybridization. Newly synthesized protein may be analyzed by radioimmunoassay, by western blotting, by immunoprecipitation following in vivo metabolic labeling, or by assays of enzymatic activity in cell extracts. For assays that involve replicate samples or treatment of transfected cells under multiple conditions or over a time course, it is desirable to avoid dish-to-dish variation in transfection efficiency. In these cases, it is best to transfect large monolayers of cells (90-mm dishes) and then to trypsinize the cells after 24 hours of incubation and distribute them among several smaller dishes.

   b. *Stable transformation:* Following 18–24 hours of incubation in nonselective medium to allow expression of the transferred gene(s) to occur, the cells are trypsinized and replated in the appropriate selective medium. This medium should be changed every 2–4 days for 2–3 weeks to remove the debris of dead cells and to allow colonies of resistant cells to grow.

   Individual colonies may be cloned and propagated for assay (for methods, see Jakoby and Pastan 1979). A permanent record of the numbers of colonies may be obtained by fixing the remaining cells with ice-cold methanol for 15 minutes and then staining them with 10% Giemsa for 15 minutes at room temperature before rinsing in tap water. The Giemsa stain should be freshly prepared in PBS or water and filtered through Whatman No. 1 filter paper.

   The dilution at which the cells should be replated to yield well-separated colonies will be determined by the efficiency of stable transformation, which can vary over several orders of magnitude (see, e.g., Spandidos and Wilkie 1984). The efficiency is dependent on (1) the recipient cell type (significant differences have been observed even between different clones or different passage numbers of the same cell line [Corsaro and Pearson 1981; Van Pel et al. 1985]), (2) the nature of the introduced gene and the efficacy of the transcriptional control signals associated with it, and (3) the amount of donor DNA used in the transfection.

DNA TRANSFECTION USING POLYBRENE

The polycation Polybrene has been used as a facilitator of DNA transfection into cells that have proved to be relatively resistant to transfection using calcium phosphate coprecipitation (Kawai and Nishizawa 1984; Chaney et al. 1986). The method outlined below works efficiently for stable transformation of CHO cells by plasmid DNA, yielding approximately 15-fold more transformants than calcium phosphate–DNA coprecipitation. However, there is no difference between the two methods in the efficiency of transformation of high-molecular-weight DNA. It is not known whether Polybrene-mediated transfection can be used for transient expression of cloned DNA or whether it can be adapted for stable transformation with cell lines other than CHO.

1. Harvest exponentially growing CHO cells by trypsinization, and replate them at a density of 5×10^5 cells per 90-mm tissue culture dish in 10 ml of alpha medium (GIBCO) containing 10% fetal calf serum. Incubate the cultures for 18–20 hours at 37°C in a humidified incubator in an atmosphere of 5–7% CO_2.

2. Replace the medium with 3 ml of serum-containing medium containing DNA (5 ng to 40 μg; no carrier DNA) and 30 μg of Polybrene (Aldrich). Mix the DNA with the medium before adding the Polybrene solution (10 mg/ml in water and sterilized by passage through a 0.22-micron filter). Return the cells to the incubator for 6 hours, gently rocking the dishes every 90 minutes to ensure even exposure of the cells to the DNA–Polybrene mixture.

3. Remove the medium containing the DNA and Polybrene by aspiration, and treat the cells with dimethyl sulfoxide (DMSO) for 4 minutes by adding 5 ml of a 30% solution of DMSO in serum-containing alpha medium. Remove the DMSO solution by aspiration, wash the cells once with prewarmed (37°C), serum-free medium, and add 10 ml of complete medium containing 10% fetal calf serum. Incubate the cultures for 48 hours at 37°C in a humidified incubator in an atmosphere of 5–7% CO_2.

4. Recover the cells by trypsinization, and replate them in appropriate selective medium. Change the medium every 2–4 days for 2–3 weeks to allow death of sensitive cells and the emergence of resistant colonies.

   The dilution at which the cells should be replated to yield well-separated colonies is determined by the efficiency of stable transformation, which can vary over several orders of magnitude (see, e.g., Spandidos and Wilkie 1984). The efficiency is dependent on (1) the recipient cell type (significant differences have been observed even between different clones or different passage numbers of the same cell line [Corsaro and Pearson 1981; Van Pel et al. 1985]), (2) the nature of the introduced gene and the efficacy of the transcriptional control signals associated with it, and (3) the amount of donor DNA used in the transfection. Typically, 1 μg of a standard plasmid such as pSV2-*neo* will transform approximately 3% of the cells to neomycin resistance.

5. Individual colonies may be cloned and propagated for assay (for methods, see Jakoby and Pastan 1979). A permanent record of the numbers of colonies may be obtained by fixing the remaining cells as described in step 7b, page 16.46.

DNA TRANSFECTION BY PROTOPLAST FUSION

Cloned DNA can be introduced into mammalian cells by fusing protoplasts, prepared from bacteria carrying the plasmid DNA of interest, with cultured cells (Schaffner 1980; Rassoulzadegan et al. 1982). The bacteria are grown in the presence of chloramphenicol to amplify the plasmid DNA and then treated with lysozyme to remove the cell wall. The resulting protoplasts are centrifuged onto a monolayer of mammalian cells, and the resulting mixture is treated with polyethylene glycol (PEG) to promote fusion. During this process, bacterial and plasmid DNAs are transferred into the mammalian cell. PEG is then removed, and the cells are incubated in fresh tissue culture medium containing kanamycin to inhibit the growth of any surviving bacteria.

Protoplast fusion has been used both for transient expression of cloned genes and for establishment of stable lines of mammalian cells. An estimate of the efficiency of these processes can be gained from the following data. When bacteria and mammalian cells were mixed in a ratio of 10,000:1, approximately 6% of the mammalian cells transiently expressed a gene carried on the amplified plasmid (Schaffner 1980). In subsequent experiments, in which a different bacterial host cell was employed, 100% of the mammalian cells were transiently transfected (Rassoulzadegan et al. 1982). The efficiency of stable transformation achieved in this experiment was 0.02% (40 transformants per 200,000 mammalian cells, with an input ratio of 10,000 protoplasts per mammalian cell).

Protoplast fusion has been used to stably introduce immunoglobulin genes into B cells (Gillies et al. 1983) and globin genes into mouse erythroleukemia cells (Charnay et al. 1984). The advantage of this method is its high efficiency. However, the manipulations are time-consuming and cotransformation is usually not possible. Thus, the gene of interest must always be carried on a plasmid containing the desired selectable marker.

Preparation of Protoplasts

1. Inoculate 100 ml of LB medium containing the appropriate antibiotic with a fresh overnight culture of bacteria containing the plasmid of interest. When the bacteria have grown to mid-log phase ($\sim 2 \times 10^8$ bacteria/ml), chloramphenicol should be added from a stock solution (34 mg/ml in absolute ethanol and sterilized by passage through a 0.22-micron filter) to a final concentration of 250 μg/ml. Continue incubation overnight.

   See Chapter 1, pages 1.21 and 1.33, for a discussion of the use of chloramphenicol in the amplification of plasmids.

2. Recover the bacterial cells from 50 ml of the culture by centrifugation at 3000g for 10 minutes at 4°C, and resuspend the pellet in 2.5 ml of an ice-cold solution of 20% (w/v) sucrose in 50 mM Tris·Cl (pH 8.0).

3. Add 0.5 ml of a fresh solution of lysozyme (5 mg/ml in 250 mM Tris·Cl [pH 8.0]). Incubate the suspension for 5 minutes at 4°C.

   Lysozyme will not work efficiently if the pH of the solution is less than 8.0.

4. Add 1 ml of ice-cold 0.25 M EDTA (pH 8.0). Incubate on ice for 5 minutes.

5. Slowly add 1 ml of ice-cold 50 mM Tris·Cl (pH 8.0), and incubate the suspension for 15 minutes at 37°C. The bacterial cell wall is digested during this incubation. The resulting protoplasts are fragile and must be treated gently in all subsequent steps.

   The efficiency of protoplast formation varies from one strain of bacteria to another and is affected by the physiological state of the cells. It is therefore essential to monitor the appearance of protoplasts under a microscope. Prior to protoplast formation, the bacteria should appear as rod-shaped particles. When the peptidoglycan layer surrounding the inner membrane has been digested by lysozyme, the cells become spherical. Digestion should be stopped when approximately 80% of the bacteria have been converted to protoplasts.

6. Slowly add 20 ml of 10% (w/v) sucrose, 10 mM MgCl$_2$ made up in Dulbecco's modified Eagle's medium (DMEM) and equilibrated at 37°C. Gently swirl the tube to mix the solution. Incubate the suspension for 15 minutes at room temperature. Check the protoplasts using the following criteria:

   a. The color should be red-orange.

   b. Upon swirling, there should be turbidity but the viscosity should be low.

   c. Debris, if present at all, should be minimal.

   The final concentration of protoplasts is approximately 5×10^8/ml.

   This step is included to eliminate the viscosity that results from the liberation of high-molecular-weight DNA from the small percentage of protoplasts that lyse during preparation.

Fusion of Protoplasts to Adherent Mammalian Cells

1. Twenty-four hours before transfection, harvest exponentially growing cells by trypsinization and seed six-well tissue culture plates with 7.5×10^5 cells/well. Incubate the cultures overnight at 37°C in a humidified incubator in an atmosphere of 5–7% CO_2.

2. Remove the medium by aspiration, rinse the wells once with DMEM, and add 1.5 ml of the suspension of bacterial protoplasts prepared in step 6, page 16.49. This corresponds to approximately 10,000 protoplasts for every mammalian cell. Deposit the protoplasts onto the cells by centrifugation at 500g for 10 minutes at room temperature in a IEC-8R centrifuge (rotor no. 216) (or its equivalent).

   The protoplast fusion can also be carried out in 60-mm dishes, but plates are more convenient to centrifuge. Use approximately 10,000 protoplasts for every mammalian cell. When centrifuging bacterial protoplasts onto the adherent layer of mammalian cells in 60-mm dishes, it is best to stack the dishes in an empty tissue culture dish and tape the stack together. The stack is then placed on the flat bottom of a swinging-bucket rotor for centrifugation. A few sheets of tissue (Kimwipes or Kleenex) should be placed at the bottom of the stack to prevent the bottom dish from cracking.

3. Immediately after centrifugation, remove the supernatant from the wells by aspiration without tilting the plate. Add 2 ml of 50% polyethylene glycol 1000 (PEG 1000) in DMEM minus fetal calf serum (equilibrated to 37°C) above the same spot used for aspiration. Incubate the plates for 1–2 minutes at room temperature.

   In a microwave oven, melt sufficient PEG 1000 to make a 50% solution by volume in DMEM (prewarmed to 37°C). Sterilize the solution by passage through a 0.22-micron filter.

   Different cell lines may require different times of incubation or different concentrations of PEG to achieve the optimal balance between toxicity and efficient fusion. For example, 42.5% PEG is used to fuse bacterial protoplasts to WI-38 cells because higher concentrations of PEG are toxic. When using protoplast fusion for the first time with a particular line of mammalian cells, preliminary experiments should be carried out to measure the amount of transient expression obtained when fusion performed under sets of different conditions.

4. Add 5 ml of prewarmed (37°C) medium to each well at the same spot used for aspiration. Remove the medium from the plate by aspiration at the same spot. Do not agitate. Wash the monolayer of cells four times with prewarmed (37°C) DMEM without fetal calf serum.

   The scummy layer of protoplasts that adheres to the monolayer of cells should not be rinsed away.

5. Add the appropriate amount of prewarmed (37°C) tissue culture medium containing serum, penicillin (100 units/ml), streptomycin (100 μg/ml), and kanamycin (100 μg/ml). Incubate the dishes for 36–60 hours at 37°C in a humidified incubator in an atmosphere of 5–7% CO_2. Depending on the experiment continue with step 6a (transient expression) or 6b (stable transformation).

6. a. *Transient expression:* Harvest the cells 48–60 hours after transfection for analysis of RNA or DNA by hybridization. Newly synthesized protein may be analyzed by radioimmunoassay, by western blotting, by immunoprecipitation following in vivo metabolic labeling, or by assays of enzymatic activity in cell extracts. For assays that involve replicate samples or treatment of transfected cells under multiple conditions or over a time course, it is desirable to avoid dish-to-dish variation in transfection efficiency. In these cases, it is best to transfect large monolayers of cells (90-mm dishes) and then to trypsinize the cells after 24 hours of incubation and distribute them among several smaller dishes.

b. *Stable transformation:* Following 18–24 hours of incubation in non-selective medium to allow expression of the transferred gene(s) to occur, the cells are trypsinized and replated in the appropriate selective medium. This medium should be changed every 2–4 days for 2–3 weeks to remove the debris of dead cells and to allow colonies of resistant cells to grow.

Individual colonies may be cloned and propagated for assay (for methods, see Jakoby and Pastan 1979). A permanent record of the numbers of colonies may be obtained by fixing the remaining cells with ice-cold methanol for 15 minutes and then staining them with 10% Giemsa for 15 minutes at room temperature before rinsing in tap water. The Giemsa stain should be freshly prepared in PBS or water and filtered through Whatman No. 1 filter paper.

The dilution at which the cells should be replated to yield well-separated colonies will be determined by the efficiency of stable transformation, which can vary over several orders of magnitude (see, e.g., Spandidos and Wilkie 1984). The efficiency is dependent on (1) the recipient cell type (significant differences have been observed even between different clones or different passage numbers of the same cell line [Corsaro and Pearson 1981; Van Pel et al. 1985]), (2) the nature of the introduced gene and the efficacy of the transcriptional control signals associated with it, and (3) the amount of donor DNA used in the transfection.

Fusion of Protoplasts to Mammalian Cells Growing in Suspension

1. Recover the mammalian cells growing in suspension by centrifugation at 1000g for 5 minutes at 4°C. Wash the cells twice in ice-cold phosphate-buffered saline (PBS; see Appendix B).

2. Add to the mammalian cell pellet the protoplasts prepared in step 6, page 16.49. Use a ratio of protoplasts to mammalian cells of 10,000:1. Gently resuspend the mammalian cells in the suspension of protoplasts by drawing the mixture up and down in a wide-bore plastic pipette.

   Important: Do not vortex! If the protoplasts lyse at this stage and the solution becomes viscous, the efficiency of transformation will almost certainly be very low.

3. Centrifuge the mixture of bacterial protoplasts and mammalian cells at 2000 rpm for 10 minutes at room temperature in a Beckman TJ6 centrifuge (or its equivalent). Remove the supernatant by aspiration.

4. Flick the side of the tube with a finger until the pellet is loosened, and then add 0.5 ml of 50% polyethylene glycol (PEG 1000) in Dulbecco's modified Eagle's medium (DMEM) (equilibrated to 37°C). Incubate for 1 minute at room temperature.

   In a microwave oven, melt sufficient PEG 1000 to make a 50% solution by volume in DMEM (prewarmed to 37°C). Sterilize the solution by passage through a 0.22-micron filter.

5. Over a period of approximately 5 minutes while still at room temperature, dilute the suspension to 20 ml with DMEM or PBS.

6. Recover the cells by centrifugation at 1000g for 5 minutes at room temperature. Gently resuspend the pellet in prewarmed (37°C) medium containing serum, penicillin (100 units/ml), streptomycin (100 μg/ml), and kanamycin (100 μg/ml). Transfer the cells to a tissue culture dish and incubate at 37°C in a humidified incubator in an atmosphere of 5–7% CO_2. Depending on the experiment continue with step 7a (transient expression) or 7b (stable transformation).

7. a. *Transient expression:* Harvest the cells 48–60 hours after transfection for analysis of RNA or DNA by hybridization. Newly synthesized protein may be analyzed by radioimmunoassay, by western blotting, by immunoprecipitation following in vivo metabolic labeling, or by assays of enzymatic activity in cell extracts. For assays that involve replicate samples or treatment of transfected cells under multiple conditions or over a time course, it is desirable to avoid dish-to-dish variation in transfection efficiency. In these cases, it is best to transfect large numbers of cells (90-mm dishes) and then to divide the cells after 24 hours of incubation and distribute them among several smaller dishes.

   b. *Stable transformation:* Following 18–24 hours of incubation in non-selective medium to allow expression of the transferred gene(s) to occur, the cells are divided and plated in the appropriate selective medium.

This medium should be changed every 2–4 days for 2–3 weeks to remove the debris of dead cells and to allow colonies of resistant cells to grow.

The dilution at which the cells should be plated will be determined by the efficiency of stable transformation, which can vary over several orders of magnitude (see, e.g., Spandidos and Wilkie 1984). The efficiency is dependent on (1) the recipient cell type (significant differences have been observed even between different clones or different passage numbers of the same cell line [Corsaro and Pearson 1981; Van Pel et al. 1985]), (2) the nature of the introduced gene and the efficacy of the transcriptional control signals associated with it, and (3) the amount of donor DNA used in the transfection.

DNA TRANSFECTION BY ELECTROPORATION

The use of pulsed electric fields to introduce DNA into cells in culture has been termed electroporation (Neumann et al. 1982). The method has been used to introduce DNA into a variety of animal cells (Neumann et al. 1982; Wong and Neumann 1982; Potter et al. 1984; Sugden et al. 1985; Toneguzzo et al. 1986; Tur-Kaspa et al. 1986), plant cells (Fromm et al. 1985, 1986; Ecker and Davis 1986) and recently into bacteria. (For a review of methods for the introduction of DNA molecules into eukaryotic cells by electroporation, see Andreason and Evans [1988]. For a discussion of the use of electroporation to introduce DNA into bacterial cells, see Chapter 1, page 1.74.) The procedure has been used for both transient expression and stable transformation, but the efficiency of transfection has varied widely in reports from different laboratories. Recently, two groups have examined the parameters that affect transfection efficiencies and have shown that different conditions must be used for different cell lines (Chu et al. 1987; Knutson and Yee 1987). It is therefore essential to carry out a series of preliminary experiments to determine the conditions that lead to acceptable levels of transient expression or stable transformation for a particular cell line.

The efficiency of transfection by electroporation is influenced by a number of factors.

- *The strength of the applied electric field*. At low voltage, the plasma membranes of cultured cells are not sufficiently altered to allow passage of DNA molecules; at higher voltage, the cells are irreversibly damaged. For most lines of mammalian cells, the maximal level of transient expression (as measured by assays of CAT activity, for example) is reached when voltages between 250 V/cm and 750 V/cm are applied. Typically, between 20% and 50% of the cells survive this treatment (as measured by exclusion of trypan blue [Patterson 1979]).

- *The length of the electric pulse*. Usually, a single electric pulse is passed through the cells. Some electroporation devices allow the experimenter to control the length and shape of the pulse; in others, the characteristics of the pulse are determined solely by the capacitance of the power supply. The available data, although scanty, indicate that the optimal length of the electric pulse required for electroporation is 20–1.0 milliseconds. The efficiency of transient expression is increased if the cells are incubated for 1–2 minutes in the electroporation chamber after exposure to the electric pulse (Rabussay et al. 1987).

- *Temperature*. Some workers report that maximal levels of transient expression are obtained when the cells are maintained at room temperature during electroporation (Chu et al. 1987); others have obtained better results when the cells are maintained at 0°C (Reiss et al. 1986). These discrepancies may result from differences in the responses of various types of mammalian cells to the passage of electric current or in the amount of heat generated during electroporation when large electrical voltages (>1000 V/cm) and/or extended electric pulses (>100 milliseconds) are used.

- *Conformation and concentration of DNA.* Although both linear and circular DNAs can be transfected by electroporation, higher levels of both transient expression and stable transformation are obtained when linear DNA is used (Neumann et al. 1982; Potter et al. 1984; Toneguzzo et al. 1986). Effective transfection has been obtained with concentrations of DNA ranging from 1 μg/ml to 40 μg/ml.

- *Ionic composition of the medium.* The efficiency of transfection is manyfold higher when the cells are suspended in buffered salt solutions (e.g., HEPES-buffered saline) rather than in buffered solutions of nonionic substances such as mannitol or sucrose (Rabussay et al. 1987).

Electroporation has one major advantage: It works well with cell lines that are refractive to other techniques, such as calcium phosphate–DNA coprecipitation. However, considerable work may be required to define optimal conditions for the particular cell line under study. Furthermore, the conditions established in one laboratory do not necessarily work well in another. A number of different electroporation instruments are commercially available, and the manufacturers provide detailed protocols for their use.

Strategies for Studying Gene Regulation

Our understanding of the ways in which mammalian genes are regulated comes chiefly from experiments in which the activity of wild-type and mutant versions of putative *cis*-acting controlling elements is measured in transfected mammalian cells.

The accuracy with which transcripts are initiated can be determined by mapping the precise location of the 5' terminus of the mRNA on its template DNA. Techniques to map the 5' termini of mRNAs, which are described in Chapter 7, include primer extension, digestion of DNA:RNA hybrids with nuclease S1, and digestion of RNA:RNA hybrids with RNAase A. The last two methods have several advantages. These techniques can easily be adapted to yield estimates of the concentration of the RNA of interest. Furthermore, because only accurately initiated RNA is detected, any expression from cryptic promoters located in the flanking DNA sequences or the vector is ignored. By analyzing the RNA, it is also possible to avoid the use of a "reporter" gene. Thus, *cis*-acting regulatory sequences located within the gene of interest or downstream from it can be detected. However, it is often necessary to discriminate between transcripts originating from the transfected gene and those originating from the homologous gene present in the cellular chromosome. This is usually accomplished by deleting or inserting a small sequence in the 5' untranslated region of the transfected DNA, so that the resulting transcript is readily distinguished from its normal counterpart. The transcript of the endogenous gene can then be used as an internal control in transfection experiments.

Direct assays of the abundance and structure of mRNAs have one major disadvantage: They quickly become tiresome, since RNAs encoded by many different mutants must be purified and analyzed by hybridization and gel electrophoresis. In most cases, therefore, the ability of the *cis*-acting elements to modulate the rate of transcription is not assayed directly. Instead, the *cis*-acting elements are joined to a reporter gene that codes for a novel enzymatic activity (e.g., chloramphenicol acetyltransferase [CAT]). The amount of the enzymatic activity that accumulates during the course of a transfection is taken as a measure of the ability of the *cis*-acting element to regulate transcription. Although this is an extremely indirect assay of the activity of *cis*-acting elements that control transcription, it is often the only one that is practicable. The major problem is that CAT activity can be generated not only from transcripts accurately initiated at the promoter of interest, but also from transcripts initiated at other sites in the vector (Langner et al. 1986). The level of such aberrant transcription may greatly dilute the contribution of the promoter being tested. Recently, Heard et al. (1987) demonstrated that the background caused by spurious transcription could be eliminated by placing a putative transcription termination signal upstream of the promoter. Such a signal can be provided by a DNA fragment located upstream of the murine c-*mos* gene (designated UMS for upstream mouse sequence). The mechanism by which UMS blocks read-through transcription is not known. Conceivably, the element contains a transcription terminator or destabilizes read-through mRNAs. To reduce the possibility of error caused by read-through mRNAs, the conclusions obtained by analysis of mutant promoters/enhancers with reporter genes should be confirmed by direct assay of the mRNAs encoded by critical mutants.

VECTORS CARRYING REPORTER GENES

A number of different prokaryotic genes have been used as reporters of the transcriptional activity of mammalian promoters. However, not all of these are suitable for use in transient expression systems. For example, the bacterial *gpt* (Mulligan and Berg 1980), *neo*[r] (Schöler and Gruss 1984), and *gal*K (galactokinase) (Schümperli et al. 1982) genes can be expressed in mammalian cells when linked to a mammalian promoter. However, the assays for all three of these enzymes are cumbersome, requiring, for example, the use of starch gel electrophoresis to separate the bacterial enzymatic activity from activities endogenous to mammalian cells.

The following considerations are important when selecting a reporter gene (Gorman et al. 1982a):

- The enzymatic activity encoded by the prokaryotic gene must be readily distinguishable from any similar activities present in the mammalian cells prior to transfection

- There should be no interference or competition from other enzymatic activities in the cells

- The assay for the encoded enzymatic activity should be rapid, sensitive, reproducible, and convenient

The enzyme CAT fulfills all of these criteria and its gene has therefore become the most widely used reporter gene for indirect assay of promoter activity in transfected mammalian cells. The *cat* gene was originally derived from the transposable element Tn*9* and confers resistance to chloramphenicol. The coding region is 1102 bp in length, which is ordinarily flanked by two 768-bp IS elements. A plasmid, pSV2CAT, has been constructed that contains the SV40 promoter/enhancer, 29 bp of 5′ untranslated sequence, the CAT coding sequence, and 8 bp of DNA 3′ to the UAA stop codon. pSV2CAT cannot confer chloramphenicol resistance on bacteria because the *cat* gene is not linked to a prokaryotic promoter. To assay putative promoters in mammalian cells, a derivative of pSV2CAT has been constructed (pSV0CAT; Gorman et al. 1982a) in which the promoter region of SV40 is replaced by the promoter being tested. CAT modifies and inactivates chloramphenicol by mono- and diacetylation, and a number of convenient assays have been developed to measure CAT activity in mammalian cells transfected with derivatives of pSV0CAT, including:

- *Incubation of extracts prepared from transfected cells with* [14]*C-labeled chloramphenicol.* The extent of modification of chloramphenicol is measured by thin-layer chromatography on silica gels (see pages 16.60–16.62), which separates the mono- and diacetylated derivatives of chloramphenicol from the unmodified compound. The silica gel is exposed to X-ray film and aligned with the resulting autoradiograph. Regions of the gel corresponding to spots on the film are scraped from the plate, and the amount of radioactivity is measured in a liquid scintillation counter. This assay can be somewhat tedious when large numbers of samples are assayed.

• *Incubation of extracts prepared from transfected cells with unlabeled chloramphenicol and ^{14}C-labeled acetyl coenzyme A.* CAT catalyzes the transfer of the ^{14}C-labeled acetyl group from acetyl coenzyme A to chloramphenicol. At the end of the reaction, the mixture is extracted with ethyl acetate. Acetylated forms of chloramphenicol partition into the organic phase, whereas the acetyl coenzyme A remains in the aqueous phase. The amount of chloramphenicol converted to the acetylated form can then be measured in a liquid scintillation counter. Before using this assay to measure CAT activity, it is necessary to heat the cell extracts for 10 minutes at 65°C to destroy an activity that consumes acetyl coenzyme A. This problem is not encountered with the silica gel assay because large amounts of unlabeled acetyl coenzyme A are present in the reaction.

When measuring the effect of promoters and enhancers on gene expression, it is essential to include an internal control that will distinguish differences in the level of transcription from differences in the efficiency of transfection or in the preparation of extracts. This is best achieved by cotransfecting the cells with two plasmids—one that carries the construct under investigation and another that constitutively expresses an activity that can be assayed in the same cell extracts prepared for measurement of CAT activity. An enzyme frequently used for this purpose is *E. coli* β-galactosidase, which is expressed in transfected mammalian cells from a promoter with a broad host range (the SV40 early promoter or the Rous sarcoma virus LTR). Extracts of most types of mammalian cells contain relatively low levels of endogenous β-galactosidase activity, and an increase in enzyme activity of up to 100-fold can usually be detected during the course of a transfection. However, β-galactosidase should not be used as an internal control in certain specialized cells (e.g., gut epithelial cells) that express high levels of this activity.

A number of different approaches can be used to normalize the CAT activity to the β-galactosidase activity. For example, the amount of protein in individual extracts prepared from a series of transfected cells is measured, and the CAT and β-galactosidase assays (see below) are then carried out using a standard amount of protein in each assay. The CAT activity is then normalized to the β-galactosidase activity. Alternatively, the β-galactosidase activity in a constant volume of extract is measured and the CAT assay is then carried out with amounts of extract that contain a defined amount of β-galactosidase activity. Finally, both enzymatic assays can be carried out in a constant volume of extract, and the results can then be normalized to a defined level of β-galactosidase activity.

The human growth hormone gene has also been used as an internal control for transfection (Selden et al. 1986). In this case, the coding region of the growth hormone gene is linked to a metallothionein or to the Rous sarcoma virus LTR—promoters that are expressed in most types of cultured mammalian cells. Rather than assaying extracts of transfected cells, the amount of growth hormone secreted into the tissue culture medium is analyzed using a commercially available radioimmunoassay. Although this assay is simple to carry out, it should be used with caution, since it is based on the assumption that the appearance of secreted protein in the medium parallels the accumulation of a cytoplasmic enzyme. Thus, the assay does not necessarily provide an accurate internal control for the measurement of enzymatic activity in cell extracts.

ASSAYS FOR CHLORAMPHENICOL ACETYLTRANSFERASE AND β-GALACTOSIDASE ACTIVITIES

As discussed above, the best method to evaluate and optimize the transfection efficiency of a particular cell line by a particular method is to use plasmids containing the gene for chloramphenicol acetyltransferase (CAT) or β-galactosidase. The time after transfection at which CAT or β-galactosidase activity is maximal must usually be determined empirically. In most cases, maximal activity is detected between 48 and 60 hours after transfection.

Preparation of Extracts

1. Remove the medium from cells growing in monolayers in 90-mm tissue culture dishes by gentle aspiration. Wash the monolayers three times with 5 ml of phosphate-buffered saline (PBS; see Appendix B) lacking magnesium and calcium ions.

2. Stand the dishes at an angle for 2–3 minutes to allow the last traces of PBS to drain to one side. Remove the last traces of PBS by aspiration. Add 1 ml of PBS to each plate, and using a policeman, scrape the cells into microfuge tubes. Store the microfuge tubes on ice until all of the plates have been processed.

3. Recover the cells by centifugation at 12,000g for 10 seconds at room temperature in a microfuge. Gently resuspend the cell pellets in 1 ml of ice-cold PBS, and again recover the cells by centrifugation. Remove the last traces of PBS from the cell pellets and from the walls of the microfuge tubes.

   The PBS can be conveniently removed with a disposable pipette tip attached to a vacuum line. Use gentle suction, and touch the tip to the surface of the liquid. Keep the tip as far away from the cell pellet as possible while the fluid is withdrawn from the tube. The pipette tip can then be used to vacuum the walls of the tube to remove any adherent droplets of fluid.

   At this stage, the cell pellet can be stored at −20°C for future analysis.

Assays for Chloramphenicol Acetyltransferase

METHOD 1: THIN-LAYER CHROMATOGRAPHY

This is a modification of methods described by Gorman et al. (1982a).

1. Resuspend the cell pellet from one 90-mm dish in 100 μl of 0.25 M Tris · Cl (pH 7.8). Vortex vigorously to break up clumps of cells.

2. Disrupt the cells by three cycles of freezing in dry ice and ethanol and thawing at 37°C. Make sure the tubes have been marked with ethanol-insoluble ink.

3. Centrifuge the suspension of disrupted cells at 12,000g for 5 minutes at 4°C in a microfuge. Transfer the supernatant to a fresh microfuge tube. Reserve 50 μl for the CAT assay, and store the remainder of the extract at −20°C.

4. Incubate the 50-μl aliquot of the extract for 10 minutes at 65°C to inactivate deacetylases. If the extract is cloudy or opaque at this stage, remove the particulate material by centrifugation at 12,000g for a further 2 minutes at 4°C in a microfuge.

   If the expression level is high, this incubation is not required (see page 16.58 and step 6 below).

5. Prepare CAT reaction mixture 1:

1 M Tris · Cl (pH 7.8)	50 μl
^{14}C-labeled chloramphenicol (60 mCi/mmole, diluted in water to 0.1 mCi/ml)	10 μl
acetyl coenzyme A (freshly prepared at a concentration of 3.5 mg/ml in H$_2$O)	20 μl

   For each 50 μl of cell extract to be tested, 80 μl of CAT reaction mixture 1 is required.

6. Mix each of the samples to be assayed with 80 μl of CAT reaction mixture 1, and incubate the reactions at 37°C. The length of the incubation depends on the concentration of CAT in the cell extract, which in turn depends on the strength of the promoter and the cell type under investigation. In most cases, incubation for 30 minutes to 2 hours is sufficient.

   The reactions can be incubated for longer periods of time (up to 16 hours) when the expression of the *cat* gene is low in the transfected cells. However, in this case, it is advisable to add another 10-μl aliquot of acetyl coenzyme A to each reaction after 2 hours of incubation. This is especially important if the extract has not been heated to 65°C to inactive enzymes that consume acetyl coenzyme A.

7. Add 1 ml of ethyl acetate to each sample, and mix thoroughly by vortexing for three periods of 10 seconds each. Centrifuge the mixtures at 12,000g for 5 minutes at room temperature in a microfuge.

The acetylated forms of chloramphenicol partition into the organic (upper) phase; unacetylated chloramphenicol remains in the aqueous phase.

8. Using an automatic pipettor, transfer exactly 900 μl of the upper phase to a fresh tube, carefully avoiding the lower phase and the interface. Discard the tube containing the lower phase in the radioactive waste.

9. Evaporate the ethyl acetate under vacuum. This is usually accomplished by placing the tubes in a rotating evaporator (e.g., Savant SpeedVac) for approximately 1 hour.

10. Redissolve the reaction products in 25 μl of ethyl acetate, carefully washing the sides of the tube.

11. Apply 10–15 μl of the dissolved reaction products to the origin of a 25-mm silica gel, thin-layer chromatography (TLC) plate. (The origin on the plate can be marked with a soft-lead pencil.) Apply 5 μl at a time, and evaporate the sample to dryness with a hair dryer after each application.

A number of different TLC plates and solvents can be used to separate acetylated forms of chloramphenicol. We routinely use chloroform:methanol (95:5) and Sybron SIL G/UV254 (Brinkmann) TLC plates.

12. Prepare a TLC chamber containing 200 ml of chloroform:methanol (95:5).

13. Place the TLC plate in the chromatography chamber, close the chamber, and allow the solvent front to move approximately 75% of the distance to the top of the plate.

14. Remove the TLC plate from the tank and allow it to dry at room temperature. Place on the TLC plate adhesive dot labels marked with radioactive ink to align the plate with the film, and then expose the plate to X-ray film.

Do not cover the TLC plate with Saran Wrap, which will block the relatively weak radiation emitted by ^{14}C.

For preparation of radioactive ink, see Appendix E.

15. Develop the X-ray film and align it with the plate. Usually, three radioactive spots are visible. The spot that has migrated the least distance from the origin consists of nonacetylated chloramphenicol that partitioned into ethyl acetate. The two faster-migrating spots are modified forms of chloramphenicol that have been acetylated at one or the other of two potential sites. Diacetylated chloramphenicol may be detected as a third, even faster migrating spot only when high concentrations of CAT are used.

16. To quantitate CAT activity, cut the radioactive spots from the TLC plate and measure the amount of radioactivity they contain in a liquid scintillation counter (see Appendix E).

Note

The sensitivity of the CAT assay can be optimized by eliminating sources of background. One source of background is due to read-through transcription—a problem that can be eliminated by placing a strong transcription termination element upstream of the promoter (Heard et al. 1987; Araki et al. 1988). "Intrinsic" acetylation of chloramphenicol in the absence of cellular extracts may also contribute to background. If this is a problem, decrease the concentration of chloramphenicol in the assay two- to fourfold (Heard et al. 1987).

METHOD 2: EXTRACTION WITH ORGANIC SOLVENTS

In this procedure (Sleigh 1986), radiolabeled acetyl coenzyme A is transferred to unlabeled chloramphenicol. The acetylated, radiolabeled forms of chloramphenicol are separated from unreacted acetyl coenzyme A by partitioning into ethyl acetate. The acetylated chloramphenicol partitions into the organic phase, and the acetyl coenzyme A remains in the aqueous phase.

Extracts of transfected cells, prepared as described on pages 16.59 and 16.60, steps 1–3 (except reserve 30 μl of the extract for the assay), are assayed as follows:

1. Prepare CAT reaction mixture 2:

8 mM chloramphenicol	20 μl
cell extract	30 μl
0.25 M Tris·Cl (pH 7.8)	30 μl
solution containing ^{14}C-labeled acetyl coenzyme A	20 μl

   8 mM chloramphenicol ($M_r = 321$) is made by dissolving 2.57 mg of chloramphenicol in 1 ml of ethanol. The solution of ^{14}C-labeled acetyl coenzyme A is prepared by dissolving ^{14}C-labeled acetyl coenzyme A (58 Ci/mmole) in water to a concentration of 50 μCi/ml. Dispense the solution into 20-μl aliquots and store at $-70°$C. Just before setting up the reaction, dilute the ^{14}C-labeled acetyl coenzyme A tenfold into a solution of unlabeled acetyl coenzyme A (0.5 mM). The concentrations of the reagents in the final reaction mixture are: 2.3 mM chloramphenicol, 0.11 M Tris·Cl, and 129 μM (1 μCi/ml) ^{14}C-labeled acetyl coenzyme A.

   For each 30 μl of cell extract to be tested, 70 μl of CAT reaction mixture 2 is required.

2. Add 30 μl of cell extract, prepared as described on page 16.59, to 70 μl of CAT reaction mixture 2. Incubate the reaction for 1 hour at 37°C and then for 10 minutes at 65°C to inactivate deacetylases.

3. Transfer the sample to an ice bath, and add 100 μl of ice-cold ethyl acetate. Mix the phases by vigorous vortexing. Centrifuge the mixtures at 12,000g for 3 minutes at room temperature in a microfuge.

4. Transfer 80 μl of the organic (upper) phase to a fresh tube. Add 100 μl of ice-cold ethyl acetate to the original tube (the aqueous phase). Vortex vigorously and centrifuge again as in step 3. Transfer 100 μl of the organic phase to the tube containing 80 μl of ethyl acetate from the first extraction.

   Take great care not to transfer any of the aqueous phase, which contains radiolabeled substrate.

5. Mix the combined organic phases with 1 ml of scintillation fluid (Instagel, Hewlett-Packard, or equivalent) in a 1.5-ml plastic tube. Measure the radioactivity by liquid scintillation counting (see Appendix E).

Note

The linear range of the assay is between 0.03 and 0.25 unit of enzyme per milliliter of incubation mixture (a unit of CAT activity is defined as the amount of enzyme that transfers 1 nmole of acetyl groups from acetyl coenzyme A to chloramphenicol in 1 minute at 37°C [pH 7.8]).

METHOD 3: DIFFUSION OF REACTION PRODUCTS INTO SCINTILLATION FLUID

This is an improved version of Method 2 that does not involve extraction of the acetylated chloramphenicol into ethyl acetate. Rather, the reaction mixture is overlaid with water-immiscible scintillation fluid. Acetylated forms of chloramphenicol produced by the activity of CAT in the aqueous phase diffuse rapidly into the scintillation fluid (Neumann et al. 1987), where they are trapped. Acetyl coenzyme A and nonacetylated chloramphenicol remain in the aqueous phase. Only the radioactivity in the scintillation fluid can be detected by liquid scintillation counting, and this can be measured at any stage during the enzymatic reaction. Because this technique is a continuous assay, rather than a single-end-point assay, it yields quantitative kinetic data that can be easily analyzed by standard mathematical means.

Extracts of transfected cells, prepared as described on pages 16.59 and 16.60, steps 1–3 (except reserve 30 μl of the extract for the assay), are assayed as follows:

1. Mix 30 μl of cell extract with 20 μl of 100 mM Tris · Cl (pH 7.8) in a 7-ml glass scintillation vial. Heat the mixture for 15 minutes at 65°C to inactivate deacetylases in the extract. Transfer the vial to a water bath set at 37°C.

2. Add 200 μl of freshly prepared CAT reaction mixture 3:

1 M Tris · Cl (pH 7.8)	25 μl
5 mM chloramphenicol	50 μl
^{3}H-labeled acetyl coenzyme A (1.4 Ci/mmole; 50 μCi/ml)	0.1 μCi
H$_2$O to 200 μl	

   5 mM chloramphenicol ($M_r = 321$) is made by dissolving 1.6 mg of chloramphenicol in 1 ml of ethanol. ^{3}H-labeled acetyl coenzyme A is prepared by dissolving ^{3}H-labeled acetyl coenzyme A (1.4 Ci/mmole) in water to a concentration of 50 μCi/ml. Dispense the solution into 20-μl aliquots and store at −70°C. The concentrations of the reagents in the final reaction mixture are: 125 mM Tris · Cl (pH 7.8), 1.25 mM chloramphenicol, and 0.35 μM acetyl coenzyme A.

   In the original description of this method, high concentrations (100 μM) of unlabeled acetyl coenzyme A were used to ensure that sufficient substrate was present to satisfy the K_m of CAT. Large amounts (5 μCi) of expensive radiolabeled acetyl coenzyme A were therefore required in each assay. However, adequate results can be obtained by omitting the unlabeled acetyl coenzyme A from the assay and reducing the amount of radioactive substrate to 0.1 μCi (Eastman 1987). Under these conditions, the reaction is linear with enzyme concentration as long as less than 0.0125 unit of CAT is present per assay; 1 unit of CAT catalyzes the acetylation of 1 nmole of chloramphenicol in 1 minute at 37°C (pH 7.8). Preparations of CAT are commercially available that are suitable as standards for the determination of enzyme activity in crude extracts of transfected cells.

3. Using an automatic pipettor, carefully overlay the reaction mixture with 5 ml of a water-immiscible scintillation fluid (e.g., Econofluor; New England Nuclear) and then incubate the mixture at 37°C. Mixing of the two phases is not necessary. As the reaction proceeds, the acetylated chloramphenicol diffuses into the organic scintillation fluid.

   Determine the amount of radioactivity in the scintillation fluid by liquid scintillation counting (see Appendix E), which does not detect radioactivity remaining in the aqueous phase. The vials containing the reaction mixtures may be counted for 0.1 minute at selected time intervals (e.g., every 20 minutes) until the enzymatic reaction is complete or at a single endpoint determined on the basis of previous experiments.

Assay for β-Galactosidase in Extracts of Mammalian Cells

The *E. coli* β-galactosidase gene has been used as an internal reference in a number of transfection studies (Hall et al. 1983; Herbomel et al. 1984; Edlund et al. 1985). Constructs are available that carry the β-galactosidase gene downstream from the SV40 early promoter/enhancer or the long terminal repeat of Rous sarcoma virus. Extracts of transfected cells that are used to assay CAT activity can also be used to measure β-galactosidase activity, but they must not be heated to 65°C prior to the assay, because β-galactosidase is sensitive to heat.

1. For each sample to be assayed, mix:

100× Mg solution	3 μl
1× ONPG (*o*-nitrophenyl-β-D-galactopyranoside)	66 μl
cell extract (see Note i)	30 μl
0.1 M sodium phosphate (pH 7.5)	201 μl

   100× Mg solution is 0.1 M $MgCl_2$, 4.5 M β-mercaptoethanol. 1× ONPG is a 4 mg/ml solution of ONPG dissolved in 0.1 M sodium phosphate (pH 7.5). 0.1 M Sodium phosphate is made by mixing 41 ml of 0.2 M $Na_2HPO_4 \cdot 2H_2O$ (m.w. = 178.05; 35.61 g/liter), 9 ml of 0.2 M $NaH_2PO_4 \cdot 2H_2O$ (m.w. = 156.01; 31.21 g/liter), and 50 ml of H_2O.

2. Incubate the reactions at 37°C for 30 minutes or until a faint yellow color has developed.

3. Stop the reactions by adding 500 μl of 1 M Na_2CO_3 to each reaction. Read the optical density of the reactions at a wavelength of 420 nm. (The linear range is 0.2–0.8 OD.)

Notes

i. The extracts are prepared as described on pages 16.59 and 16.60, steps 1–3 (except reserve 30 μl of the extract for the β-galactosidase assay). The extracts that are to be used for assay of CAT activity should be heated to 65°C for 10 minutes after the β-galactosidase assays are completed.

ii. It is essential to include positive and negative controls of the following type in β-galactosidase assays:

	Positive control (μl)	Negative control (μl)
Mg buffer	3	3
1× ONPG	66	66
β-Galactosidase (50 units/ml)	1	0
Extract from mock-transfected cells	30	30
0.1 M Sodium phosphate	201	201

E. coli β-galactosidase, which can be purchased commercially, should be dissolved at a concentration of 3000 units/ml in 0.1 M sodium phosphate (pH 7.5). Just before use, transfer 1 μl of the stock solution of β-galactosidase into 60 μl of 0.1 M sodium phosphate. Use 1 μl of the diluted stock (50 units/ml) in the positive control. One unit of *E. coli* β-galactosidase is defined as the amount of enzyme that will hydrolyze 1 μmole of ONPG in 1 minute at 37°C.

Cloning by Expression in Mammalian Cells

In addition to the uses of expression in mammalian cells described above, mammalian expression can be an important tool in the identification and isolation of eukaryotic genes. In contrast to bacterial cells, which are very useful in expressing immunologically active mammalian proteins, mammalian cells are, in principle, capable of expressing most mammalian genes in a biologically active form. Thus, using mammalian expression, genes and gene products can be identified by (1) the appearance of a selectable or identifiable new phenotype in the cells, (2) the production of a biologically active molecule that can be isolated and assayed, or (3) the presentation on the cell surface of novel proteins that can be identified either immunologically or through binding to specific ligands. This approach to gene identification effectively complements many of the biochemical approaches (protein sequencing and construction of oligonucleotide probes, in vitro translation and immunoprecipitation, bacterial expression and immunological screening) and permits the isolation of genes coding for proteins for which there is no structural information and only, for example, a crude biological assay.

Overall, the isolation of genes by expression has been successfully applied to secreted, biologically active gene products, to cell surface antigens and receptors, to intracellullar gene products for which selection systems are present, and to intracellular gene products that change the phenotype of the cell in easily identifiable ways. There are some theoretical limitations that have not yet been approached. For example, some gene products may require specific processing that is only likely to occur in special host-cell types. In other cases, the desired gene product may be part of a larger structure within or outside of the cell, and expression of multiple components may be necessary to get detectable activity. Unique solutions, such as cell-cell hybridization or chromosome transfer to appropriate recipient cells, may be necessary to circumvent these types of problems.

Below, we discuss the two broad categories of mammalian expression cloning, which are based on whether a genomic gene or a cDNA copy is being expressed.

Cloning by Expression of cDNA Clones

One approach to cloning by expression is to prepare cDNA from mRNA containing the expressed gene of interest and to insert this cDNA into an expression vector for a given target cell type. The most commonly used recipient cell type has been COS cells (see page 16.19). The expression vector contains a strong promoter sequence with an enhancer and the SV40 origin to allow replication to high copy number following transfection. For genes whose products are normally secreted, the culture supernatant is then examined for the presence of the gene product. This approach has been used most commonly with molecules that are biologically active (Lee et al. 1985; Wong et al. 1985; Yang et al. 1986), although immunological assays could be performed if the proper reagent sensitivity was available.

Isolation of particular genes is usually accomplished by dividing the cDNA library into pools prior to transfection. Any pool that demonstrates the proper activity is further subdivided and retransformed until a single clone is obtained. A matrix approach (see Wong et al. 1985) can significantly speed the process and reduce the overall amount of work. For cell surface molecules, Seed and Aruffo (1987) have described a novel approach based on COS cell expression and adherence of cells expressing the antigen to antibody-coated dishes. The cells binding to the antibody-coated dishes express primarily the desired cell surface protein, and plasmids can be rescued from these cells by simple lysis and retransformation of bacteria. Thus, the approach of expression of cDNA clones has been used effectively for biologically active secreted molecules and for cell surface molecules. A unique prerequisite of this approach is the necessity of full-length cDNA for efficient production. Furthermore, the vector should be constructed to yield maximal expression.

It is possible to use a stable transformation system, rather than a transient system, for isolation of expressing cDNA clones. This type of approach might be useful when the desired gene will change the phenotype of the cell and a selection or cell culture screening approach can be utilized.

More recently, the genes for the receptors for the growth factors interleukin-1 (IL-1) (Sims et al. 1988), IL-6 (Yamasaki et al. 1988), and erythropoietin (D'Andrea et al. 1989) have been isolated by expression cloning in COS cells. In these cases, radioactive growth factor was used as a screen to identify cultures of cells expressing the receptor following transfection.

Expression Cloning of Genomic DNA

In principle, the genome of a mammalian species contains all the information necessary for expression of a given gene. Furthermore, because the cellular machinery for transcription, translation, and processing are very uniform across all mammalian species, a single gene transferred to cells from a different species could be expected to be expressed in a biologically active form if the proper cellular environment is present. The first example of transfer of genetic information via naked DNA in mammalian cells came from experiments of Szybalska and Szybalski (1962), who transferred hypoxanthine-guanine phosphoribosyl transferase activity from cells that expressed active enzyme to cells that were deficient in such activity via naked DNA from the active cells. Later, many investigators demonstrated that for many DNA viruses, the naked viral DNA was sufficient to establish an infection if transferred to an appropriate host cell. Although these experiments were not directly aimed at the isolation of genes, they provided the experimental support for the concept that in many cases a given gene introduced into the appropriate cellular environment could be expressed and the gene product detected.

The first genes to be isolated by expression in mammalian cells were several examples of activated oncogenes (Goldfarb et al. 1982; Shih and Weinberg 1982; Goubin et al. 1983). In these experiments, DNAs from certain types of transformed cells were transfected into NIH-3T3 cells. NIH-3T3 cells were selected as target cells based on their maintenance of a flat appearance even during prolonged tissue culture growth. Expression of "transforming" genes (e.g., from oncogenic retroviruses) resulted in a distinct morphological change in these cells, leading to piling up and disorganized growth patterns. This easily recognizable phenotype served as the assay for successful transferal of an activated oncogene from a transformed cell to the NIH-3T3 cells. Precise identification of the actual gene involved in transmitting this phenotype is discussed below.

A similar approach has been used to isolate the gene for thymidine kinase (Perucho et al. 1980). In this case, cells deficient in thymidine kinase activity were used as recipients for DNA from cells containing an active *tk* gene. Growth of the cells in media containing hypoxanthine, aminopterin, and thymidine (HAT selection) results in selection against cells that do not contain an active *tk* gene (aminopterin inhibits the endogenous pyrimidine biosynthesis and cells must depend on thymidine kinase to scavenge thymidine from the medium for pyrimidine biosynthesis). Only those cells that contain an active *tk* gene can survive, and they form readily isolatable colonies in tissue culture. The hamster adenosine phosphoribosyl transferase (*aprt*) gene has been isolated by an analogous selection of *aprt*-deficient cells in media containing azaserine and adenine to select for cells capable of utilizing adenosine as a purine source (Lowy et al. 1980).

Finally, the third type of genomic cloning by expression was described initially by Kavathas and Herzenberg (1983) and Kavathas et al. (1984) for the Leu-2 gene and Kühn et al. (1984) for the transferrin receptor. Kavathas et al. (1984) identified the gene for the Leu-2 surface antigen by transferring DNA from cells containing the mouse Leu-2 gene into cells that did not express detectable amounts of Leu-2 surface antigen. By staining the recipient cells with fluorescein-labeled anti-Leu-2 antibody and sorting through a fluorescence-activated cell sorter, a population of cells that expressed the Leu-2 antigen was selected and used to isolate the Leu-2 gene. Very frequently, the sorted cells contained amplified copies of the transfected DNA, since cells with the highest levels of expression were repetitively isolated. The amplification of the DNA facilitated the cloning of the desired gene.

The above three examples illustrate the type of selection or screening that is essential for the initial identification of cells that have taken up and express the desired gene. Several strategies have been used to actually identify and isolate the desired gene. Transformation with genomic DNA results in uptake by the recipient cells of considerably more DNA than the DNA of the desired gene. Frequently, 0.01–0.1% of the genome equivalent of DNA may be taken up by recipient cells, depending on the cell type. Obviously, the purification of the desired gene is insufficient at this point, and sorting through the DNA of the primary transformant for the desired gene is not very efficient. The primary solution to this problem has been the retransformation of fresh cells with DNA from the primary transformant (secondary transformation). The original donor DNA within the primary transformant is distributed randomly throughout the genome of the recipient. Retransformation with DNA from the primary transformant and selection of a secondary transformant eliminates many of the segments of DNA that do not contain the desired gene and enhances the purity of the desired gene relative to the other remaining donor DNA. By the second or third retransformation, often the only remaining donor DNA in the cell line is the desired gene. The key is then to define a method to identify specifically the original donor DNA.

When human DNA is used as donor into a mouse or other rodent cell line, one common approach is to use human-specific, highly repetitive sequences (e.g., members of the Alu family of highly repeated sequences) as a probe to identify the human-specific DNA (Jolly et al. 1982; Pulciani et al. 1982; Shih and Weinberg 1982; Lin et al. 1983). Since such sequences are distributed, on average, every 3000 bp, most human genes would be expected to have such an adjacent sequence near enough to be retained in the transformants and to serve as a target for hybridization detection. A second approach is to utilize a marker DNA that is specifically associated with the original donor DNA prior to transfection. Such a sequence (e.g., a portion of a plasmid such a pBR322) serves as a clear marker for the donor DNA (Lowy et al. 1980). Finally, the third method for marking the donor DNA is to associate it with a selectable marker such as a bacterial suppressor tRNA gene (Goldfarb et al. 1982; Lewis et al. 1983). If such a sequence stays associated with the desired gene during the subsequent retransformations, a bacteriophage λ genomic DNA library of the final stage retransformant should contain a clone with the desired gene and the selectable marker. By applying the appropriate selection technique (e.g., plating on a nonsuppressing strain of bacteria if the selectable marker is a suppressor tRNA gene), the correct clone should be readily obtained. None of the above approaches has been entirely successful. Often, a repeated sequence is not located near the desired gene or the marker fragment is lost from the gene region during retransformation.

Obviously, successful transformation with genomic DNA requires that the gene be maintained intact and that the promoter region be active in the recipient cell. When total genomic DNA is used, efforts are made to keep the DNA as large as possible (usually > 100 kb). For some genes (the 200-kb Factor VIII gene), even this large size may not be sufficient. One useful approach for many types of genes has been the use of restriction endonuclease digestion to determine sensitivity of the target gene to digestion with a particular enzyme. If the gene is not large, it is sometimes possible to prepare a genomic DNA library in a bacteriophage λ or cosmid vector and use purified library DNA for the transformation. Three advantages of this approach are:

1. The library can be divided into sublibraries and individual sublibraries retransformed. Once an active sublibrary is identified, that sublibrary is further subdivided and retransformed again. The process is repeated until a single clone is obtained.

2. It may be possible to enhance the expression of transfected genes by placing one or more enhancer-type elements into the vector. Such elements may improve the likelihood of a successful expression (see page 16.6).

3. In some cases, the vector sequences can be used to "rescue" a clone from the transfected cell line, if the vector sequences are maintained intact (Lindenmaier et al. 1982). Such an approach significantly speeds the process and eliminates the need to construct libraries from transfected cell lines.

The factors that determine whether a given promoter will be active in a given recipient cell line are not completely understood and are the subject of considerable research in their own right. To date, the source of the genomic DNA has been a factor in determining the transfectability of a given marker gene, except for trophoblast DNA, perhaps due to the high degree of methylation of such DNA. At some point, it may be possible to define what types of sequence elements (such as recipient-cell-specific enhancers) might improve the likelihood or level of expression of a foreign gene and thereby improve the success of cloning by genomic expression approaches.

References

Akusjärvi, G., C. Svensson, and O. Nygård. 1987. A mechanism by which adenovirus virus-associated RNA$_I$ controls translation in a transient expression assay. *Mol. Cell. Biol.* **7:** 549.

Andreason, G.L. and G.A. Evans. 1988. Introduction and expression of DNA molecules in eukaryotic cells by electroporation. *BioTechniques* **6:** 650.

Araki, E., F. Shimada, M. Shichiri, M. Mori, and Y. Ebina. 1988. pSV00CAT: Low background CAT plasmid. *Nucleic Acids Res.* **16:** 1627.

Aruffo, A. and B. Seed. 1987. Molecular cloning of a CD28 cDNA by a high-efficiency COS cell expression system. *Proc. Natl. Acad. Sci.* **84:** 8573.

Bacchetti, S. and F.L. Graham. 1977. Transfer of the gene for thymidine kinase to thymidine kinase-deficient human cells by purified herpes simplex viral DNA. *Proc. Natl. Acad. Sci.* **74:** 1590.

Berkner, K.L. 1988. Development of adenovirus vectors for the expression of heterologous genes. *BioTechniques* **6:** 616.

Birnstiel, M.L., M. Busslinger, and K. Strub. 1985. Transcription termination and 3′ processing: The end is in site! *Cell* **41:** 349.

Boggs, S.S., R.G. Gregg, N. Borenstein, and O. Smithies. 1986. Efficient transformation and frequent single-site, single-copy insertion of DNA can be obtained in mouse erythroleukemia cells transformed by electroporation. *Exp. Hematol.* **14:** 988.

Bonthron, D.T., R.I. Handin, R.J. Kaufman, L.C. Wasley, E.C. Orr, L.M. Mitsock, B. Ewenstein, J. Loscalzo, D. Ginsburg, and S.H. Orkin. 1986. Structure of pre-pro-von Willebrand factor and its expression in heterologous cells. *Nature* **324:** 270.

Boshart, M., F. Weber, G. Jahn, K. Dorsch-Häsler, B. Fleckenstein, and W. Schaffner. 1985. A very strong enhancer is located upstream of an immediate early gene of human cytomegalovirus. *Cell* **41:** 521.

Broker, T.R. and M. Botchan. 1986. Papillomaviruses: Retrospectives and prospectives. *Cancer Cells* **4:** 17.

Campo, M.S. 1985. Bovine papillomavirus DNA: A eukaryotic cloning vector. In *DNA cloning: A practical approach* (ed. D.M. Glover), vol. 2, p. 213. IRL Press, Oxford.

Capecchi, M.R. 1980. High efficiency transformation by direct microinjection of DNA into cultured mammalian cells. *Cell* **22:** 479.

Chaney, W.G., D.R. Howard, J.W. Pollard, S. Sallustio, and P. Stanley. 1986. High-frequency transfection of CHO cells using Polybrene. *Somatic Cell Mol. Genet.* **12:** 237.

Charnay, P., R. Treisman, P. Mellon, M. Chao, R. Axel, and T. Maniatis. 1984. Differences in human α- and β-globin gene expression in mouse erythroleukemia cells: The role of intragenic sequences. *Cell* **38:** 251.

Chen, C. and H. Okayama. 1987. High-efficiency transformation of mammalian cells by plasmid DNA. *Mol. Cell. Biol.* **7:** 2745.

———. 1988. Calcium phosphate-mediated gene transfer: A highly efficient transfection system for stably transforming cells with plasmid DNA. *BioTechniques* **6:** 632.

Chen, E.Y., P.M. Howley, A.D. Levinson, and P.H. Seeburg. 1982. The primary structure and genetic organization of the bovine papillomavirus type 1 genome. *Nature* **299:** 529.

Chia, W., M.R.D. Scott, and P.W.J. Rigby. 1982. The construction of cosmid libraries of eukaryotic DNA using the Homer series of vectors. *Nucleic Acids Res.* **10:** 2503.

Chu, G. and P.A. Sharp. 1981. SV40 DNA transfection of cells in suspension: Analysis of the efficiency of transcription and translation of T-antigen. *Gene* **13:** 197.

Chu, G., H. Hayakawa, and P. Berg. 1987. Electroporation for the efficient transfection of mammalian cells with DNA. *Nucleic Acids Res.* **15:** 1311.

Colbère-Garapin, F., F. Horodniceanu, P. Kourilsky, and A.-C. Garapin. 1981. A new dominant hybrid selective marker for higher eukaryotic cells. *J. Mol. Biol.* **150:** 1.

Colbère-Garapin, F., S. Chousterman, F. Horodniceanu, P. Kourilsky, and A.-C. Garapin. 1979. Cloning of the active thymidine kinase gene of herpes simplex virus type 1 in *Escherichia coli* K-12. *Proc. Natl. Acad. Sci.* **76:** 3755.

Corsaro, C.M. and M.L. Pearson. 1981. Enhancing the efficiency of DNA-mediated gene transfer in mammalian cells. *Somatic Cell Mol. Genet.* **7:** 603.

Cudd, A. and C. Nicolau. 1984. Entrapment of recombinant DNA in liposomes and its transfer and expression in eukaryotic cells. In *Liposome technology: Incorporation of drugs, proteins and genetic material* (ed. G. Gregoriadis), vol. 2, p. 207. CRC Press, Boca Raton, Florida.

D'Andrea, A.D., H.F. Lodish, and G.G. Wong. 1989. Expression cloning of the murine erythropoietin receptor. *Cell* **57:** 277.

Dick, J.E., M.C. Magli, R.A. Phillips, and A. Bernstein. 1986. Genetic manipulation of hematopoietic stem cells with retrovirus vectors. *Trends Genet.* **2:** 165.

Dijkema, R., P.H. van der Meide, P.H. Pouwels, M. Caspers, M. Dubbeld, and H. Schellekens. 1985. Cloning and expression of the chromosomal immune interferon gene of the rat. *EMBO J.* **4:** 761.

DiMaio, D. 1987. Papillomavirus cloning vectors. In *The papovaviridae: The papillomaviruses* (ed. N.P. Salzman and P.M. Howley), vol. 2, p. 293. Plenum Publishing, New York.

DiMaio, D., R. Treisman, and T. Maniatis. 1982. Bovine papillomavirus vector that propagates as a plasmid in both mouse and bacterial cells. *Proc. Natl. Acad. Sci.* **79:** 4030.

DiMaio, D., V. Corbin, E. Sibley, and T. Maniatis. 1984. High-level expression of a cloned HLA heavy chain gene introduced into mouse cells on a bovine papillomavirus vector. *Mol. Cell. Biol.* **4:** 340.

Doyle, C., M.G. Roth, J. Sambrook, and M.-J. Gething. 1985. Mutations in the cytoplasmic domain of the influenza virus hemagglutinin affect different stages of intracellular transport. *J. Cell Biol.* **100:** 704.

Dynan, W.S. and R. Tjian. 1985. Control of eukaryotic messenger RNA synthesis by sequence-specific DNA-binding proteins. *Nature* **316:** 774.

Eastman, A. 1987. An improvement to the novel rapid assay for chloramphenicol acetyltransferase gene expression. *BioTechniques* **5:** 730.

Ecker, J.R. and R.W. Davis. 1986. Inhibition of gene expression in plant cells by expression of antisense RNA. *Proc. Natl. Acad. Sci.* **83:** 5372.

Edlund, T., M.D. Walker, P.J. Barr, and W.J. Rutter. 1985. Cell-specific expression of the rat insulin gene: Evidence for role of two distinct 5′ flanking elements. *Science* **230:** 912.

Eglitis, M.A. and W.F. Anderson. 1988. Retroviral vectors for introduction of genes into mammalian cells. *BioTechniques* **6:** 608.

Elder, J.T., R.A. Spritz, and S.M. Weissman. 1981. Simian virus 40 as a eukaryotic cloning vehicle. *Annu. Rev. Genet.* **15:** 295.

Felgner, P.L. and M. Holm. 1989. Cationic liposome-mediated transfection. *Bethesda Res. Lab. Focus* **11(2):** 21.

Felgner, P.L., T.R. Gadek, M. Holm, R. Roman, H.W. Chan, M. Wenz, J.P. Northrop, G.M. Ringold, and M. Danielsen. 1987. Lipofection: A highly efficient, lipid-mediated DNA-transfection procedure. *Proc. Natl. Acad. Sci.* **84:** 7413.

Fromm, M., L.P. Taylor, and V. Walbot. 1985. Expression of genes transferred into monocot and dicot plant cells by electroporation. *Proc. Natl. Acad. Sci.* **82:** 5824.

———. 1986. Stable transformation of maize after gene transfer by electroporation. *Nature* **319:** 791.

Fuerst, T.R., P.L. Earl, and B. Moss. 1987. Use of a hybrid vaccinia virus-T7 RNA polymerase system for expression of target genes. *Mol. Cell. Biol.* **7:** 2538.

Fuerst, T.R., E.G. Niles, F.W. Studier, and B. Moss. 1986. Eukaryotic transient-expression system based on recombinant vaccinia virus that synthesizes bacteriophage T7 RNA polymerase. *Proc. Natl. Acad. Sci.* **83:** 8122.

Gerard, R.D. and Y. Gluzman. 1985. New host cell system for regulated simian virus 40 DNA replication. *Mol. Cell. Biol.* **5:** 3231.

Gething, M.-J. and J. Sambrook. 1981. Cell-surface expression of influenza haemagglutinin from a cloned DNA copy of the RNA gene. *Nature* **293:** 620.

Gilboa, E., M.A. Eglitis, P.W. Kantoff, and W.F. Anderson. 1986. Transfer and expression of cloned genes using retroviral vectors. *BioTechniques* **4:** 504.

Gillies, S.D., S.L. Morrison, V.T. Oi, and S. Tonegawa. 1983. A tissue-specific transcription enhancer element is located in the major intron of a rearranged immunoglobulin heavy chain gene. *Cell* **33:** 717.

Gluzman, Y. 1981. SV40-transformed simian cells support the replication of early SV40 mutants. *Cell* **23:** 175.

Goldfarb, M., K. Shimizu, M. Perucho, and M. Wigler. 1982. Isolation and preliminary characterization of a human transforming gene from T24 bladder carcinoma cells. *Nature* **296:** 404.

Gorman, C.M., B.H. Howard, and R. Reeves. 1983a. Expression of recombinant plasmids in mammalian cells is enhanced by sodium butyrate. *Nucleic Acids Res.* **11:** 7631.

Gorman, C.M., L.F. Moffat, and B.H. Howard. 1982a. Recombinant genomes which express chloramphenicol acetyltransferase in mammalian cells. *Mol. Cell. Biol.* **2:** 1044.

Gorman, C., R. Padmanabhan, and B.H. Howard. 1983b. High efficiency DNA-mediated transformation of primate cells. *Science* **221:** 551.

Gorman, C.M., G.T. Merlino, M.C. Willingham, I. Pastan, and B.H. Howard. 1982b. The Rous sarcoma virus long terminal repeat is a strong promoter when introduced into a variety of eukaryotic cells by DNA-mediated transfection. *Proc. Natl. Acad. Sci.* **79:** 6777.

Goubin, G., D.S. Goldman, J. Luce, P.E. Neiman, and G.M. Cooper. 1983. Molecular cloning and nucleotide sequence of a transforming gene detected by transfection of chicken B-cell lymphoma DNA. *Nature* **302:** 114.

Graham, F.L. and A.J. van der Eb. 1973. A new technique for the assay of infectivity of human adenovirus 5 DNA. *Virology* **52:** 456.

Green, M.R. 1986. Pre-mRNA splicing. *Annu. Rev. Genet.* **20:** 671.

Gritz, L. and J. Davies. 1983. Plasmid-encoded hygromycin B resistance: The sequence of hygromycin B phosphotransferase gene and its expression in *Escherichia coli* and *Saccharomyces cerevisiae*. *Gene* **25:** 179.

Gruss, P. and G. Khoury. 1980. Rescue of a splicing defective mutant by insertion of an heterologous intron. *Nature* **286:** 634.

Gruss, P., C.-J. Lai, R. Dhar, and G. Khoury. 1979. Splicing as a requirement for biogenesis of functional 16S mRNA of simian virus 40. *Proc. Natl. Acad. Sci.* **76:** 4317.

Hall, C.V., P.E. Jacob, G.M. Ringold, and F. Lee. 1983. Expression and regulation of *Escherichia coli lac*Z gene fusions in mammalian cells. *J. Mol. Appl. Genet.* **2:** 101.

Hamer, D.H. and P. Leder. 1979a. Expression of the chromosomal mouse β^{maj}-globin gene cloned in SV40. *Nature* **281:** 35.

———. 1979b. SV40 recombinants carrying a functional RNA splice junction and polyadenylation site from the chromosomal mouse β^{maj} globin gene. *Cell* **17:** 737.

———. 1979c. Splicing and the formation of stable RNA. *Cell* **18:** 1299.

Heard, J.-M., P. Herbomel, M.-O. Ott, A. Mottura-Rollier, M. Weiss, and M. Yaniv. 1987. Determinants of rat albumin promoter tissue specificity analyzed by an improved transient expression system. *Mol. Cell. Biol.* **7:** 2425.

Herbomel, P., B. Bourachot, and M. Yaniv. 1984. Two distinct enhancers with different cell specificities coexist in the regulatory region of polyoma. *Cell* **39:** 653.

Howley, P.M., M.-F. Law, C. Heilman, L. Engel, M.C. Alonso, M.A. Israel, D.R. Lowy, and W.D. Lancaster. 1980. Molecular characterization of papilloma virus genomes. *Cold Spring Harbor Conf. Cell Proliferation* **7:** 233.

Hughes, S., K. Mellstrom, E. Kosik, F. Tamanoi, and J. Brugge. 1984. Mutation of a termination codon affects *src* initiation. *Mol. Cell. Biol.* **4:** 1738.

Ingolia, D.E., C.-Y. Yeung, I.F. Orengo, M.L. Harrison, E.G. Frayne, F.B. Rudolph, and R.E. Kellems. 1985. Purification and characterization of adenosine deaminase from a genetically enriched mouse cell line. *J. Biol. Chem.* **260:** 13261.

Jakoby, W.B. and I.H. Pastan, eds. 1979. *Cell culture. Methods in enzymology*, vol. 58. Academic Press, New York.

Jimenez, A. and J. Davies. 1980. Expression of a transposable antibiotic resistance element in *Saccharomyces*. *Nature* **287:** 869.

Jolly, D.J., A.C. Esty, H.U. Bernard, and T. Friedmann. 1982. Isolation of a genomic clone partially encoding human hypoxanthine phosphoribosyltransferase. *Proc. Natl. Acad. Sci.* **79:** 5038.

Karin, M., G. Cathala, and M.C. Nguyen-Huu. 1983. Expression and regulation of a human metallothionein gene carried on an autonomously replicating shuttle vector. *Proc. Natl. Acad. Sci.* **80:** 4040.

Karlsson, S., K. Van Doren, S.G. Schweiger, A.W. Nienhuis, and Y. Gluzman. 1986. Stable gene transfer and tissue-specific expression of a human globin gene using adenoviral vectors. *EMBO J.* **5:** 2377.

Kaufman, R.J. 1987. High level production of proteins in mammalian cells. In *Genetic engineering: Principles and methods* (ed. J.K. Setlow), vol. 9, p. 155. Plenum Publishing, New York.

Kaufman, R.J. and P. Murtha. 1987. Translational control mediated by eucaryotic initiation factor-2 is restricted to specific mRNAs in transfected cells. *Mol. Cell. Biol.* **7:** 1568.

Kaufman, R.J. and P.A. Sharp. 1982a. Amplification and expression of sequences cotransfected with a modular dihydrofolate reductase complementary DNA gene. *J. Mol. Biol.* **159:** 601.

———. 1982b. Construction of a modular dihydrofolate reductase cDNA gene: Analysis of signals utilized for efficient expression. *Mol. Cell. Biol.* **2:** 1304.

Kaufman, R.J., P. Murtha, and M.V. Davies. 1987. Translational efficiency of polycistronic mRNAs and their utilization to express heterologous genes in mammalian cells. *EMBO J.* **6:** 187.

Kaufman, R.J., M.V. Davies, V.K. Pathak, and J.W.B. Hershey. 1989. The phosphorylation state of eucaryotic initiation factor 2 alters translational efficiency of specific mRNAs. *Mol. Cell. Biol.* **9:** 946.

Kaufman, R.J., P. Murtha, D.E. Ingolia, C.-Y. Yeung, and R.E. Kellems. 1986a. Selection and amplification of heterologous genes encoding adenosine deaminase in mammalian cells. *Proc. Natl. Acad. Sci.* **83:** 3136.

Kaufman, R.J., L.C. Wasley, B.C. Furie, B. Furie, and C.B. Shoemaker. 1986b. Expression, purification, and characterization of recombinant γ-carboxylated Factor IX synthesized in Chinese hamster ovary cells. *J. Biol. Chem.* **261:** 9622.

Kaufman, R.J., L.C. Wasley, A.J. Spiliotes, S.D. Gossels, S.A. Latt, G.R. Larsen, and R.M. Kay. 1985. Coamplification and coexpression of human tissue-type plasminogen activator and murine dihydrofolate reductase sequences in Chinese hamster ovary cells. *Mol. Cell. Biol.* **5:** 1750.

Kavathas, P. and L.A. Herzenberg. 1983. Stable transformation of mouse L cells for human membrane T-cell differentiation antigens, HLA and β_2-microglobulin: Selection by fluorescence-activated cell sorting. *Proc. Natl. Acad. Sci.* **80:** 524.

Kavathas, P., V.P. Sukhatme, L.A. Herzenberg, and J.R. Parnes. 1984. Isolation of the gene encoding the human T-lymphocyte differentiation antigen Leu-2 (T8) by gene transfer and cDNA subtraction. *Proc. Natl. Acad. Sci.* **81:** 7688.

Kawai, S. and M. Nishizawa. 1984. New procedure for DNA transfection with polycation and dimethyl sulfoxide. *Mol. Cell. Biol.* **4:** 1172.

Kempe, T.D., E.A. Swyryd, M. Bruist, and G.R. Stark. 1976. Stable mutants of mammalian cells that overproduce the first three enzymes of pyrimidine nucleotide biosynthesis. *Cell* **9:** 541.

Kim, S.K. and B.J. Wold. 1985. Stable reduction of thymidine kinase activity in cells expressing high levels of antisense RNA. *Cell* **42:** 129.

Kioussis, D., F. Wilson, C. Daniels, C. Leveton, J. Taverne, and J.H.L. Playfair. 1987. Expression and rescuing of a cloned human tumour necrosis factor gene using an EBV-based shuttle cosmid vector. *EMBO J.* **6:** 355.

Kit, S., D.R. Dubbs, L.J. Piekarski, and T.C. Hsu. 1963. Deletion of thymidine kinase activity from L cells resistant to bromodeoxyuridine. *Exp. Cell Res.* **31:** 297.

Kitajewski, J., R.J. Schneider, B. Safer, S.M. Munemitsu, C.E. Samuel, B. Thimmappaya, and T. Shenk. 1986. Adenovirus VAI RNA antagonizes the antiviral action of interferon by preventing activation of the interferon-induced eIF-2α kinase. *Cell* **45:** 195.

Knutson, J.C. and D. Yee. 1987. Electroporation: Parameters affecting transfer of DNA into mammalian cells. *Anal. Biochem.* **164:** 44.

Kozak, M. 1989. The scanning model for translation: An update. *J. Cell Biol.* **108:** 229.

Krainer, A.R. and T. Maniatis. 1988. RNA splicing. In *Transcription and splicing* (ed. B.D. Hames and D.M. Glover), p. 131. IRL Press, Oxford.

Krystal, M., R. Li, D. Lyles, G. Pavlakis, and P. Palese. 1986. Expression of the three influenza virus polymerase proteins in a single cell allows growth complementation of viral mutants. *Proc. Natl. Acad. Sci.* **83:** 2709.

Kühn, L.C., A. McClelland, and F.H. Ruddle. 1984. Gene transfer, expression, and molecular cloning of the human transferrin receptor gene. *Cell* **37:** 95.

Langner, K.-D., U. Weyer, and W. Doerfler. 1986. *Trans* effect of the E1 region of adenoviruses on the expression of a prokaryotic gene in mammalian cells: Resistance to 5'-CCGG-3' methylation. *Proc. Natl. Acad. Sci.* **83:** 1598.

Law, M.-F., J.C. Byrne, and P.M. Howley. 1983. A stable bovine papillomavirus hybrid plasmid that expresses a dominant selective trait. *Mol. Cell. Biol.* **3:** 2110.

Lee, F., T. Yokota, T. Otsuka, L. Gemmell, N. Larson, J. Luh, K. Arai, and D. Rennick. 1985. Isolation of cDNA for a human granulocyte–macrophage colony-stimulating factor by functional expression in mammalian cells. *Proc. Natl. Acad. Sci.* **82:** 4360.

Lewis, J.A., K. Shimizu, and D. Zipser. 1983. Isolation and preliminary characterization of the Chinese hamster thymidine kinase gene. *Mol. Cell. Biol.* **3:** 1815.

Lin, P.-F., S.-Y. Zhao, and F.H. Ruddle. 1983. Genomic cloning and preliminary characterization of the human thymidine kinase gene. *Proc. Natl. Acad. Sci.* **80:** 6528.

Lindenmaier, W., H. Hauser, I. Greiser de Wilke, and G. Schütz. 1982. Gene shuttling: Moving of cloned DNA into and out of eukaryotic cells. *Nucleic Acids Res.* **10:** 1243.

Littlefield, J.W. 1964. Selection of hybrids from matings of fibroblasts in vitro and their presumed recombinants. *Science* **145:** 709.

———. 1966. The use of drug-resistant markers to study the hybridization of mouse fibroblasts. *Exp. Cell Res.* **41:** 190.

Liu, C.-C., C.C. Simonsen, and A.D. Levinson. 1984. Initiation of translation at internal AUG codons in mammalian cells. *Nature* **309:** 82.

Lowy, I., A. Pellicer, J.F. Jackson, G.-K. Sim, S. Silverstein, and R. Axel. 1980. Isolation of transforming DNA: Cloning the hamster *aprt* gene. *Cell* **22:** 817.

Luckow, V.A. and M.D. Summers. 1988. Trends in the development of baculovirus expression vectors. *Bio/ Technology* **6:** 47.

Lupton, S. and A.J. Levine. 1985. Mapping genetic elements of Epstein-Barr virus that facilitate extrachromosomal persistence of Epstein-Barr virus-derived plasmids in human cells. *Mol. Cell. Biol.* **5:** 2533.

Lusky, M. and M. Botchan. 1981. Inhibition of SV40 replication in simian cells by specific pBR322 DNA sequences. *Nature* **293:** 79.

Luthman, H. and G. Magnusson. 1983. High efficiency polyoma DNA transfection of chloroquine treated cells. *Nucleic Acids Res.* **11:** 1295.

Mackett, M., G.L. Smith, and B. Moss. 1985. The construction and characterisation of vaccinia virus recombinants expressing foreign genes. In *DNA cloning: A practical approach* (ed. D.M. Glover), vol. 2, p. 191. IRL Press, Oxford.

Maniatis, T., S. Goodbourn, and J.A. Fischer. 1987. Regulation of inducible and tissue-specific gene expression. *Science* **236:** 1237.

Mannino, R.J. and S. Gould-Fogerite. 1988. Liposome mediated gene transfer. *BioTechniques* **6:** 682.

Mansour, S.L., T. Grodzicker, and R. Tjian. 1985. An adenovirus vector system used to express polyoma virus tumor antigens. *Proc. Natl. Acad. Sci.* **82:** 1359.

Mason, P.J., J.A. Elkington, M.M. Lloyd, M.B. Jones, and J.G. Williams. 1986. Mutations downstream of the polyadenylation site of a *Xenopus* β-globin mRNA affect the position but not the efficiency of 3' processing. *Cell* **46:** 263.

Maurer, R.A. 1989. Cationic liposome-mediated transfection of primary cultures of rat pituitary cells. *Bethesda Res. Lab. Focus* **11(2):** 25.

McCutchan, J.H. and J.S. Pagano. 1968. Enhancement of the infectivity of simian virus 40 deoxyribonucleic acid with diethyl aminoethyl-dextran. *J. Natl. Cancer Inst.* **41:** 351.

McKnight, S. and R. Tjian. 1986. Transcriptional selectivity of viral genes in mammalian cells. *Cell* **46:** 795.

Mellon, P., V. Parker, Y. Gluzman, and T. Maniatis. 1981. Identification of DNA sequences required for transcription of the human α1-globin gene in a new SV40 host–vector system. *Cell* **27:** 279.

Mishina, M., T. Kurosaki, T. Tobimatsu, Y. Morimoto, M. Noda, T. Yamamoto, M. Terao, J. Lindstrom, T. Takahashi, M. Kuno, and S. Numa. 1984. Expression of functional acetylcholine receptor from cloned cDNAs. *Nature* **307:** 604.

Moss, B. 1985. Replication of poxviruses. In *Virology* (ed. B.N. Fields et al.), p. 685. Raven Press, New York.

Mulligan, R.C. and P. Berg. 1980. Expression of a bacterial gene in mammalian cells. *Science* **209:** 1422.

———. 1981a. Selection for animal cells that express the *Escherichia coli* gene coding for xanthine-guanine phosphoribosyltransferase. *Proc. Natl. Acad. Sci.* **78:** 2072.

———. 1981b. Factors governing the expression of a bacterial gene in mammalian cells. *Mol. Cell. Biol.* **1:** 449.

Murray, M.J., R.J. Kaufman, S.A. Latt, and R.A. Weinberg. 1983. Construction and use of a dominant, selectable marker: A Harvey sarcoma virus-dihydrofolate reductase chimera. *Mol. Cell. Biol.* **3:** 32.

Neumann, E., M. Schaefer-Ridder, Y. Wang, and P.H. Hofschneider. 1982. Gene transfer into mouse lyoma cells by electroporation in high electric fields. *EMBO J.* **1:** 841.

Neumann, J.R., C.A. Morency, and K.O. Russian. 1987. A novel rapid assay for chloramphenicol acetyltransferase gene expression. *BioTechniques* **5:** 444.

Nevins, J.R. 1983. The pathway of eukaryotic mRNA formation. *Annu. Rev. Biochem.* **52:** 441.

O'Hare, K., C. Benoist, and R. Breathnach. 1981. Transformation of mouse fibroblasts to methotrexate resistance by a recombinant plasmid expressing a prokaryotic dihydrofolate reductase. *Proc. Natl. Acad. Sci.* **78:** 1527.

Okayama, H. and P. Berg. 1982. High-efficiency cloning of full-length cDNA. *Mol. Cell. Biol.* **2:** 161.

———. 1983. A cDNA cloning vector that permits expression of cDNA inserts in mammalian cells. *Mol. Cell. Biol.* **3:** 280.

———. 1985. Bacteriophage lambda vector for transducing a cDNA clone library into mammalian cells. *Mol. Cell. Biol.* **5:** 1136.

O'Malley, R.P., T.M. Mariano, J. Siekierka, and M.B. Mathews. 1986. A mechanism for the control of protein synthesis by adenovirus VA RNA$_I$. *Cell* **44:** 391.

Orkin, S.H., S.C. Goff, W.N. Kelley, and P.E. Daddona. 1985. Transient expression of human adenosine deaminase cDNAs: Identification of a nonfunctional clone resulting from a single amino acid substitution. *Mol. Cell. Biol.* **5:** 762.

Orth, G. and M. Favre. 1985. Human papillomaviruses: Biochemical and biologic properties. *Clin. Dermatol.* **3:** 27.

Padgett, R.A., P.J. Grabowski, M.M. Konarska, S. Seiler, and P.A. Sharp. 1986. Splicing of messenger RNA precursors. *Annu. Rev. Biochem.* **55:** 1119.

Parker, B.A. and G.R. Stark. 1979. Regulation of simian virus 40 transcription: Sensitive analysis of the RNA species present early in infections by virus or viral DNA. *J. Virol.* **31:** 360.

Patterson, M.K., Jr. 1979. Measurement of growth and viability of cells in culture. *Methods Enzymol.* **58:** 141.

Pavlakis, G.N. and D.H. Hamer. 1983. Regulation of a metallothionein–growth hormone hybrid gene in bovine papilloma virus. *Proc. Natl. Acad. Sci.* **80:** 397.

Perez, L., J.W. Wills, and E. Hunter. 1987. Expression of the Rous sarcoma virus *env* gene from a simian virus 40 late-region replacement vector: Effects of upstream initiation codons. *J. Virol.* **61:** 1276.

Perucho, M., D. Hanahan, L. Lipsich, and M. Wigler. 1980. Isolation of the chicken thymidine kinase gene by plasmid rescue. *Nature* **285:** 207.

Potter, H., L. Weir, and P. Leder. 1984. Enhancer-dependent expression of human κ immunoglobulin genes introduced into mouse pre-B lymphocytes by electroporation. *Proc. Natl. Acad. Sci.* **81:** 7161.

Proudfoot, N.J. 1989. How RNA polymerase II terminates transcription in higher eukaryotes. *Trends Biochem. Sci.* **14:** 105.

Proudfoot, N.J. and E. Whitelaw. 1988. Termination and 3′ end processing of eukaryotic RNA. In *Transcription and splicing* (ed. B.D. Hames and D.M. Glover), p. 97. IRL Press, Oxford.

Pulciani, S., E. Santos, A.V. Lauver, L.K. Long, K.C. Robbins, and M. Barbacid. 1982. Oncogenes in human tumor cell lines: Molecular cloning of a transforming gene from human bladder carcinoma cells. *Proc. Natl. Acad. Sci.* **79:** 2845.

Rabussay, D., L. Uher, G. Bates, and W. Piastuch. 1987. Electroporation of mammalian and plant cells. *Bethesda Res. Lab. Focus* **9(3):** 1.

Rassoulzadegan, M., B. Binetruy, and F. Cuzin. 1982. High frequency of gene transfer after fusion between bacteria and eukaryotic cells. *Nature* **295:** 257.

Reed, R. and T. Maniatis. 1986. A role for exon sequences and splice-site proximity in splice-site selection. *Cell* **46:** 681.

Reiss, M., M.M. Jastreboff, J.R. Bertino, and R. Narayanan. 1986. DNA-mediated gene transfer into epidermal cells using electroporation. *Biochem. Biophys. Res. Commun.* **137:** 244.

Rigby, P.W.J. 1982. Expression of cloned genes in eukaryotic cells using vector systems derived from viral replicons. In *Genetic engineering* (ed. R. Williamson), vol. 3, p. 83. Academic Press, London.

―――. 1983. Cloning vectors derived from animal viruses. *J. Gen. Virol.* **64:** 255.

Rio, D.C., S.G. Clark, and R. Tjian. 1985. A mammalian host-vector system that regulates expression and amplification of transfected genes by temperature induction. *Science* **227:** 23.

Robert de Saint Vincent, B., S. Delbrück, W. Eckhart, J. Meinkoth, L. Vitto, and G. Wahl. 1981. The cloning and reintroduction into animal cells of a functional CAD gene, a dominant amplifiable genetic marker. *Cell* **27:** 267.

Ruiz, J.C. and G.M. Wahl. 1986. *Escherichia coli* aspartate transcarbamylase: A novel marker for studies of gene amplification and expression in mammalian cells. *Mol. Cell. Biol.* **6:** 3050.

Sambrook, J., D. Hanahan, L. Rodgers, and M.-J. Gething. 1986. Expression of human tissue-type plasminogen activator from lytic virus vectors and in established cell lines. *Mol. Biol. Med.* **3:** 459.

Sambrook, J., L. Rodgers, J. White, and M.-J. Gething. 1985. Lines of BPV-transformed murine cells that constitutively express influenza virus hemagglutinin. *EMBO J.* **4:** 91.

Sarver, N., J.C. Byrne, and P.M. Howley. 1982. Transformation and replication in mouse cells of a bovine papillomavirus–pML2 plasmid vector that can be rescued in bacteria. *Proc. Natl. Acad. Sci.* **79:** 7147.

Sarver, N., P. Gruss, M.-F. Law, G. Khoury, and P.M. Howley. 1981. Bovine papilloma virus deoxyribonucleic acid: A novel eucaryotic cloning vector. *Mol. Cell. Biol.* **1:** 486.

Sassone-Corsi, P. and E. Borrelli. 1986. Transcriptional regulation by trans-acting factors. *Trends Genet.* **2:** 215.

Schaffner, W. 1980. Direct transfer of cloned genes from bacteria to mammalian cells. *Proc. Natl. Acad. Sci.* **77:** 2163.

Schimke, R.T., ed. 1982. *Gene amplification.* Cold Spring Harbor Laboratory, Cold Spring Harbor, New York.

Schimke, R.T. 1984. Gene amplification in cultured animal cells. *Cell* **37:** 705.

―――. 1988. Gene amplification in cultured cells. *J. Biol. Chem.* **263:** 5989.

Schöler, H.R. and P. Gruss. 1984. Specific interaction between enhancer-containing molecules and cellular components. *Cell* **36:** 403.

Schümperli, D., B.H. Howard, and M. Rosenberg. 1982. Efficient expression of *Escherichia coli* galactokinase gene in mammalian cells. *Proc. Natl. Acad. Sci.* **79:** 257.

Seed, B. and A. Aruffo. 1987. Molecular cloning of the CD2 antigen, the T-cell erythrocyte receptor, by a rapid immunoselection procedure. *Proc. Natl. Acad. Sci.* **84:** 3365.

Selden, R.F., K. Burke Howie, M.E. Rowe, H.M. Goodman, and D.D. Moore. 1986. Human growth hormone as a reporter gene in regulation studies employing transient gene expression. *Mol. Cell. Biol.* **6:** 3173.

Serfling, E., M. Jasin, and W. Schaffner. 1985. Enhancers and eukaryotic gene transcription. *Trends Genet.* **1:** 224.

Shaw, G. and R. Kamen. 1986. A conserved AU sequence from the 3′ untranslated region of GM-CSF mRNA mediates selective mRNA degradation. *Cell* **46:** 659.

Shih, C. and R.A. Weinberg. 1982. Isolation of a transforming sequence from a human bladder carcinoma cell line. *Cell* **29:** 161.

Shimizu, Y., B. Koller, D. Geraghty, H. Orr, S. Shaw, P. Kavathas, and R. DeMars. 1986. Transfer of cloned human class I major histocompatibility complex genes into HLA mutant human lymphoblastoid cells. *Mol. Cell. Biol.* **6:** 1074.

Simonsen, C.C. and A.D. Levinson. 1983. Isolation and expression of an altered mouse dihydrofolate reductase cDNA. *Proc. Natl. Acad. Sci.* **80:** 2495.

Simonsen, C.C., H.M. Shepard, P.W. Gray, D.W. Leung, D. Pennica, E. Yelverton, R. Derynck, P.J. Sherwood, A.D. Levinson, and D.V. Goeddel. 1982. Plasmid-directed synthesis of human immune interferon in *E. coli* and monkey cells. *UCLA Symp. Mol. Cell. Biol.* **25:** 1.

Sims, J.E., C.J. March, D. Cosman, M.B. Widmer, H.R. MacDonald, C.J. McMahan, C.E. Grubin, J.M. Wignall, J.L. Jackson, S.M. Call, D. Friend, A.R. Alpert, S. Gillis, D.L. Urdal, and S.K. Dower. 1988. cDNA expression cloning of the IL-1 receptor, a member of the immunoglobulin superfamily. *Science* **241:** 585.

Sleigh, M.J. 1986. A nonchromatographic assay for expression of the chloramphenicol acetyltransferase gene in eucaryotic cells. *Anal. Biochem.* **156:** 251.

Solnick, D. 1981. Construction of an adenovirus–SV40 recombinant producing SV40 T antigen from an adenovirus late promoter. *Cell* **24:** 135.

Southern, P.J. and P. Berg. 1982. Transformation of mammalian cells to antibiotic resistance with a bacterial gene under control of the SV40 early region promoter. *J. Mol. Appl. Genet.* **1:** 327.

Spandidos, D.A. and L. Siminovitch. 1977. Transfer of codominant markers by isolated metaphase chromosomes in Chinese hamster ovary cells. *Proc. Natl. Acad. Sci.* **74:** 3480.

Spandidos, D.A. and N.M. Wilkie. 1984. Expression of exogenous DNA in mammalian cells. In *Transcription and translation: A practical approach* (ed. B.D. Hames and S.J. Higgins), p. 1. IRL Press, Oxford.

Stark, G.R. and G.M. Wahl. 1984. Gene amplification. *Annu. Rev. Biochem.* **53:** 447.

Subramani, S., R. Mulligan, and P. Berg. 1981. Expression of the mouse dihydrofolate reductase complementary deoxyribonucleic acid in simian virus 40 vectors. *Mol. Cell. Biol.* **1:** 854.

Sugden, B. 1982. Epstein-Barr virus: A human pathogen inducing lymphoproliferation *in vivo* and *in vitro. Rev. Infect. Dis.* **4:** 1048.

Sugden, B., K. Marsh, and J. Yates. 1985. A vector that replicates as a plasmid and can be efficiently selected in B-lymphoblasts transformed by Epstein-Barr virus. *Mol. Cell. Biol.* **5:** 410.

Summers, M.D. and G.E. Smith. 1987. A manual of methods for baculovirus vectors and insect cell culture procedures. *Tex. Agric. Exp. Stn. Bull.* No. 1555.

Szybalska, E.H. and W. Szybalski. 1962. Genetics of human cell lines, IV. DNA-mediated heritable transformation of a biochemical trait. *Proc. Natl. Acad. Sci.* **48:** 2026.

Thummel, C., R. Tjian, and T. Grodzicker. 1981. Expression of SV40 T antigen under control of adenovirus promoters. *Cell* **23:** 825.

————. 1982. Construction of adenovirus expression vectors by site-directed *in vivo* recombination. *J. Mol. Appl. Genet.* **1:** 435.

Thummel, C., R. Tjian, S.-L. Hu, and T. Grodzicker. 1983. Translational control of SV40 T antigen expressed from the adenovirus late promoter. *Cell* **33:** 455.

Toneguzzo, F., A.C. Hayday, and A. Keating. 1986. Electric field-mediated DNA transfer: Transient and stable gene expression in human and mouse lymphoid cells. *Mol. Cell. Biol.* **6:** 703.

Toole, J.J., J.L. Knopf, J.M. Wozney, L.A. Sultzman, J.L. Buecker, D.D. Pittman, R.J. Kaufman, E. Brown, C. Shoemaker, E.C. Orr, G.W. Amphlett, W.B. Foster, M.L. Coe, G.J. Knutson, D.N. Fass, and R.M. Hewick. 1984. Molecular cloning of a cDNA encoding human antihaemophilic factor. *Nature* **312:** 342.

Tooze, J., ed. 1980. *Molecular biology of tumor viruses,* 2nd edition: *DNA tumor viruses,* part 2. Cold Spring Harbor Laboratory, Cold Spring Harbor, New York.

Topp, W.C. 1981. Normal rat cell lines deficient in nuclear thymidine kinase. *Virology* **113:** 408.

Treisman, R., S.H. Orkin, and T. Maniatis. 1983. Specific transcription and RNA splicing defects in five cloned β-thalassaemia genes. *Nature* **302**: 591.

Treisman, R., U. Novak, J. Favaloro, and R. Kamen. 1981. Transformation of rat cells by an altered polyoma virus genome expressing only the middle-T protein. *Nature* **292**: 595.

Tur-Kaspa, R., L. Teicher, B.J. Levine, A.I. Skoultchi, and D.A. Shafritz. 1986. Use of electroporation to introduce biologically active foreign genes into primary rat hepatocytes. *Mol. Cell. Biol.* **6**: 716.

Urlaub, G. and L.A. Chasin. 1980. Isolation of Chinese hamster cell mutants deficient in dihydrofolate reductase activity. *Proc. Natl. Acad. Sci.* **77**: 4216.

Vaheri, A. and J.S. Pagano. 1965. Infectious poliovirus RNA: A sensitive method of assay. *Virology* **27**: 434.

Van Doren, K., D. Hanahan, and Y. Gluzman. 1984. Infection of eucaryotic cells by helper-independent recombinant adenoviruses: Early region 1 is not obligatory for integration of viral DNA. *J. Virol.* **50**: 606.

Van Pel, A., E. De Plaen, and T. Boon. 1985. Selection of highly transfectable variant from mouse mastocytoma P815. *Somatic Cell Mol. Genet.* **11**: 467.

Voss, S.D., U. Schlokat, and P. Gruss. 1986. The role of enhancers in the regulation of cell-type-specific transcriptional control. *Trends Biochem. Sci.* **11**: 287.

Wahl, G.M., B. Robert de Saint Vincent, and M.L. DeRose. 1984. Effect of chromosomal position on amplification of transfected genes in animal cells. *Nature* **307**: 516.

Warden, D. and H.V. Thorne. 1968. Infectivity of polyoma virus DNA for mouse embryo cells in presence of diethylaminoethyl-dextran. *J. Gen. Virol.* **3**: 371.

Wieringa, B., F. Meyer, J. Reiser, and C. Weissmann. 1983. Unusual splice sites revealed by mutagenic inactivation of an authentic splice site of the rabbit β-globin gene. *Nature* **301**: 38.

Wigler, M., S. Silverstein, L.-S. Lee, A. Pellicer, Y.-C. Cheng, and R. Axel. 1977. Transfer of purified herpes virus thymidine kinase gene to cultured mouse cells. *Cell* **11**: 223.

Wigler, M., R. Sweet, G.K. Sim, B. Wold, A. Pellicer, E. Lacy, T. Maniatis, S. Silverstein, and R. Axel. 1979. Transformation of mammalian cells with genes from procaryotes and eucaryotes. *Cell* **16**: 777.

Wong, G.G., J.S. Witek, P.A. Temple, K.M. Wilkens, A.C. Leary, D.P. Luxenberg, S.S. Jones, E.L. Brown, R.M. Kay, E.C. Orr, C. Shoemaker, D.W. Golde, R.J. Kaufman, R.M. Hewick, E.A. Wang, and S.C. Clark. 1985. Human GM-CSF: Molecular cloning of the complementary DNA and purification of the natural and recombinant proteins. *Science* **228**: 810.

Wong, T.-K. and E. Neumann. 1982. Electric field mediated gene transfer. *Biochem. Biophys. Res. Commun.* **107**: 584.

Wood, W.I., D.J. Capon, C.C. Simonsen, D.L. Eaton, J. Gitschier, B. Keyt, P.H. Seeburg, D.H. Smith, P. Hollingshead, K.L. Wion, E. Delwart, E.G.D. Tuddenham, G.A. Vehar, and R.M. Lawn. 1984. Expression of active human factor VIII from recombinant DNA clones. *Nature* **312**: 330.

Yamasaki, K., T. Taga, Y. Hirata, H. Yawata, Y. Kawanishi, B. Seed, T. Taniguchi, T. Hirano, and T. Kishimoto. 1988. Cloning and expression of the human interleukin-6 (BSF-2/IFNβ 2) receptor. *Science* **241**: 825.

Yang, Y.-C., A.B. Ciarletta, P.A. Temple, M.P. Chung, S. Kovacic, J.S. Witek-Giannotti, A.C. Leary, R. Kriz, R.E. Donahue, G.G. Wong, and S.C. Clark. 1986. Human IL-3 (multi-CSF): Identification by expression cloning of a novel hematopoietic growth factor related to murine IL-3. *Cell* **47**: 3.

Yates, J.L., N. Warren, and B. Sugden. 1985. Stable replication of plasmids derived from Epstein-Barr virus in various mammalian cells. *Nature* **313**: 812.

Yeung, C.-Y., D.E. Ingolia, D.B. Roth, C. Shoemaker, M.R. Al-Ubaidi, J.-Y. Yen, C. Ching, C. Bobonis, R.J. Kaufman, and R.E. Kellems. 1985. Identification of functional murine adenosine deaminase cDNA clones by complementation in *Escherichia coli*. *J. Biol. Chem.* **260**: 10299.

Zimmermann, U. 1982. Electric field-mediated fusion and related electrical phenomena. *Biochim. Biophys. Acta* **694**: 227.

Zinn, K., D. DiMaio, and T. Maniatis. 1983. Identification of two distinct regulatory regions adjacent to the human β-interferon gene. *Cell* **34**: 865.

17
Expression of Cloned Genes in Escherichia coli

Methods for expressing large amounts of protein from a cloned gene introduced into *Escherichia coli* have proven invaluable in the purification, localization, and functional analysis of proteins. For example, fusion proteins consisting of amino-terminal peptides encoded by a portion of the *E. coli lacZ* or *trpE* gene linked to eukaryotic proteins have been used to prepare polyclonal and monoclonal antibodies against these eukaryotic proteins. These antibodies have been used (1) to purify proteins by immunoaffinity chromatography; (2) in diagnostic assays to quantitate the levels of protein; and (3) to localize the proteins in organisms, tissues, and individual cells by immunofluorescence. Intact native proteins have also been produced in *E. coli* in large amounts for functional studies. For example, both prokaryotic and eukaryotic DNA-binding proteins produced using *E. coli* expression vectors have been used to study the role of these proteins in gene expression. In addition, proteins encoded by viral and cellular oncogenes have been produced in bacteria and then shown to be biologically active after microinjection into mammalian cells in culture. Finally, protein engineering studies are based on methods for introducing mutations into genes that encode proteins of interest and for producing the mutant proteins in large amounts in bacteria.

In this chapter, we describe methods and plasmid vectors for producing fusion proteins and intact native proteins in bacteria. Fusion proteins can be made in large amounts, are easy to purify, and can be used to elicit an antibody response. (Procedures for raising antibodies against fusion proteins in rabbits are given in Chapter 18, pages 18.5–18.6.) Native proteins can be produced in bacteria by placing a strong, regulated promoter and an efficient ribosome-binding site upstream of the cloned gene. If low levels of protein are produced, additional steps may be taken to increase protein production; if high levels of protein are produced, purification is relatively easy. Often, proteins expressed at high levels are found in insoluble inclusion bodies. Methods for extracting proteins from these aggregates are described on pages 17.37–17.38.

Many of the procedures utilized in this chapter (e.g., ligation, transformation, screening, and plasmid minipreps) are described in detail in Chapter 1. Readers who are not familiar with these procedures should refer to the relevant sections there.

Production of Fusion Proteins

If antibodies to the protein encoded by the cloned DNA are not available, it is often useful to produce a fusion protein for use as an antigen. Gene fusions in which DNA encoding part of the cloned gene is inserted near the 3′ terminus of the *lacZ* gene have been particularly useful for the following reasons:

- The fusion protein is usually made at high levels because transcription and translation initiation are directed by normal *E. coli* sequences.

- Fusion of foreign sequences to *E. coli* genes often results in products that are more stable than the native foreign proteins.

- The fusion protein is larger than most *E. coli* proteins and is therefore easy to identify in a protein gel. The fusion protein band can be cut out of the gel, lyophilized, ground into a powder, and used as an antigen.

The procedures presented below are designed to allow production of large amounts of a readily isolatable antigen.

Vector Systems for the Expression of *lacZ* Fusion Genes

Several vector systems have been developed for the expression of *lacZ* fusion genes. One such system, the pUR series of vectors (Rüther and Müller-Hill 1983), has cloning sites for *Bam*HI, *Sal*I, *Pst*I, *Xba*I, *Hin*dIII, and *Cla*I in all three reading frames at the 3′ terminus of the *lacZ* gene (see Figure 17.1 for a description of which vectors contain which cloning sites). By the appropriate choice of vector and restriction site, it should be possible to construct an in-frame fusion for virtually any cloned gene. In some cases, the fusions can be made by removing protruding termini or filling recessed termini prior to ligation.

A similar vector system developed for the production of fusion proteins uses bacterial expression vectors pEX1–3 (Stanley and Luzio 1984) in which the bacteriophage λ p_R promoter directs the expression of large amounts of a Cro–β-galactosidase fusion protein (see Figure 17.2). The bacteriophage λ p_R promoter is regulated and induced in the same manner as the bacteriophage λ p_L promoter (see page 17.11). The vectors pEX1–3 each contain a polycloning site in a different translational reading frame with *Eco*RI, *Sma*I, *Bam*HI, *Sal*I, and *Pst*I sites located in the 3′ end of the *lacZ* gene. Translation termination codons and transcription stop signals have been placed downstream from the polycloning site.

An alternative approach to the production of fusion proteins employs open reading frame (ORF) vectors (Gray et al. 1982; Weinstock et al. 1983) (see Figure 17.3). The ORF vectors are designed so that *lacZ* is not in-frame with sequences encoding the amino terminus of the protein and they therefore express very low levels of functional β-galactosidase. Insertion of a DNA fragment that contains an open reading frame and that places the *lacZ* sequences in-frame results in a plasmid expressing a protein that has β-galactosidase activity. After cDNA fragments are inserted into the ORF vector, *E. coli* strain LG90 (in which *lacZ* has been deleted) is transformed and plated on MacConkey lactose indicator plates. Cells expressing high levels of β-galactosidase are screened for the proteins synthesized.

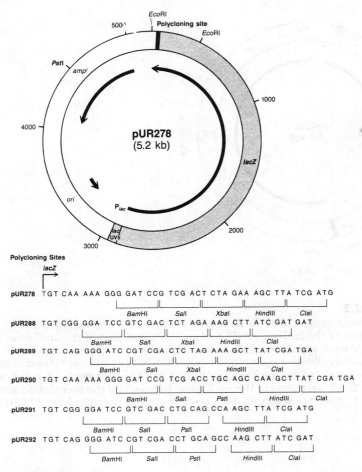

Polycloning Sites

```
             lacZ
             ┌──→

pUR278   TGT CAA AAA GGG GAT CCG TCG ACT CTA GAA AGC TTA TCG ATG
                         └──┘  └──┘  └──┘  └──┘  └──┘
                        BamHI   SalI  XbaI  HindIII  ClaI

pUR288   TGT CGG GGA TCC GTC GAC TCT AGA AAG CTT ATC GAT GAT
                     └──┘  └──┘  └──┘  └──┘  └──┘
                    BamHI   SalI  XbaI  HindIII  ClaI

pUR289   TGT CAG GGG ATC CGT CGA CTC TAG AAA GCT TAT CGA TGA
                     └──┘  └──┘  └──┘  └──┘  └──┘
                    BamHI   SalI  XbaI  HindIII  ClaI

pUR290   TGT CAA AAA GGG GAT CCG TCG ACC TGC AGC CAA GCT TAT CGA TGA
                         └──┘  └──┘  └──┘  └──┘  └──┘
                        BamHI   SalI  PstI  HindIII  ClaI

pUR291   TGT CGG GGA TCC GTC GAC CTG CAG CCA AGC TTA TCG ATG
                     └──┘  └──┘  └──┘  └──┘  └──┘
                    BamHI   SalI  PstI  HindIII  ClaI

pUR292   TGT CAG GGG ATC CGT CGA CCT GCA GCC AAG CTT ATC GAT
                     └──┘  └──┘  └──┘  └──┘  └──┘
                    BamHI   SalI  PstI  HindIII  ClaI
```

FIGURE 17.1

Vectors pUR278, pUR288, pUR289, pUR290, pUR291, and pUR292. These *lacZ* fusion vectors have the cloning sites *Bam*HI, *Sal*I, *Pst*I, *Xba*I, *Hin*dIII, and *Cla*I in all three reading frames at the 3′ terminus of the *lacZ* gene. Insertion of cDNA sequences in the appropriate cloning sites allows the production of a fusion protein of active β-galactosidase and the peptide encoded by the cDNA. pUR278, pUR288, and pUR289 plasmids contain polycloning sites that allow fusions in each of the three reading frames, and they contain a *Pst*I site in the *amp*[r] gene. pUR290, pUR291, and pUR292 plasmids contain a unique *Pst*I site within the polycloning site because the *Pst*I site in *amp*[r] has been destroyed. These plasmids are useful if the cDNA of interest was cloned into a *Pst*I site using the GC-tailing method.

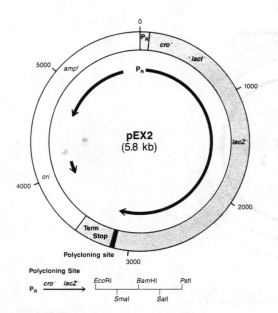

FIGURE 17.2

pEX2, a plasmid about 5.8 kb in length, is designed for expression of cDNA fused at the 3′ terminus of the *lacZ* gene. The amino-terminal part of the *lacZ* gene has been replaced with some sequences from the bacteriophage λ *cro* gene and the *E. coli lacI* gene. The bacteriophage λ p_R promoter is used, which allows expression of the fusion protein to be regulated by the bacteriophage λ *c*Its857 repressor. A polycloning site is present at the 3′ terminus of the *lacZ* gene, followed by the translation stop codons (Stop) and a transcription terminator (Term) from the bacteriophage fd (Stanley and Luzio 1984).

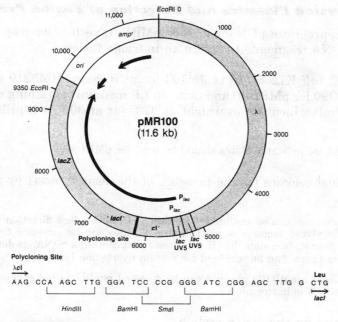

Polycloning Site

λcI →

AAG CCA AGC TTG GGA TCC CCG GGG ATC CGG AGC TTG G | Leu CTG
                                                    → lacI

HindIII   BamHI   SmaI   BamHI

FIGURE 17.3

pMR100 is an ORF vector that carries a *lac* promoter, which directs the transcription of part of the *c*I gene of bacteriophage λ, followed by a polycloning site and an out-of-frame *lacZ* gene deleted at its 5′ terminus. When foreign DNA is inserted into the polycloning site such that the frameshift is corrected, a fusion protein is made that has the amino terminus of the bacteriophage λ *c*I gene, a middle segment of foreign protein sequence, and β-galactosidase at the carboxyl terminus. Cells making a fusion protein will express a high level of β-galactosidase, which can be detected by plating on MacConkey lactose plates (Gray et al. 1982).

1. Ligate the appropriate pUR (or pEX or pMR100) vector (see pages 17.4–17.7) and cDNA fragments to create an in-frame fusion.

2. Transform *E. coli* K12 71/18 or JM103 for pUR vectors (M5219 for pEX vectors or LG90 for pMR100) and plate on LB medium containing ampicillin (100 μg/ml). Incubate overnight at 37°C (or at 30°C if a pEX vector was used).

   MacConkey lactose indicator plates should be used for pMR100.

3. Test individual colonies for the presence of the desired insert by plasmid minipreps.

   If most of the colonies can be assumed to contain a cDNA (because directional cloning or a dephosphorylated vector was used in step 1), they can be screened for protein production in parallel (see step 4b). If not, clones that contain a cDNA, as determined by plasmid minipreps, can be screened for protein expression later.

   cDNA inserts into a pMR100 plasmid can be detected readily as red colonies on the MacConkey lactose indicator plates.

4. Screen colonies for the fusion protein.

   a. Grow small cultures from 5–10 colonies in LB medium containing ampicillin (100 μg/ml). Incubate overnight at 37°C (or at 30°C for pEX).

   b. Inoculate 5 ml of LB medium containing ampicillin (100 μg/ml) with 50 μl of each overnight culture. Incubate for 2 hours at 37°C (or at 30°C for pEX) with aeration. Remove 1 ml of uninduced culture, place it in a microfuge tube, and process as described in steps d and e. If screening for protein production is being done in parallel, prepare plasmid minipreps from 1-ml aliquots of the overnight cultures.

   c. Induce each culture as follows: For pUR or pMR100 vectors, add isopropylthio-β-D-galactoside (IPTG) to a final concentration of 1 mM and continue incubation at 37°C with aeration. For pEX vectors, transfer the culture to 40°C and continue incubating with aeration.

   For preparation of a stock solution of IPTG, see Appendix B.

   d. At various time points during the incubation (i.e., 1, 2, 3, and 4 hours), transfer 1 ml of each culture to a microfuge tube, and centrifuge at 12,000g for 1 minute at room temperature in a microfuge. Remove the supernatant by aspiration.

   The kinetics of induction varies with different proteins, so it is necessary to determine the time at which the maximum amount of product is produced.

   e. Resuspend each pellet in 100 μl of 1 $\times$ SDS gel-loading buffer, heat to 100°C for 3 minutes, and then centrifuge at 12,000g for 1 minute at room temperature. Load 15 μl of each suspension on a 6% SDS-polyacrylamide gel (see Chapter 18, pages 18.47–18.54). Use suspensions of cells containing the vector alone as a control. (For pEX and

ORF vectors, also use β-galactosidase [Sigma] as a control.) The fusion protein should appear as a novel band migrating more slowly than the intense β-galactosidase band in the control. It is not uncommon for a protein the size of β-galactosidase to be present along with the fusion protein.

> *1 × SDS gel-loading buffer*
>
> 50 mM Tris · Cl (pH 6.8)
> 100 mM dithiothreitol
> 2% SDS (electrophoresis grade)
> 0.1% bromophenol blue
> 10% glycerol
>
> 1× SDS gel-loading buffer lacking dithiothreitol can be stored at room temperature. Dithiothreitol should then be added, just before the buffer is used, from a 1 M stock (see Appendix B).

Preparation of Fusion Proteins for Antibody Production

The fusion protein can be prepared for injection in a number of ways, including urea extraction of insoluble aggregates (Schoner et al. 1985) (see page 17.40), aminophenylthiogalactoside affinity chromatography (Germino and Bastia 1984), preparative SDS-polyacrylamide gel electrophoresis, or any combination of these. One simple approach is to scale up the screening procedure.

1. Based on the optimal induction time determined in the screening procedure (step 4, page 17.8), grow and induce a scaled-up version (e.g., 200 ml) of the culture as described on page 17.8, step 4a–c. Transfer the culture to centrifuge tubes, and centrifuge at 5000 rpm for 15 minutes at 4°C. Remove the supernatant, and resuspend the pellet in a volume of 1× SDS gel-loading buffer equivalent to one tenth of the original volume of the cell culture.

2. Using a 5% SDS-polyacrylamide gel with a single large well, load the maximum amount of protein that still forms a reasonably tight band. (For preparation of SDS-polyacrylamide gels, see Chapter 18, pages 18.47–18.54.) Be certain from control experiments (page 17.8, step 4e) that the band of interest is clearly identifiable.

3. Stain the gel by soaking it in cold 0.25 M KCl (Hager and Burgess 1980) for 10–20 minutes with gentle agitation.

4. Excise the band from the gel, lyophilize for about 2 days, and then grind it into a fine powder. This powder is then used for injection into rabbits as described in Chapter 18, pages 18.5–18.6.

Production of Intact Native Proteins

Intact native proteins can be made in *E. coli* by providing a strong, regulated promoter and an efficient ribosome-binding site. To express a prokaryotic gene that has a strong ribosome-binding site, only a promoter must be supplied (see pages 17.11–17.16). To express a eukaryotic gene (or a prokaryotic gene with a weak ribosome-binding site), both a promoter and a ribosome-binding site must be provided (see pages 17.17–17.25). Levels of expression may vary from less than 1% to more than 30% of total cell protein. Measures to improve expression are discussed on page 17.36.

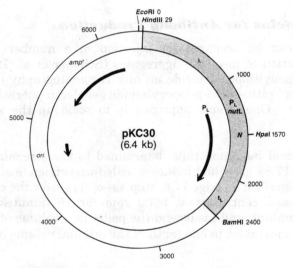

FIGURE 17.4

pKC30, a plasmid approximately 6.4 kb in length, carries the p_L promoter of bacteriophage λ and a *Hpa*I recognition site located 321 nucleotides downstream from the p_L transcriptional start site. The plasmid is a derivative of pBR322 and contains a *Hin*dIII-*Bam*HI fragment (gray box) derived from bacteriophage λ inserted between the *Hin*dIII and *Bam*HI sites within the tetracycline resistance gene (*tet*^{r}). The insertion contains the promoter signal, p_L, a site recognized by the *N* gene product (*nut*L), the *N* gene itself, and the strong *rho*-dependent transcription-termination signal t_L. The *Hpa*I recognition site lies within the coding region of the *N* gene. Sequences inserted into the *Hpa*I site can be regulated by introducing the recombinant plasmid into a temperature-sensitive bacteriophage λ lysogen (*c*I*ts*857). The cells are grown to mid-log phase at 30°C and then shifted to 40°C to inactivate the *c*I gene product and to turn on the p_L promoter. This vector has been used to express the bacteriophage λ *c*II protein at a level such that the protein comprises 4% of the total protein of the cell (Shimatake and Rosenberg 1981).

EXPRESSION OF PROKARYOTIC GENES: PROMOTERS

The first step in expressing eukaryotic proteins in bacteria is to choose an expression vector that carries a strong, regulated prokaryotic promoter. Here we describe the use of expression vectors that contain a bacteriophage λ p_L promoter, a hybrid *trp-lac* promoter, or a bacteriophage T7 promoter.

The Bacteriophage λ p_L Promoter

The bacteriophage λ p_L promoter is regulated by a temperature-sensitive repressor, cIts857, which represses p_L transcription at low temperatures but not at elevated temperatures. *E. coli* strain M5219 contains a defective bacteriophage λ prophage, which encodes the cIts857 repressor, and the bacteriophage λ N protein, an antagonist of transcription termination. This system is particularly suitable if the product of the gene to be expressed is toxic to *E. coli*, since cIts857 strongly represses transcription. In addition, the antitermination function of the N gene may allow RNA polymerase to read through potential termination sites within the gene. One drawback of this system is that the temperature shift (step 5) induces not only the p_L promoter, but also the heat-shock genes, some of which encode proteases (Buell et al. 1985). This problem can be alleviated by using a bacteriophage λ cI$^+$ lysogen and inducing with mitomycin C or nalidixic acid (Shatzman and Rosenberg 1987). pKC30 (see Figure 17.4) (Shimatake and Rosenberg 1981) is one of many p_L vectors available.

To provide a p_L promoter to direct transcription of a cloned gene that has an *E. coli* ribosome-binding site:

1. Digest pKC30 with *Hpa*I.

2. Digest the cloned DNA with appropriate restriction enzymes at a position 5′ of the initiation codon and at a site 3′ of the terminus of the cloned gene.

3. Insert the cDNA into the plasmid, ligate the DNA, and transform *E. coli* strain M5219. Plate transformants on LB medium containing ampicillin (100 μg/ml) and incubate overnight at 30°C.

4. Screen transformants for the presence of the desired insert by colony hybridization and/or by plasmid minipreps and restriction enzyme analysis.

5. To obtain high levels of transcription of the cloned gene, grow strain M5219 containing the expression plasmid to mid-log phase at 30°C, and then shift the temperature of the culture to 40°C. Continue to incubate for several hours at 40°C. Remove small aliquots at various times and analyze them by one of the methods discussed on pages 17.34–17.35. The kinetics of induction varies with different proteins, so it is necessary to determine the time at which the maximum amount of product is present.

Although 42–45°C is used elsewhere in this manual to inactivate bacteriophage λ cIts857, 40°C is used here in order to reduce the induction of heat-shock proteins and to allow the cells to continue growth.

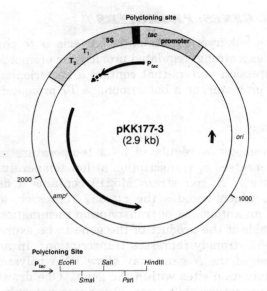

FIGURE 17.5

pKK177-3 is a *tac* vector containing multiple sites downstream from the *tac* promoter into which a gene can be cloned. Downstream from these sites is *rrn*B, which contains an *E. coli* 5S gene and the T₁ and T₂ terminators (Amann and Brosius 1985).

The *trp-lac* **Promoter**

Another promoter that has been used successfully to produce large amounts of proteins in *E. coli* is the *tac* promoter, a hybrid *trp-lac* promoter that is regulated by *lac* repressor (Amann et al. 1983; de Boer et al. 1983). Transcription is repressed in *E. coli* strains such as RB791, a *lacI*^{q} strain that makes high levels of *lac* repressor. If the protein to be expressed is toxic to *E. coli*, then a *lacI*^{q} gene should be cloned into the expression plasmid to make higher levels of *lac* repressor. One factor to consider in choosing this vector system is that the *tac* promoter is induced by adding isopropylthio-β-D-galactoside (IPTG), a relatively expensive compound. A useful *tac* promoter expression plasmid, pKK177-3 (Amann and Brosius 1985), is illustrated in Figure 17.5.

The steps used in cloning and expression in *tac* vectors (or *lac* vectors) are similar to those for other promoter systems and are described below.

1. Clone the cDNA to be expressed, along with its ribosome-binding site, into the polycloning site of pKK177-3.

2. Transform *E. coli* strain RB791, and screen transformants by colony hybridization and/or plasmid minipreps and restriction enzyme analysis.

3. To obtain high levels of transcription, grow cells to mid-log phase at 37°C, add IPTG to a final concentration of 1 mM, continue growth, and monitor the amount of protein made at various times after induction using one of the methods discussed on pages 17.34–17.35.

For preparation of a stock solution of IPTG, see Appendix B.

Note

Hasan and Szybalski (1987) have designed *tac* vectors in which expression of a cloned gene is controlled by promoter inversion in vivo. The promoter in the vector is directed away from the gene to be expressed so that transcription of the gene is kept extremely low until production is desired. The promoter is flanked by *att*P and *att*B and can be inverted efficiently by the bacteriophage λ Int protein. These vectors also utilize bacteriophage λ *N*-mediated antitermination as described on page 17.11 for pKC30.

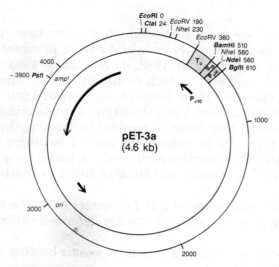

EcoRI 0
ClaI 24
EcoRV 190
NheI 230
EcoRV 380
BamHI 510
NheI 560
NdeI 560
BglII 610

T_φ

P_φ10

4000

~3900 _PstI_

amp^r

pET-3a
(4.6 kb)

1000

3000

ori

2000

FIGURE 17.6

pET-3 carries the bacteriophage T7 ϕ10 promoter (P$_{\phi 10}$) and ϕ terminator (T$_\phi$). The terminator may make the transcripts more resistant to exonucleolytic degradation (Studier and Moffatt 1986). pET-3a is a derivative of pET-3 into which the translation start (S$_{10}$) of bacteriophage T7 ϕ10 (the major capsid protein of bacteriophage T7) with a _Bam_HI site at codon 11 has been inserted. The _Nde_I site (CATATG) is located at the translation start site and can be used to construct a plasmid that directs the expression of native proteins.

The Bacteriophage T7 Promoter

A novel expression system has been developed by Studier and Moffatt (1986) and Tabor and Richardson (1985) using a bacteriophage T7 RNA polymerase/promoter system. This system is designed for the exclusive expression of the cloned gene. Bacteriophage T7 RNA polymerase recognizes solely bacteriophage T7 promoters, will transcribe around a plasmid several times, and may transcribe sequences that are not efficiently transcribed by *E. coli* RNA polymerase. This system allows high levels of expression of some genes that are not expressed efficiently in other systems.

Two components are required for the bacteriophage T7 expression system:

- The first component is bacteriophage T7 RNA polymerase. This polymerase is the product of bacteriophage T7 gene *1* and can be provided on an infecting bacteriophage λ vector or produced from a gene copy inserted into the *E. coli* chromosome (Tabor and Richardson 1985; Studier and Moffatt 1986). If expression of the cloned gene product is toxic, then the level of bacteriophage T7 RNA polymerase must be kept low during cell growth. One way to accomplish this is to use the lysogen BL21(DE3), in which bacteriophage T7 gene *1* is expressed from a *lac*UV5 promoter. In some cases, it is necessary to use cells in which no bacteriophage T7 RNA polymerase is present until expression is desired. This is accomplished by infecting the host cells (e.g., HMS174) that carry the expression plasmid with bacteriophage CE6 (λcIts857 Sam7) carrying bacteriophage T7 gene *1* (Studier and Moffatt 1986).

- The second component is a plasmid vector with a bacteriophage T7 promoter upstream of the gene to be expressed. pET-3 is a derivative of pBR322 that carries the bacteriophage T7 gene *10* promoter ($P_{\phi 10}$), a *Bam*HI cloning site, and the bacteriophage T7 transcription terminator (T_ϕ) (Rosenberg et al. 1987). Derivatives of pET-3 (e.g., pET-3a; see Figure 17.6) have been constructed to provide a bacteriophage T7 gene *10* translation start (S_{10}) through codon 11. DNA can be inserted in each of the three reading frames at this codon to express fusion proteins. The pET-3 derivatives can also be used for production of intact native proteins as described on pages 17.17–17.24 by fusing coding sequences at the *Nde*I site immediately preceding the ATG. Another bacteriophage T7 expression plasmid, pT7-1, has a polycloning site downstream from the bacteriophage T7 $\phi 10$ promoter (Tabor and Richardson 1985).

To express a gene in *E. coli* using bacteriophage T7 RNA polymerase:

1. Clone the gene of interest into a bacteriophage T7 promoter expression plasmid. Identify the correct plasmids in a standard *E. coli* strain by miniprep analysis.

2. Transform *E. coli* strain BL21(DE3), and select for ampicillin-resistant transformants. BL21(DE3) is a lysogen bearing the bacteriophage T7 polymerase gene under the control of the *lac*UV5 promoter. (See precautions recommended by Studier and Moffatt [1986].)

3. Inoculate NZCYM medium with one colony, and incubate overnight at 37°C to obtain a saturated culture.

4. Induce the culture and determine the amount of protein produced as follows:

   a. Inoculate 5 ml of NZCYM medium containing ampicillin (100 μg/ml) with 50 μl of saturated culture. Incubate the culture for 2 hours at 37°C.

   b. Remove 1 ml of the uninduced culture and place in a microfuge tube. Process as described in steps c and d below. Induce the remaining culture by adding isopropylthio-β-D-galactoside (IPTG) to a final concentration of 1 mM.

      For preparation of a stock solution of IPTG, see Appendix B.

   c. Remove 1-ml aliquots of the induced culture at 0.5, 1, 2, and 3 hours after induction. Immediately centrifuge these samples at 12,000g for 1 minute at room temperature in a microfuge. Remove the supernatants by aspiration.

   d. Resuspend each pellet in 100 μl of 1× SDS gel-loading buffer, heat to 100°C for 3 minutes, and store at 0°C until all of the samples are collected and ready to load on a gel.

   1× SDS gel-loading buffer

   50 mM Tris · Cl (pH 6.8)
   100 mM dithiothreitol
   2% SDS (electrophoresis grade)
   0.1% bromophenol blue
   10% glycerol

   1× SDS gel-loading buffer lacking dithiothreitol can be stored at room temperature. Dithiothreitol should then be added, just before the buffer is used, from a 1 M stock (see Appendix B).

   e. Thaw the samples, and then centrifuge them at 12,000g for 1 minute at room temperature. Load 15 μl of each suspension on an SDS-polyacrylamide gel of the appropriate concentration (see Chapter 18, pages 18.47–18.54). Use as a control a suspension of cells containing the vector alone.

5. Stain the gel with Coomassie Brilliant Blue or silver stain, or carry out a western blot (if antibody already exists), to identify induced protein (see Chapter 18, pages 18.55–18.57 or 18.60–18.75, respectively).

Note

For expression of some proteins, it is advisable to increase the ampicillin concentration to levels as high as 200 μg/ml to select for cells that retain the expression plasmid.

EXPRESSION OF EUKARYOTIC GENES: PROMOTERS AND RIBOSOME-BINDING SITES

In addition to the bacterial promoter, the second major factor required to express cloned DNA in *E. coli* is an efficient ribosome-binding site. In *E. coli*, the ribosome-binding site includes an initiation codon (ATG) and a sequence 3–9 nucleotides in length located 3–11 nucleotides upstream of the initiation codon (Shine and Dalgarno 1975; Steitz 1979). This sequence, which is called the Shine-Dalgarno (SD) sequence, is complementary to the 3' terminus of *E. coli* 16S rRNA. Binding of the ribosome to mRNA is thought to be promoted by base pairing between the SD sequence in the mRNA and the sequence at the 3' terminus of the 16S rRNA (Steitz 1979).

When expressing a prokaryotic gene, the ribosome-binding site of that gene may be sufficient. In this case, the protocols presented earlier, which place the gene downstream from a strong, regulated promoter and employ a restriction site 5' of the ribosome-binding site, may allow high levels of expression. Eukaryotic genes and prokaryotic genes with weak ribosome-binding sites require that an efficient ribosome-binding site be provided. This can be accomplished in several ways, as described below.

Preparation of a DNA Fragment for Placement Adjacent to a Functional Ribosome-binding Site

To insert a cloned gene into a vector that carries a functional ribosome-binding site (an initiator ATG an appropriate distance from an SD sequence), the second codon of the cloned gene must be placed adjacent to the initiator ATG. A DNA fragment extending from the second codon to a restriction site 3' of the cloned gene is made by one of the methods given below.

SYNTHESIS OF DNA ENCODING THE AMINO TERMINUS

1. Synthesize a double-stranded DNA adapter extending from the second codon of the gene (blunt end) to a restriction site within the gene (sticky end).

2. Ligate this adapter to a fragment of the gene extending from the restriction site used in step 1 to a site downstream from the end of the gene.

3. Depending on the choice of vector and availability of compatible restriction sites, it may not be necessary to create a blunt end at the downstream site or it may be possible to use a restriction enzyme that leaves a blunt end. If necessary, repair the DNA from step 2 with the Klenow fragment of *E. coli* DNA polymerase I to create a blunt end at the downstream site (see page 17.23, step 5a–c).

4. Isolate the DNA fragment containing the gene following gel electrophoresis to obtain a DNA fragment that can be inserted after the ATG of plasmid pAS1 (see Figure 17.8) or pKK 240-11 (see Figure 17.9).

PRIMER REPAIR

1. Synthesize a DNA primer whose 5' terminus encodes the second amino acid of the protein to be expressed.

2. Phosphorylate the 5' terminus of the primer with bacteriophage T4 polynucleotide kinase.

   a. Mix:

primer	100 pmoles
H_2O	42 μl
10 × bacteriophage T4 polynucleotide kinase buffer	5 μl
10 mM ATP	1 μl
bacteriophage T4 polynucleotide kinase (10 units)	1 μl

   10 × Bacteriophage T4 polynucleotide kinase buffer

   500 mM Tris · Cl (pH 7.6)
   100 mM $MgCl_2$
   50 mM dithiothreitol
   1 mM spermidine HCl
   1 mM EDTA (pH 8.0)

b. Incubate the reaction for 30 minutes at 37°C.

c. Heat the reaction for 5 minutes at 68°C to inactivate the bacteriophage T4 polynucleotide kinase. Place on dry ice and store at −20°C.

3. Hybridize the primer to template bacteriophage M13 single-stranded DNA carrying the cloned gene, and extend the primer with the Klenow fragment of *E. coli* DNA polymerase I.

a. Mix:

phosphorylated primer	1 μl
DNA template	1 μl
0.5 M NaCl	1 μl
H_2O	6 μl

b. Incubate the mixture for 5 minutes at 70°C to denature the DNAs.

c. Slowly cool to 18°C.

d. Add 10 μl of 2 × elongation mix and 2 μl of the Klenow fragment of *E. coli* DNA polymerase I. Incubate for 30 minutes at room temperature.

> *2 × Elongation mix*
>
> 20 mM Tris · Cl (pH 7.5)
> 10 mM $MgCl_2$
> 10 mM dithiothreitol
> a mixture of all four dNTPs, each at a concentration of 1 mM

4. Treat the mixture with nuclease S1 to remove single-stranded DNA.

a. Add 2 μl of 10 × nuclease-S1 buffer and 1 unit of nuclease S1.

> *10 × Nuclease-S1 buffer*
>
> 2 M NaCl
> 0.5 M sodium acetate (pH 4.5)
> 10 mM $ZnSO_4$
> 5% glycerol

b. Incubate the reaction for 30 minutes at 37°C.

c. Extract the reaction once with phenol:chloroform. Transfer the aqueous phase to a fresh microfuge tube, and precipitate the DNA with 2 volumes of ethanol at 0°C. Recover the DNA by centrifugation at 12,000*g* for 10 minutes at 0°C in a microfuge. Remove the supernatant by aspiration, and resuspend the DNA in the appropriate 1 × restriction enzyme buffer.

5. Digest the DNA with an enzyme that cleaves at a downstream site.

6. Continue with steps 3 and 4, page 17.18.

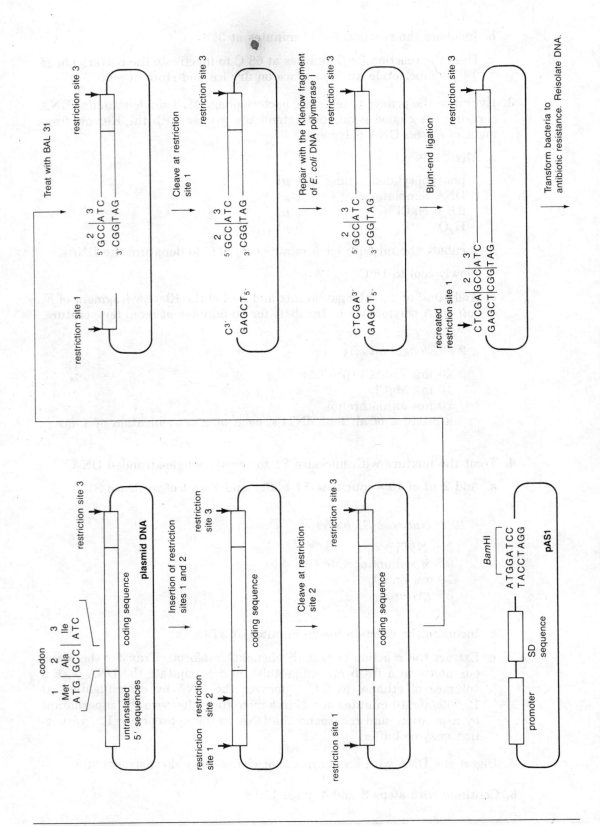

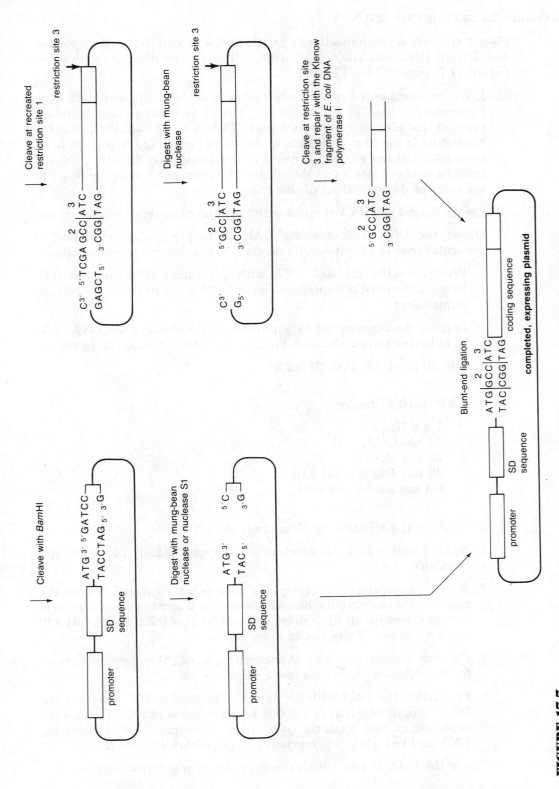

FIGURE 17.7

A method to create a blunt end immediately before a particular nucleotide in a segment of cloned DNA (see text for details).

ENGINEERING A RESTRICTION SITE

Place a restriction site immediately preceding the second codon (Panayotatos and Truong 1981; Shatzman et al. 1983). This can be done as follows (see Figure 17.7, pages 17.20–17.21):

1. Insert two unique restriction sites upstream of the coding sequence to be expressed by inserting linkers or adapters or by cloning the gene into a plasmid that already has the two sites. Choose site 1 such that the sixth nucleotide is the first nucleotide of the second codon of the protein to be expressed and the enzyme leaves a 5′ overhang upon digestion. Site 2 must lie between site 1 and the start of the gene, and it must be closer to the start of the gene than to site 1.

2. Cleave 5 μg of the DNA at site 2 with the appropriate restriction enzyme.

3. Digest the DNA with nuclease BAL 31 to produce a population of molecules that randomly terminate close to the desired codon as follows:

   a. Precipitate the digested DNA with 2 volumes of ethanol at 0°C. Recover the DNA by centrifugation at 12,000g for 10 minutes at 0°C in a microfuge.

   b. Remove the supernatant by aspiration, and redissolve the DNA in 50 μl of bovine serum albumin (Fraction V; Sigma) (500 μg/ml in water).

   c. Add 50 μl of 2 × BAL 31 buffer.

   > *2 × BAL 31 buffer*
   >
   > 1.2 M NaCl
   > 24 mM $CaCl_2$
   > 24 mM $MgCl_2$
   > 40 mM Tris · Cl (pH 8.0)
   > 0.4 mM EDTA (pH 8.0)

   d. Incubate the mixture for 30 minutes at 30°C.

   e. Add 0.5 unit of BAL 31, which removes approximately 10 bp/minute/terminus.

   f. At five appropriate, closely spaced time points centered around the time needed to delete the bases between site 2 and the second codon of the gene, remove 10-μl samples and add 10 μl of 0.2 M EGTA (pH 8.0) to each sample. Store the samples on ice.

   g. Combine the samples, extract once with phenol:chloroform, and transfer the aqueous phase to a fresh microfuge tube.

   h. Precipitate the DNA with 2 volumes of ethanol at 0°C. Recover the DNA by centrifugation at 12,000g for 10 minutes at 0°C. Remove the supernatant, and wash the pellet with 70% ethanol. Redissolve the DNA in 20 μl of the appropriate 1 × restriction enzyme buffer.

4. Cleave the DNA at site 1 with the appropriate restriction enzyme.

---

5. Repair the DNA with the Klenow fragment of *E. coli* DNA polymerase I to create a blunt end containing 5 of the 6 nucleotides of restriction site 1.

a. Mix the DNA with:

10 × nick-translation buffer	2.5 μl
a mixture of all four dNTPs, each at a concentration of 2 mM	1 μl
Klenow fragment of *E. coli* DNA polymerase I	1 μl
H$_2$O to 25 μl	

> *10 × Nick-translation buffer*
>
> 0.5 M Tris · Cl (pH 7.5)
> 0.1 M MgSO$_4$
> 1 mM dithiothreitol
> 500 μg/ml bovine serum albumin (Fraction V; Sigma) (optional)
>
> Divide the stock solution into small aliquots and store them at −20°C.

b. Incubate the reaction for 30 minutes at room temperature.

c. Stop the reaction by adding 1 μl of 0.5 M EDTA (pH 8.0).

d. Extract once with phenol:chloroform.

e. Separate the DNA from unincorporated dNTPs by chromatography on, or centrifugation through, small columns of Sephadex G-50 (see Appendix E).

6. Recircularize the DNA by ligation, and use the plasmid to transform bacteria to antibiotic resistance.

7. Screen individual colonies by plasmid minipreps and restriction enzyme analysis for the presence of plasmids in which restriction site 1 has been regenerated. Such regeneration occurs when digestion with BAL 31 generates, by chance, a DNA molecule whose terminus carries the particular base pair required to complete site 1. Limited DNA sequence (see Chapter 13) may be required to identify the plasmid within the subpopulation in which the last nucleotide of site 1 is the first nucleotide of the second codon of the gene to be expressed.

8. Cleave 5 μg of the desired plasmid at site 1. Repeat step 3g–h, except resuspend the DNA in 10 μl of 10 mM Tris · Cl (pH 7.5). Digest with mung-bean nuclease to generate a blunt end directly preceding the first nucleotide of the second codon as follows:

a. Mix the 10 μl of DNA with:

10 × mung-bean nuclease buffer	2 μl
mung-bean nuclease (1.5 units/μl)	2.5 μl
H$_2$O	5.5 μl

10 × Mung-bean nuclease buffer

    300 mM sodium acetate (pH 4.5)
    500 mM NaCl
    10 mM $ZnCl_2$
    50% glycerol

    b. Incubate the reaction for 1 hour at 37°C.

    c. Extract the reaction once with phenol:chloroform, and transfer the aqueous phase to a fresh microfuge tube.

    d. Precipitate the DNA with 2 volumes of ethanol at 0°C. Recover the DNA by centrifugation at 12,000g for 10 minutes at 0°C in a microfuge. Remove the supernatant, and wash the pellet with 70% ethanol. Redissolve the DNA in 20 μl of the appropriate 1 × restriction enzyme buffer.

  9. Cleave the DNA with a restriction enzyme that cuts downstream from the gene (site 3 in Figure 17.7).

 10. Depending on the choice of vector and availability of compatible restriction sites, it may not be necessary to create a blunt end at the downstream site or it may be possible to use a restriction enzyme that leaves a blunt end. If necessary, repair the DNA from step 9 with the Klenow fragment of *E. coli* DNA polymerase I to create a blunt end at the downstream site (see step 5a–c).

 11. Isolate the DNA fragment containing the gene following gel electrophoresis to obtain a DNA fragment that can be inserted after the ATG of plasmid pAS1 (see Figure 17.8) or pKK240-11 (see Figure 17.9).

Expression of a Gene from the Prepared Fragments

To express any gene:

1. Digest pAS1 (Figure 17.8) with *Bam*HI and remove protruding termini with mung-bean nuclease or nuclease S1 (see step 8, page 17.23, or step 4, page 17.19, respectively). pAS1 provides a bacteriophage λ promoter, p_L, and the ribosome-binding site of the bacteriophage λ *c*II gene. Alternatively, digest pKK240-11 (Figure 17.9) with *Nco*I and fill in recessed termini with the Klenow fragment of *E. coli* DNA polymerase I (see step 5, page 17.23). pKK240-11 provides a *tac* promoter and a *lacZ* ribosome-binding site.

2. If necessary, cleave the vector with a restriction enzyme that cuts at a downstream site compatible with the DNA fragment prepared by one of the methods described above (see pages 17.18–17.24).

   In most cases, a blunt-ended fragment will be inserted and this step will not be required. In a few cases, the gene fragment prepared as described on pages 17.18–17.24 will contain a compatible sticky end with a site within the vector.

3. Ligate the purified DNA fragment (see pages 17.18–17.24) to the vector prepared in steps 1 and 2, and transform the appropriate strain (see Chapter 1). For pKK240-11, use strain RB791; for pAS1, use M5219.

4. Screen transformants for colonies that contain the cloned gene by plasmid minipreps and restriction enzyme analysis. Sequence the junction between the ATG and the second codon (see Chapter 13). Alternatively, synthesize an oligonucleotide complementary to the correct sequence extending from the ribosome-binding site through the first part of the gene. Screen transformants by colony hybridization with the oligonucleotide that bridges the junction between the ATG and the second codon (see Chapter 11), and sequence this region in the positive clones (see Chapter 13).

5. To obtain high levels of transcription, follow the procedure described in step 5, page 17.11, or step 3, page 17.13, for induction of the p_L or *tac* promoter, respectively. Test the correct clones for protein production (see pages 17.34–17.35).

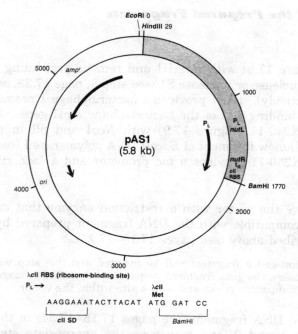

FIGURE 17.8

pAS1, a plasmid approximately 5.8 kb in length, carries the bacteriophage λ p_L promoter and a unique *Bam*HI site located at the ATG of the bacteriophage λ *c*II gene. This plasmid is a derivative of pKC30 (see Figure 17.4) into which the bacteriophage λ *c*II gene was inserted at the *Hpa*I site. The *c*II gene was then resected by exonuclease digestion until only the initiation codon ATG remained (the G of the ATG is the first nucleotide of a *Bam*HI site). To express a gene lacking an initiation codon, pAS1 is digested with *Bam*HI and then treated with mung-bean nuclease or nuclease S1 to remove the protruding, single-stranded termini. Ligation of this blunt-ended DNA to a blunt-ended DNA fragment that begins with the second codon of the gene to be expressed places that gene in-frame with the ATG. Genes inserted in this manner are regulated by introducing the recombinant plasmid into a temperature-sensitive bacteriophage λ lysogen (*c*I*ts*857). The cells are grown to late log phase at 30°C and then shifted to 40°C to inactivate the repressor and to turn on the p_L promoter. The inserted gene can also be regulated by the action of the N protein at *nut*L and *nut*R to antiterminate at t_{R1} (Shatzman et al. 1983).

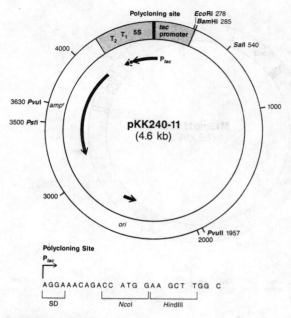

FIGURE 17.9

pKK240-11, a plasmid approximately 4.6 kb in length, carries a *tac* promoter, which is an *E. coli* hybrid promoter composed of *trp* and *lac* promoter sequences. A unique *Nco*I site, located at the ATG, can be filled in using the Klenow fragment of *E. coli* DNA polymerase I to form a blunt end to which the second codon of the gene to be expressed can be fused. Downstream from the polycloning site is *rrn*B, which contains an *E. coli* 5S gene and the T_1 and T_2 terminators (Amann and Brosius 1985).

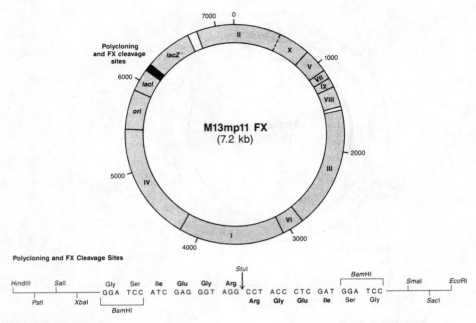

FIGURE 17.10

M13mp11 FX is an M13mp11 derivative that has the polycloning site oriented with *Eco*RI adjacent to *lacZ'*. DNA encoding the recognition sequence for blood coagulation factor X_a (Ile Glu Gly Arg [shown in bold]) is cloned into the *Bam*HI site. Any coding sequence with blunt ends can be cloned into the *Stu*I site to place the FX_a cleavage site adjacent to the desired protein sequence. The entire DNA sequence can then be inserted into an appropriate expression vector (e.g., pLcII) (Nagai and Thøgersen 1984).

ALTERNATIVE EXPRESSION SYSTEMS

In some cases, the methods described above do not result in the production of the desired protein. In particular, it may be important that the protein not have an amino-terminal methionine, which is coded for by the ATG in these vectors. The amino-terminal methionine is removed in *E. coli* to different extents for different proteins. Another common problem is that some proteins produced intracellularly are inactive, perhaps due to incorrect folding. Some foreign proteins are degraded by intracellular proteases. Two other expression systems may solve some of these problems.

Expression of a Cloned Gene as Part of a Fusion Protein That Can Be Cleaved by a Protease or Cyanogen Bromide

The cloned gene can be expressed as part of a fusion protein composed of a prokaryotic amino terminus connected to the protein of interest by a sequence that can be cleaved by a protease or by cyanogen bromide (Itakura et al. 1977; Goeddel et al. 1979; Nagai and Thøgersen 1984; Sung et al. 1986). The fusion protein is often stable compared to the foreign protein, and a protein can be made that does not contain an amino-terminal methionine.

Nagai and Thøgersen (1984) have developed a vector designed to produce hybrid proteins that can be cleaved with the blood coagulation factor X_a to liberate the foreign protein. Factor X_a cleaves specifically after the arginine of the tetrapeptide Ile-Glu-Gly-Arg, which connects the 31 amino-terminal residues of the bacteriophage λ *c*II protein to the protein of interest (Figure 17.10).

Other proteases have been similarly employed. For example, Jia et al. (1987) used λgt11 to make a β-galactosidase–carbohydrate-binding protein 35 (CBP35) fusion protein, which was then cleaved with V8 protease to release CBP35.

Proinsulin is readily degraded in *E. coli*. However, expression of a fusion protein composed of two molecules of proinsulin has been shown to stabilize the protein, perhaps because the protein product is larger than proinsulin. This fusion protein can then be cleaved by cyanogen bromide (Shen 1984).

In each case, it is important that the target amino acids for the protease or cyanogen bromide (which cleaves after methionine residues) does not occur at other locations within the protein of interest.

PRODUCTION OF HYBRID PROTEINS THAT CAN BE CLEAVED WITH FACTOR X$_a$

The expression plasmid of Nagai and Thøgersen is constructed as follows:

1. Using one of the three methods described on pages 17.18–17.24, prepare a cDNA fragment with a blunt 5′ terminus immediately preceding the first codon of the sequence to be expressed.

2. Clone the blunt-ended coding sequence into the *Stu*I site of M13mp11 FX. This *Stu*I site is located at the 3′ terminus of the sequence encoding the factor X$_a$ recognition site.

3. Clone a *Bam*HI fragment encoding the factor X$_a$ site and the foreign gene into the *Bam*HI site of pLcII, a plasmid encoding the bacteriophage λ p_L promoter that directs transcription of bacteriophage λ *c*II.

4. Transform *E. coli* strain MZ-1, which carries a defective bacteriophage λ *c*I*ts*857 prophage, and plate on LB medium containing ampicillin (100 μg/ml). Incubate overnight at 30°C.

5. Test individual colonies for the presence of the desired insert by plasmid minipreps.

6. To obtain high levels of transcription, follow the procedure in step 5, page 17.11, for induction of the p_L promoter. Test correct clones for protein production (see pages 17.34–17.35).

7. Purify the fusion protein and cleave it with factor X$_a$ as follows:

   a. Grow 500 ml of MZ-1 cells carrying the plasmid encoding the *c*II-Ile-Glu-Gly-Arg-foreign protein at 30°C in LB medium containing ampicillin (100 μg/ml) to an A$_{600}$ of 0.7.

   b. Add 500 ml of LB medium preheated to 65°C, and continue incubation for 2 hours at 40°C.

   c. Collect the cells and purify the fusion protein (see Nagai and Thøgersen 1984). Store at −20°C until needed.

   d. Activate bovine blood coagulation factor X to factor X$_a$ with Russell's viper venom (Fujikawa et al. 1972).

   e. Dissolve the purified fusion protein in 100 mM NaCl, 50 mM Tris · Cl (pH 8.0), 1 mM CaCl$_2$, and add factor X$_a$ at a 1:100 (factor X$_a$: protein) molar ratio. Incubate for 2 hours at 25°C. The cleavage reaction should be tested by SDS-polyacrylamide gel electrophoresis. Depending on the use, the cleaved product should be either used directly or purified further.

Expression of Secreted Foreign Proteins

Secretion of the foreign protein is an alternative to intracellular expression. This is accomplished by fusing the coding sequence to DNA encoding a signal peptide that is cleaved by signal peptidase when the protein is secreted into the periplasm located between the inner and outer membranes of *E. coli*. The two major problems encountered in secretion of foreign proteins are that yields are often low and cleavage of the signal peptide may not occur or may occur at an inappropriate position. However, there are a number of advantages:

- Some proteins that are degraded by intracellular proteases are stable in the periplasm (Talmadge et al. 1980).

- Some proteins that are inactive when produced intracellularly are active when secreted; secretion may allow them to be properly folded (Gray et al. 1985).

- Proteins are produced that do not have an amino-terminal methionine, since cleavage occurs between the signal peptide and the coding sequence.

Several expression vectors have been designed for the secretion of foreign proteins (Talmadge et al. 1980; Oka et al. 1985; Takahara et al. 1985). It may be necessary to try more than one secretion system to find a vector and an *E. coli* strain that will allow secretion of a given foreign protein. Even then, processing may not occur in the correct position and protein levels may be low. One system that may allow secretion and processing takes advantage of the alkaline phosphatase promoter (*phoA*) and signal sequence. Upon phosphate starvation, this promoter and signal direct the synthesis of large amounts of alkaline phosphatase, which is secreted into the periplasm. This system has been used for secretion of human epidermal growth factor (hEGF) into the periplasm of *E. coli* (Oka et al. 1985). The secreted hEGF has the correct amino terminus.

Plasmid pTA1529 (Figure 17.11) is a convenient vector for *phoA*-mediated expression and secretion.

1. Using one of the three methods described on pages 17.18–17.24, prepare a DNA fragment encoding the foreign protein such that there is a blunt 5′ terminus immediately preceding the first codon of the sequence to be expressed.

2. Digest pTA1529 with *Hin*dIII and fill in recessed termini with the Klenow fragment of *E. coli* DNA polymerase I (see page 17.23, step 5a and b, and page 17.22, step 3g–h). Cleave the vector at an appropriate downstream site (*Eco*RI, *Sma*I, *Bam*HI).

3. Clone the fragment into the vector, transform *E. coli* strain YK537, and plate on LB medium containing ampicillin (100 μg/ml). Screen transformants with a junction oligonucleotide or by plasmid minipreps, and sequence the junction region of the potentially correct clones to verify (see Chapter 13).

4. To obtain high levels of transcription from the *phoA* promoter, grow an overnight culture of YK537 transformants in high-phosphate medium (640 μM KH_2PO_4). Dilute the culture 1:50 in low-phosphate medium (32 μM KH_2PO_4). Phosphate will gradually become limiting, and the *phoA* promoter will be induced. Test for protein production at various times (2, 4, 8, and 24 hours) after induction using one of the methods described on pages 17.34–17.35.

5. Determine the cellular location of the protein by cell fractionation.

   a. Place 1 ml each of the induced and uninduced culture in separate microfuge tubes. Centrifuge at 12,000g for 1 minute at 4°C in a microfuge to pellet the cells. Discard the supernatant.

   b. Suspend the cell pellets in 100 μl of a freshly prepared solution of lysozyme (1 mg/ml), 20% w/v sucrose, 30 mM Tris·Cl (pH 8.0), 1 mM EDTA (pH 8.0), and place on ice for 10 minutes.

   Lysozyme will not work efficiently if the pH of the solution is less than 8.0.

   c. Recover the cells by centrifugation as in step a. The supernatant is the periplasmic fraction, which should be stored at 4°C until step g.

   d. Resuspend the cell pellets in 0.1 M Tris·Cl (pH 8.0), and break open the cells by freezing and thawing (i.e., place the cells on dry ice, thaw at 37°C, and repeat).

   e. Centrifuge the suspension at 12,000g for 5 minutes at 4°C in a microfuge. The cytoplasmic proteins are mainly in the supernatant, which should be stored at 4°C until step g. The pellet contains the membrane fraction and insoluble inclusion bodies.

   f. Solubilize the membrane proteins by incubating the pellet in 1% Triton X-100 for 10 minutes at 4°C.

g. Analyze the periplasmic, cytoplasmic, and membrane proteins by SDS-polyacrylamide gel electrophoresis (see Chapter 18, pages 18.47–18.54), potentially coupled with western blotting (see Chapter 18), to determine how effectively the protein of interest is secreted into the periplasm.

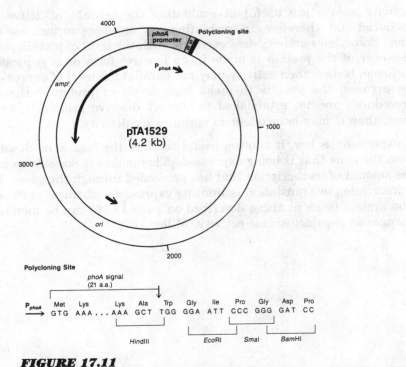

Polycloning Site

phoA signal
(21 a.a.)

P$_{phoA}$ Met Lys   Lys Ala Trp Gly Ile Pro Gly Asp Pro
→ GTG AAA . . . AAA GCT TGG GGA ATT CCC GGG GAT CC

HindIII   EcoRI SmaI BamHI

FIGURE 17.11

pTA1529 contains the promoter and signal sequence (SS) of the _E. coli phoA_ gene. The first codon of the signal sequence is GTG (Kikuchi et al. 1981). A polycloning site is present near the signal sequence cleavage site (↓) (Oka et al. 1985).

QUANTITATING THE LEVELS OF EXPRESSION OF CLONED GENES

After plasmids containing the transcription and translation initiation signals that allow the protein of interest to be expressed in *E. coli* are constructed, the amount of protein made is measured. Generally, four approaches are used:

1. SDS-polyacrylamide gels are run to determine whether a protein of the appropriate size is made at increased levels only after induction in cells that contain the plasmid expressing the protein of interest (Laemmli 1970). Often the protein of interest can be visualized by staining the gel with Coomassie Brilliant Blue or by silver staining (Oakley et al. 1980) (see Chapter 18, pages 18.55–18.57). If no new protein band is seen using either of these stains, metabolic labeling with 100 μCi of [^{35}S]Met or [^{35}S]Cys per milliliter of culture for 5 minutes following induction (see Chapter 18, pages 18.27–18.28), followed by SDS-polyacrylamide gel electrophoresis and autoradiography, may allow detection of the protein of interest.

2. Western gel analysis (Towbin et al. 1979) (see Chapter 18, pages 18.60–18.75) using antibodies that bind specifically to the protein of interest is usually necessary to confirm the identity of the gene product detected by SDS-polyacrylamide gel electrophoresis. In addition, sometimes proteins that are not detectable in Laemmli gels are detected by western gels.

3. Activity assays are useful in evaluating the amount of active protein produced and, therefore, in evaluating any attempts to increase production. Sometimes activity does correlate with the level of protein produced. However, if the protein is made in an inactive form or is sequestered in inclusion bodies, then activity may not parallel the level of expression. In some cases, the goal is to make high levels of protein. However, if procedures are not established to convert inactive protein to an active form, then it may be prudent to optimize for activity.

4. If expression is low, it is often useful to place the *lacZ* gene downstream from the gene that is being expressed. This makes it possible to measure the amount of transcription that has proceeded through the gene. Thus, if transcription or translation is limiting expression, changes in the expression system (such as those described on page 17.36) can be monitored by changes in β-galactosidase activity (Miller 1972).

Monitoring Expression by β-Galactosidase Activity

1. Grow an overnight culture in A medium containing 0.4% glucose, 1 μg/ml vitamin B1, 1 mM $MgSO_4$, and the appropriate antibiotic. Dilute the culture 1:50 in the same medium and grow to an A_{600} of 0.4. Place the culture on ice.

   To prepare A medium, mix:

K_2HPO_4	10.5 g
KH_2PO_4	4.5 g
$(NH_4)_2SO_4$	1.0 g
Na citrate $\cdot$ $2H_2O$	0.5 g
H_2O to 1 liter	

2. Mix 0.1 ml of the cell culture, 0.9 ml of Z buffer, 2 drops of chloroform, and 1 drop of 0.1% SDS. Vortex for 10 seconds, and then equilibrate to 28°C.

   Z buffer

   0.06 M $Na_2HPO_4 \cdot 7H_2O$
   0.04 M $NaH_2PO_4 \cdot H_2O$
   0.01 M KCl
   0.001 M $MgSO_4 \cdot 7H_2O$
   0.05 M β-mercaptoethanol

   The pH of the buffer should be 7.0. Do not autoclave.

3. Add 0.2 ml of *o*-nitrophenyl-β-D-galactopyranoside in A medium (4 mg/ml) to the lysed culture prepared in step 2 at 1-minute intervals. When a yellow color develops, stop the reaction by adding 0.5 ml of 1 M Na_2CO_3. Record the time at which the yellow color develops.

4. Determine the A_{600} of the cultures. Read the A_{550} and A_{420} of the reaction samples. Units of β-galactosidase activity =

$$1000 \times \frac{A_{420} - (1.75 \times A_{550})}{t \times 0.1 \times A_{600}}$$

where t = time in minutes. Improvements in transcription through the gene should be reflected by increases in the amount of β-galactosidase activity.

INCREASING EXPRESSION OF CLONED GENES

The level of expression of the cloned gene may be low due to RNA instability, premature termination, inefficient translation, or protein instability. Protein instability is distinguished from the other potential problems by pulse-chase experiments. A culture of cells carrying the expression plasmid is grown to mid-log phase, induced, and labeled for 10 minutes with [^{35}S]Met or [^{35}S]Cys (see Chapter 18, pages 18.27–18.28), followed by the addition of excess methionine or cysteine. Samples taken every 10 minutes from immediately before induction to 1 hour after induction are run on an SDS-polyacrylamide gel. The gel is dried and subjected to autoradiography to determine whether the protein is unstable. If the protein is unstable, then the use of protease-deficient strains (*lon*, *hfl*, or *htpR*) (Buell et al. 1985) or protease inhibitors during growth and harvesting may increase yields (1 mM phenylmethylsulfonyl fluoride [PMSF], 1–5 mM EDTA, 20 mM benzamidine). In many cases, problems of protein instability have been overcome by increasing the amount of foreign protein synthesized.

Several strategies have been successful in increasing the synthesis of foreign proteins. These include:

- *Use of site-directed mutagenesis to increase translation initiation.* The presence of secondary structure involving the ribosome-binding site can reduce the efficiency of translation initiation. Thus, removal of potential secondary structure involving the ribosome-binding site can increase translation initiation (Iserentant and Fiers 1980; Queen and Rosenberg 1981; Hall et al. 1982; Coleman et al. 1985). The SD sequence and the initiation codon are the major factors in ribosome binding. However, conserved base pairs in the region from −20 to +13 have been reported (Gold et al. 1981), and mutations in base pairs other than those in the SD sequence or ATG have been shown to alter translation initiation (Hui et al. 1984; Wood et al. 1984). Changing the DNA sequence near the SD and ATG sequences may increase expression.

- *Use of mutant strains defective in termination of transcription (*rho*) or altered in RNA metabolism (*pnp, rna*) to increase the amount of functional RNA.* Also try several different strains of *E. coli* because expression levels vary in different strains for reasons that are not clear. If the p_L promoter is used, then the strains must be lysogenized with bacteriophage λ *c*I*ts*857. If the *tac* promoter is used, then F′ *lacI*^{q} must be transferred to the host or *lacI* must be cloned into the expression plasmid.

- *Placing sequences that may stabilize the mRNA or make it more translatable downstream from the inserted DNA (i.e., transcription terminators or RNAase III sites [Panayotatos and Truong 1981; Studier and Moffatt 1986; Rosenberg et al. 1987]).* Studies of sib mutants of bacteriophage λ demonstrate the effect of secondary structure on the stability of the mRNA in *E. coli.*

- *Altering the kinetics of synthesis by changing temperature, ribosome-binding site (Marquis et al. 1986), promoter strength, plasmid copy number, or amount of inducer.* The kinetics of synthesis of foreign proteins may change cell physiology, protein folding, or susceptibility to proteolysis.

PROTEIN PURIFICATION

Inclusion Bodies

A high level of expression of proteins in *E. coli* often results in cytoplasmic granules that can be seen with a phase-contrast microscope and that can be separated from crude cell lysates by centrifugation. Cells expressing high levels of foreign protein are concentrated by centrifugation and lysed by mechanical techniques, sonication, or lysozyme plus detergents. The inclusion bodies are pelleted by centrifugation and washed with Triton X-100 and EDTA (Marston et al. 1984) or urea (Schoner et al. 1985).

To obtain soluble, active protein, the washed inclusion bodies must be solubilized and then refolded. Each protein may require a different procedure, which must be determined empirically. Various conditions (e.g., guanidine HCl [5–8 M], urea [6–8 M], SDS, alkaline pH, or acetonitrile/propanol) may be used to solubilize the inclusion bodies. A number of strategies to identify effective solubilization reagents have been described by Marston (1987). The procedure given below has been used to solubilize prorennin inclusion bodies and is adapted from Marston et al. (1984).

After successful solubilization, various refolding methods involving dilution or dialysis may be tried. The yield of active protein or of protein with the same disulfide bonds as the original protein depends on the concentration, purity, and size of the polypeptide; the pH and ionic strength of the solvent; and the rate of refolding. Other factors include the number of disulfide bonds and the nature of the protein itself. A number of successful refolding protocols are detailed in Marston (1987).

CELL LYSIS

The following procedure utilizes lysozyme plus detergent for lysis and is adapted from Marston (1987).

Important: Steps 2–4 should be performed in a coldroom.

1. Centrifuge one liter of cell culture at 500*g* for 15 minutes at 4°C.

2. Remove the supernatant and weigh the *E. coli* pellet. For each gram (wet weight) of *E. coli*, add 3 ml of lysis buffer. Resuspend the pellet.

> *Lysis buffer*
>
> 50 mM Tris·Cl (pH 8.0)
> 1 mM EDTA
> 100 mM NaCl

3. For each gram of *E. coli*, add 8 μl of 50 mM phenylmethylsulfonylfluoride (PMSF) and then 80 μl of lysozyme (10 mg/ml). Stir occasionally for 20 minutes.

   Caution: PMSF is extremely destructive to mucous membranes and the tissues of the respiratory tract, eyes, and skin. PMSF may be fatal if inhaled, swallowed, or absorbed through the skin. In case of contact, immediately flush eyes or skin with copious amounts of water. Discard contaminated clothing.

   PMSF is inactivated in aqueous solutions. The rate of inactivation increases with pH and is faster at 25°C than at 4°C. The half-life of a 20 μM aqueous solution of PMSF is about 35 minutes at pH 8.0 (James 1978). This means that aqueous solutions of PMSF can be safely discarded after they have been rendered alkaline (pH > 8.6) and stored for several hours at room temperature. PMSF is usually stored as a 10 mM or 100 mM stock solution (1.74 or 17.4 mg/ml in isopropanol) at −20°C.

   Lysozyme will not work efficiently if the pH of the solution is less than 8.0.

4. Add 4 mg of deoxycholic acid per gram of *E. coli* while stirring continuously.

5. Place at 37°C and stir with a glass rod. When the lysate becomes viscous, add 20 μl of DNAase I (1 mg/ml) per gram of *E. coli*.

6. Place the lysate at room temperature until it is no longer viscous (about 30 minutes).

PURIFICATION AND WASHING OF INCLUSION BODIES: METHOD 1

The following procedure is adapted from Marston et al. (1984).

Important: Steps 2 and 5 should be performed in a coldroom.

1. Centrifuge the cell lysate at 12,000g for 15 minutes at 4°C in a microfuge.

2. Decant the supernatant. Resuspend the pellet in 9 volumes of lysis buffer containing 0.5% Triton X-100 and 10 mM EDTA (pH 8.0).

3. Store at room temperature for 5 minutes.

4. Centrifuge at 12,000g for 15 minutes at 4°C in a microfuge.

5. Decant the supernatant and set aside for the next step. Resuspend the pellet in 100 μl of H$_2$O.

6. Remove 10-μl samples of the supernatant and the resuspended pellet. Mix each with 10 μl of 2× SDS gel-loading buffer and analyze by SDS-polyacrylamide gel electrophoresis to determine if most of the protein of interest is in the pellet.

> *2 × SDS gel-loading buffer*
>
> 100 mM Tris · Cl (pH 6.8)
> 200 mM dithiothreitol
> 4% SDS (electrophoresis grade)
> 0.2% bromophenol blue
> 20% glycerol
>
> 2× SDS gel-loading buffer lacking dithiothreitol can be stored at room temperature. Dithiothreitol should then be added, just before the buffer is used, from a 1 M stock (see Appendix B).

PURIFICATION AND WASHING OF INCLUSION BODIES: METHOD 2

The following procedure is adapted from Schoner et al. (1985).

Important: Steps 2, 4, and 6 should be performed in a coldroom.

1. Centrifuge the cell lysate at 12,000*g* for 15 minutes at 4°C in a microfuge.

2. Decant the supernatant. Resuspend the pellet in 1 ml of H_2O per gram of *E. coli*. Transfer 100-μl aliquots to four microfuge tubes and store the remainder.

3. Centrifuge at 12,000*g* for 15 minutes at 4°C in a microfuge.

4. Discard the supernatants. Resuspend each pellet in 100 μl of 0.1 M Tris · Cl (pH 8.5) containing a different concentration of urea (e.g., 0.5, 1, 2, and 5 M).

5. Centrifuge at 12,000*g* for 15 minutes at 4°C in a microfuge.

6. Decant the supernatants and set aside for the next step. Resuspend each pellet in 100 μl of H_2O.

7. Remove 10-μl samples of each supernatant and each resuspended pellet. Mix each with 10 μl of 2× SDS gel-loading buffer and analyze by SDS-polyacrylamide gel electrophoresis to determine the appropriate washing conditions for the protein.

SOLUBILIZATION OF INCLUSION BODIES

1. Suspend the washed pellet in 100 μl of lysis buffer (see page 17.37) containing 0.1 mM PMSF (added fresh), 8 M urea (deionized).

   Caution: PMSF is extremely destructive to mucous membranes and the tissues of the respiratory tract, eyes, and skin. PMSF may be fatal if inhaled, swallowed, or absorbed through the skin. In case of contact, immediately flush eyes or skin with copious amounts of water. Discard contaminated clothing.

   PMSF is inactivated in aqueous solutions. The rate of inactivation increases with pH and is faster at 25°C than at 4°C. The half-life of a 20 μM aqueous solution of PMSF is about 35 minutes at pH 8.0 (James 1978). This means that aqueous solutions of PMSF can be safely discarded after they have been rendered alkaline (pH > 8.6) and stored for several hours at room temperature. PMSF is usually stored as a 10 mM or 100 mM stock solution (1.74 or 17.4 mg/ml in isopropanol) at −20°C.

2. Store for 1 hour at room temperature.

3. Add this solution to 9 volumes of 50 mM KH_2PO_4 (pH 10.7), 1 mM EDTA (pH 8.0), 50 mM NaCl, and store for 30 minutes at room temperature. Maintain the pH at 10.7 with KOH.

   Caution: Handle concentrated KOH and HCl solutions with great care. Use gloves and a face protector.

4. Adjust the pH to 8.0 with HCl, and store for at least 30 minutes at room temperature.

5. Centrifuge at 12,000g for 15 minutes at room temperature in a microfuge.

6. Decant the supernatant and set aside for the next step. Resuspend the pellet in 100 μl of 1× SDS gel-loading buffer.

7. Remove 10-μl samples of the supernatant and resuspended pellet. Mix the supernatant sample with 10 μl of 2× SDS gel-loading buffer. Mix the pellet sample with 1× SDS gel-loading buffer. Analyze both samples by SDS-polyacrylamide gel electrophoresis to determine the degree of solubilization.

References

Amann, E. and J. Brosius. 1985. 'ATG vectors' for regulated high-level expression of cloned genes in *Escherichia coli*. *Gene* **40**: 183.

Amann, E., J. Brosius, and M. Ptashne. 1983. Vectors bearing a hybrid *trp-lac* promoter useful for regulated expression of cloned genes in *Escherichia coli*. *Gene* **25**: 167.

Buell, G., M.-F. Schulz, G. Selzer, A. Chollet, N.R. Movva, D. Semon, S. Escanez, and E. Kawashima. 1985. Optimizing the expression in *E. coli* of a synthetic gene encoding somatomedin-C (IGF-I). *Nucleic Acids Res.* **13**: 1923.

Coleman, J., M. Inouye, and K. Nakamura. 1985. Mutations upstream of the ribosome-binding site affect translational efficiency. *J. Mol. Biol.* **181**: 139.

de Boer, H.A., L.J. Comstock, and M. Vasser. 1983. The *tac* promoter: A functional hybrid derived from the *trp* and *lac* promoters. *Proc. Natl. Acad. Sci.* **80**: 21.

Fujikawa, K., M.E. Legaz, and E.W. Davie. 1972. Bovine factors X_1 and X_2 (Stuart factor). Isolation and characterization. *Biochemistry* **11**: 4882.

Germino, J. and D. Bastia. 1984. Rapid purification of a cloned gene product by genetic fusion and site-specific proteolysis. *Proc. Natl. Acad. Sci.* **81**: 4692.

Goeddel, D.V., D.G. Kleid, F. Bolivar, H.L. Heyneker, D.G. Yansura, R. Crea, T. Hirose, A. Kraszewski, K. Itakura, and A.D. Riggs. 1979. Expression in *Escherichia coli* of chemically synthesized genes for human insulin. *Proc. Natl. Acad. Sci.* **76**: 106.

Gold, L., D. Pribnow, T. Schneider, S. Shinedling, B.S. Singer, and G. Stormo. 1981. Translational initiation in prokaryotes. *Annu. Rev. Microbiol.* **35**: 365.

Gray, G.L., J.S. Baldridge, K.S. McKeown, H.L. Heyneker, and C.N. Chang. 1985. Periplasmic production of correctly processed human growth hormone in *Escherichia coli*: Natural and bacterial signal sequences are interchangeable. *Gene* **39**: 247.

Gray, M.R., H.V. Colot, L. Guarente, and M. Rosbash. 1982. Open reading frame cloning: Identification, cloning, and expression of open reading frame DNA. *Proc. Natl. Acad. Sci.* **79**: 6598.

Hager, D.A. and R.R. Burgess. 1980. Elution of proteins from sodium dodecyl sulfate–polyacrylamide gels, removal of sodium dodecyl sulfate, and renaturation of enzymatic activity: Results with sigma subunit of *Escherichia coli* RNA polymerase, wheat germ DNA topoisomerase, and other enzymes. *Anal. Biochem.* **109**: 76.

Hall, M.N., J. Gabay, M. Débarbouillé, and M. Schwartz. 1982. A role for mRNA secondary structure in the control of translation initiation. *Nature* **295**: 616.

Hasan, N. and W. Szybalski. 1987. Control of cloned gene expression by promoter inversion in vivo: Construction of improved vectors with a multiple cloning site and the p_{tac} promoter. *Gene* **56**: 145.

Hui, A., J. Hayflick, K. Dinkelspiel, and H.A. de Boer. 1984. Mutagenesis of the three bases preceding the start codon of the β-galactosidase mRNA and its effect on translation in *Escherichia coli*. *EMBO J.* **3**: 623.

Iserentant, D. and W. Fiers. 1980. Secondary structure of mRNA and efficiency of translation initiation. *Gene* **9**: 1.

Itakura, K., T. Hirose, R. Crea, A.D. Riggs, H.L. Heyneker, F. Bolivar, and H.W. Boyer. 1977. Expression in *Escherichia coli* of a chemically synthesized gene for the hormone somatostatin. *Science* **198**: 1056.

James, G.T. 1978. Inactivation of the protease inhibitor phenylmethylsulfonyl fluoride in buffers. *Anal. Biochem.* **86**: 574.

Jia, S., R.P. Mee, G. Morford, N. Agrwal, P.G. Voss, I.K. Moutsatsos, and J.L. Wang. 1987. Carbohydrate-binding protein 35: Molecular cloning and expression of a recombinant polypeptide with lectin activity in *Escherichia coli*. *Gene* **60**: 197.

Kikuchi, Y., K. Yoda, M. Yamasaki, and G. Tamura. 1981. The nucleotide sequence of the promoter and the amino-terminal region of alkaline phosphatase structural gene (*phoA*) of *Escherichia coli*. *Nucleic Acids Res.* **9:** 5671.

Laemmli, U.K. 1970. Cleavage of structural proteins during the assembly of the head of bacteriophage T4. *Nature* **227:** 680.

Marquis, D.M., J.M. Smolec, and D.H. Katz. 1986. Use of a portable ribosome-binding site for maximizing expression of a eukaryotic gene in *Escherichia coli*. *Gene* **42:** 175.

Marston, F.A.O. 1987. The purification of eukaryotic polypeptides expressed in *Escherichia coli*. In *DNA cloning: A practical approach* (ed. D.M. Glover), vol. 3, p. 59. IRL Press, Oxford.

Marston, F.A.O., P.A. Lowe, M.T. Doel, J.M. Schoemaker, S. White, and S. Angal. 1984. Purification of calf prochymosin (prorennin) synthesized in *Escherichia coli*. *Bio/Technology* **2:** 800.

Miller, J.H. 1972. *Experiments in molecular genetics*. Cold Spring Harbor Laboratory, Cold Spring Harbor, New York.

Nagai, K. and H.C. Thøgersen. 1984. Generation of β-globin by sequence-specific proteolysis of a hybrid protein produced in *Escherichia coli*. *Nature* **309:** 810.

Oakley, B.R., D.R. Kirsch, and N.R. Morris. 1980. A simplified ultrasensitive silver stain for detecting proteins in polyacrylamide gels. *Anal. Biochem.* **105:** 361.

Oka, T., S. Sakamoto, K. Miyoshi, T. Fuwa, K. Yoda, M. Yamasaki, G. Tamura, and T. Miyake. 1985. Synthesis and secretion of human epidermal growth factor by *Escherichia coli*. *Proc. Natl. Acad. Sci.* **82:** 7212.

Panayotatos, N. and K. Truong. 1981. Specific deletion of DNA sequences between preselected bases. *Nucleic Acids Res.* **9:** 5679.

Queen, C. and M. Rosenberg. 1981. Differential translation efficiency explains discoordinate expression of the galactose operon. *Cell* **25:** 241.

Rosenberg, A.H., B.N. Lade, D. Chui, S.-W. Lin, J.J. Dunn, and F.W. Studier. 1987. Vectors for selective expression of cloned DNAs by T7 RNA polymerase. *Gene* **56:** 125.

Rüther, U. and B. Müller-Hill. 1983. Easy identification of cDNA clones. *EMBO J.* **2:** 1791.

Schoner, R.G., L.F. Ellis, and B.E. Schoner. 1985. Isolation and purification of protein granules from *Escherichia coli* cells overproducing bovine growth hormone. *Bio/Technology* **3:** 151.

Shatzman, A.R. and M. Rosenberg. 1987. Expression, identification, and characterization of recombinant gene products in *Escherichia coli*. *Methods Enzymol.* **152:** 661.

Shatzman, A., Y.-S. Ho, and M. Rosenberg. 1983. Use of phage λ regulatory signals to obtain efficient expression of genes in *Escherichia coli*. In *Experimental manipulation of gene expression* (ed. M. Inouye), p. 1. Academic Press, New York.

Shen, S.-H. 1984. Multiple joined genes prevent product degradation in *Escherichia coli*. *Proc. Natl. Acad. Sci.* **81:** 4627.

Shimatake, H. and M. Rosenberg. 1981. Purified λ regulatory protein *c*II positively activates promoters for lysogenic development. *Nature* **292:** 128.

Shine, J. and L. Dalgarno. 1975. Determinant of cistron specificity in bacterial ribosomes. *Nature* **254:** 34.

Stanley, K.K. and J.P. Luzio. 1984. Construction of a new family of high efficiency bacterial expression vectors: Identification of cDNA clones coding for human liver proteins. *EMBO J.* **3:** 1429.

Steitz, J.A. 1979. Genetic signals and nucleotide sequences in messenger RNA. In *Biological regulation and development: Gene expression* (ed. R.F. Goldberger), vol. 1, p. 349. Plenum Publishing, New York.

Studier, F.W. and B.A. Moffatt. 1986. Use of bacteriophage T7 RNA polymerase to direct selective high-level expression of cloned genes. *J. Mol. Biol.* **189:** 113.

Sung, W.L., F.-L. Yao, D.M. Zahab, and S.A. Narang. 1986. Short synthetic oligodeoxyribonucleotide leader sequences enhance accumulation of human proinsulin synthesized in *Escherichia coli*. *Proc. Natl. Acad. Sci.* **83:** 561.

Tabor, S. and C.C. Richardson. 1985. A bacteriophage T7 RNA polymerase/promoter system for controlled exclusive expression of specific genes. *Proc. Natl. Acad. Sci.* **82:** 1074.

Takahara, M., D.W. Hibler, P.J. Barr, J.A. Gerlt, and M. Inouye. 1985. The *ompA* signal peptide directed secretion of staphylococcal nuclease A by *Escherichia coli*. *J. Biol. Chem.* **260:** 2670.

Talmadge, K., J. Kaufman, and W. Gilbert. 1980. Bacteria mature preproinsulin to proinsulin. *Proc. Natl. Acad. Sci.* **77:** 3988.

Towbin, H., T. Staehelin, and J. Gordon. 1979. Electrophoretic transfer of proteins from polyacrylamide gels to nitrocellulose sheets: Procedure and some applications. *Proc. Natl. Acad. Sci.* **76:** 4350.

Weinstock, G.M., C. ap Rhys, M.L. Berman, B. Hampar, D. Jackson, T.J. Silhavy, J. Weisemann, and M. Zweig. 1983. Open reading frame expression vectors: A general method for antigen production in *Escherichia coli* using protein fusions to β-galactosidase. *Proc. Natl. Acad. Sci.* **80:** 4432.

Wood, C.R., M.A. Boss, T.P. Patel, and J.S. Emtage. 1984. The influence of messenger RNA secondary structure on expression of an immunoglobulin heavy chain in *Escherichia coli*. *Nucleic Acids Res.* **12:** 3937.

18

Detection and Analysis of Proteins Expressed from Cloned Genes

18

The scope of recombinant DNA research has undergone a dramatic expansion. In its early days, the field was concerned mainly with describing the structure of eukaryotic cDNAs, establishing the topography of the corresponding genomic sequences, and defining the location of controlling elements. Although this kind of cataloging remains an essential part of molecular cloning, the field has now broadened to include expression of cloned genes and analysis of the proteins they encode. Methods commonly used to express cloned genes in prokaryotic and eukaryotic systems are discussed elsewhere in this manual. In this chapter, we describe techniques to detect and quantitate foreign proteins synthesized in these systems. Although we emphasize methods to analyze proteins expressed in mammalian cells, the same techniques are applicable to proteins synthesized in yeast and prokaryotic systems. We also describe methods to synthesize proteins in vitro using the rabbit reticulocyte system and natural mRNAs or synthetic RNAs.

Proteins synthesized in heterologous systems can be detected either by assaying for a particular biological activity or by employing assays that are independent of such activity. In a few cases, the protein of interest carries an enzymatic or other biological activity (e.g., the ability to bind a specific ligand) that can be assayed in intact cells or in extracts in vitro. Although such assays can be extremely useful, they frequently suffer from one of several practical limitations: (1) They may not be sufficiently sensitive to detect the small amounts of protein that are synthesized in small-scale mammalian cultures. (2) The host cells may themselves express an endogenous protein that either displays the same biological activity as the foreign protein or interferes with it. (3) The detection of biological activity in an extract says nothing about the specific activity of the protein of interest (i.e., the activity displayed by a given amount of the protein). This point becomes especially important when measuring the activity of mutated forms of the protein that have been generated by in vitro mutagenesis or when expressing the wild-type protein in heterologous cells that may not allow the protein to fold correctly or may not carry out the correct posttranslational modifications. For these and other reasons, it is essential to develop assays that are independent of biological activity and sensitive enough to measure very small amounts of the protein. The reagents of choice for these assays are antibodies that react specifically with the foreign protein.

Production of Antibodies

It is important to bear in mind from the outset that antibodies that react specifically with the foreign protein fall into three general classes:

- *Antibodies that react with the foreign protein independently of its conformation.* Antibodies of this type are particularly useful for measuring the total amount of the target protein present in crude preparations or cell extracts. They are usually raised by immunizing animals with partially denatured protein or with a peptide whose sequence corresponds to part of the intact protein. However, monoclonal antibodies that are pan-specific are not uncommon.

- *Antibodies that react only with epitopes specific to the native form of the target protein.* Antibodies of this type are typically monoclonal and recognize a given sequence of amino acids only when it occurs in the particular three-dimensional conformation that is characteristic of the native form of the target protein.

- *Antibodies that react only with denatured forms of the protein.* These are raised against fully denatured antigens and can be either monoclonal or polyclonal.

Factors Affecting Immune Response

Although there is no way to guarantee the production of particular types of antibodies, it is advisable to choose an immunization regimen that will favor antibodies with the desired characteristics. It is then necessary to screen several independent antisera or a series of monoclonal antibodies to identify those best suited to the tasks at hand. A number of factors that affect the strength and specificity of the immune response should be considered by workers planning to raise antisera. These are described below.

SPECIES OF ANIMAL

To raise antisera that react with as many potential epitopes as possible, choose a species of animal that is phylogenetically different from the species from which the target protein was isolated. Inoculation of a native protein into an animal of the same species is likely to yield antibodies directed only against minor epitopes that vary from individual to individual. For most proteins, rabbits are the animals of choice: They are easy to bleed, yielding several milliliters of serum; they are sufficiently robust to withstand multiple bleedings; and they are reasonably inexpensive. Chickens have many of the same virtues; however, they are clearly a second choice because their antibodies, unlike those of rabbits, do not bind to protein A of *Staphylococcus aureus*. Mice and guinea pigs are less desirable because their small size makes repeated drawing of blood difficult.

GENETIC FACTORS

Outbred animals are preferred to inbred strains, which may be genetically incapable of responding vigorously to certain antigens. Even outbred animals, however, exhibit a broad range of intensity of response to the same immunogen. It is therefore essential to immunize several animals, especially when attempting to raise antisera against poor immunogens. In some cases, it is possible to improve the response to such antigens by physically coupling them to a carrier protein such as keyhole limpet hemocyanin (KLH) or bovine serum albumin.

PHYSICAL STATE OF THE ANTIGEN

Aggregated protein is a much better immunogen than soluble, dispersed protein. Partial denaturation may help to increase immunogenicity but may yield antisera that fail to react with native protein. Severely denatured proteins (e.g., proteins that have been extensively boiled, reduced, and alkylated) often yield antisera that react exclusively or preferentially with denatured protein. Many immunology textbooks state that denatured proteins are poor antigens. This is a fallacy that almost certainly stems from the failure to assay the antisera against the appropriately denatured form of the antigen.

When the amount of antigen is limiting, adjuvants are used to enhance the immune response. These substances prolong the half-life of the antigen by protecting it from degradation. In addition, they minimize any direct toxic effects of the antigen and, by allowing a slow, sustained release from the site of immunization, increase the efficiency of uptake of the antigen by macrophages. A large variety of substances can be used as adjuvants, including alum, mineral oils, charcoal, and Freund's adjuvant (complete or incomplete). Nitrocellulose filters and polyacrylamide are particularly useful adjuvants when only small amounts of antigen are available. Bands of polyacrylamide containing specific polypeptides can be excised from polyacrylamide gels or electrophoretically transferred to nitrocellulose filters and used directly for immunization (see discussion below on immunizing with small amounts of antigen).

Note: If the antisera will be used to detect proteins expressed in *E. coli,* Freund's adjuvant should not be used due to the presence of bacterial components in the Freund's adjuvant that will generate cross-reactive antibodies to *E. coli* proteins.

AMOUNT OF ANTIGEN

Injection of large amounts of a soluble immunogen can lead to the establishment of a tolerant state. Although the use of adjuvants greatly reduces the chances that tolerance will be induced, it is important not to use more antigen than necessary, since this often leads to the generation of antisera of low avidity. Although there is no foolproof method to predict what the lowest effective dose of an unknown antigen will be, the immunization regimens described below usually yield antisera of high titer and avidity. If sufficient animals and antigen are available, it is best to inoculate batches of animals with different amounts of antigen. Those that receive too little antigen to elicit a response can always be given a larger dose subsequently. Booster injections are essential to obtain antisera of high titer and avidity. The first of these booster injections should be given when the primary response is past its peak (about 4–6 weeks after the primary injection of antigen in the case of rabbits). About 7–10 days after the first booster injection, test bleeds should be taken, and further booster injections should be given at regular intervals to those animals that show the maximum response to the antigen. The length of these intervals varies from antigen to antigen and cannot be predicted in advance. The best procedure is to boost the animal when the titer of circulating antibody begins to drop. The amount of antigen used for boosting is usually two- to threefold less than that used for primary immunization. Such booster injections are usually given to rabbits by the same route (although not necessarily at the same site) as the primary immunization. However, mice are more frequently boosted by intravenous injection of soluble antigen even if the primary immunization was by another route.

ROUTE OF INJECTION

Antigens stimulate the strongest immune responses when they are injected directly into the animal's popliteal lymph nodes. However, this requires special skill, and the much simpler methods of injection of antigen mixed with adjuvant into multiple intradermal or subcutaneous sites or into a single intramuscular site are used more commonly. In multiple intradermal injections, where the antigen-adjuvant complex is injected at 20–40 sites spread over the body of the animal, the primary response is produced faster and the titer of antibody is generally higher than when the antigen is injected into a single intramuscular site. Furthermore, fewer booster injections are required. However, since an ulcer forms at the site of each injection of adjuvant, the multiple-injection method may cause more distress to the animal.

IMMUNIZATION SCHEDULES

The injection of antigen should be carried out by someone experienced in handling animals according to the procedures approved by your animal care facility. Some facilities prefer not to use certain types of adjuvants (e.g., Freund's complete adjuvant or crushed polyacrylamide gels), and it is therefore important to agree on an immunization regimen before animals are ordered. *Make sure that a sample of blood is drawn from each animal before immunization is begun.* Although the immunization schedule will vary according to the nature of the antigen, the amount of antigen available, and its immunogenicity, a reasonable general plan for raising antisera against an antigen that is readily available is described below.

Rabbits

Dissolve 100 μg of antigen in 0.5 ml of a buffer in which the antigen is soluble, and emulsify the solution with an equal volume of Freund's complete adjuvant. Use the emulsion immediately. An emulsion of the desired consistency is readily obtained by passing the solution through a double-canula microemulsifying needle (Thomas Scientific 345OD46).

- *For intramuscular injection:* Inject 0.5 ml of emulsified antigen-adjuvant into each of two limbs of the animal. Four to 6 weeks later, inject 0.25 ml of freshly prepared emulsion into another limb. Limbs should be used in rotation every 4–6 weeks for subsequent booster injections.

- *For intradermal injection:* Inject 0.025 ml of emulsified antigen at 24–36 sites distributed over the back of a shaved rabbit. Four to 6 weeks later, boost the animal by intramuscular injection of 0.25 ml of freshly prepared emulsified antigen-adjuvant into each of two limbs of the animal.

Arrange for blood (20–40 ml) to be drawn 7–10 days after each booster injection. The serum should be tested for the presence of antibody by one of the standard methods, such as radioimmunoassay (RIA), enzyme-linked immunosorbent assay (ELISA), or immunoprecipitation. Animals that respond poorly to the antigen after two booster immunizations should be sacrificed. Responsive animals should be boosted at regular intervals (2–3 months) until a high titer of antibody is attained. Blood (40 ml) can then be drawn weekly until the titer drops. The animal should then be allowed to rest for 2–3 months and reboosted before bleeding is resumed.

Mice

Dissolve 20–50 μg of antigen in 100 μl of a buffer in which the antigen is soluble, and emulsify the solution with an equal volume of Freund's complete adjuvant. Use the emulsion immediately. Inject 50 μl of the emulsified antigen-adjuvant subcutaneously at each of four sites. Four to 6 weeks later, boost the mice with an intraperitoneal injection of 100 μl of the appropriate buffer containing 5–10 μg of soluble antigen. The mice should be bled from the median tail vein 7–10 days after the booster injection, and the serum should be tested for the presence of antibody by one of the standard methods. The animals can be reboosted if necessary, and blood can be withdrawn every 3–4 days until the antibody titer drops.

Immunizing with Small Amounts of Antigen

Frequently, the molecular biologist is faced with the problem of raising antibodies to proteins that are available in extremely small amounts. To maximize the chances of obtaining an immune response, the soluble antigen (up to 10 μg) emulsified in Freund's complete adjuvant should be injected into the popliteal lymph node of a rabbit. However, immune responses can also be obtained when animals are immunized by conventional routes with small amounts of polypeptides (100 ng to 5 μg) that have been recovered from polyacrylamide gels. After electrophoresis, the polypeptides are transferred from the gel to a nitrocellulose filter by western blotting, and the relevant segment of the nitrocellulose filter is then either implanted subcutaneously in a rabbit or dissolved in a small volume of dimethyl sulfoxide (DMSO), mixed with Freund's complete adjuvant, and injected intradermally as described above. Alternatively, the band of polyacrylamide containing the polypeptide of interest is cut from the gel. The polyacrylamide is then broken into small fragments by freezing and thawing and passage through a narrow-gauge hypodermic needle. The fragments of gel are then emulsified in an equal volume of Freund's complete adjuvant and injected subcutaneously into mice or intradermally at many sites into a rabbit.

Monoclonal Antibodies

The introduction of cell hybridization to generate myeloma-like cell lines that secrete monoclonal antibodies with desired specificities has greatly expanded the range, number, and reproducibility of immunological reagents (Köhler and Milstein 1975, 1976). Although the protocols to develop hybridomas are nearly as numerous as the number of investigators using them, almost all of them involve fusing spleen cells from a hyperimmunized donor with cells from a genetically marked myeloma that has been adapted to grow in tissue culture. Mixed populations of the resulting hybrid cells are tested for their ability to secrete antibody of the desired specificity, and clones derived from individual cells are then established. In the best cases, the supernatant of the resulting cultures contains up to 50 μg/ml of immunoglobulin. However, even greater amounts of antibody (5–20 mg/ml) can be obtained by growing the cloned hybridoma as an ascites tumor in the peritoneal cavity of mice.

Monoclonal antibodies are particularly useful when an immune reagent is required that reacts specifically with a single component in a complex mixture of antigens, for example, an enzyme that is contaminated with other proteins. Although an immunized animal may develop antibodies against all of the proteins in the mixture, an individual B lymphocyte will produce antibody specific for only one of them. The establishment of permanent cell lines from such individual B cells provides a method to obtain an antibody that is specific for the enzyme of interest. When using a monoclonal antibody, it is important to determine its subclass, using commercially available anti-idiotypic reagents, since not all subclasses of immunoglobulins bind efficiently to commonly used immunoabsorbents such as staphylococcal protein A.

A description of methods to generate monoclonal antibodies is beyond the scope of this manual. However, a number of excellent reviews are available that describe the techniques in detail (see, e.g., Campbell 1984; Goding 1986; Kipps and Herzenberg 1986; Harlow and Lane 1988).

Raising Antisera against Synthetic Peptides

If the amino acid sequence of the protein of interest is known or can be deduced from the corresponding nucleic acid sequence, it is possible to raise specific antisera by immunizing animals with short synthetic peptides (for review, see Lerner 1984). If the experimenter wishes to generate antibodies that react with specific sites in the target protein, there may be little or no choice in the sequence of the peptide that is used as an immunogen. However, if the antibody is to be used merely as a reagent to identify the protein, there may be an opportunity to select a sequence of amino acids that is likely to be highly immunogenic. If possible, choose a sequence of 12–15 amino acids that contains no more than four adjacent hydrophobic residues and no more than six hydrophobic residues in total—the greater the number of charged amino acids in the sequence the better. Peptides with these properties are likely to be soluble in aqueous solvents and therefore easy to couple to carrier proteins. In addition, the cognate sequence is more likely to be present on the surface of the target protein and therefore accessible to the antibody.

Usually, the synthetic peptide is coupled to a carrier protein such as keyhole limpet hemocyanin (KLH) through an amino- or carboxy-terminal cysteine residue. Ideally, therefore, the sequence of the peptide should end with a "natural" cysteine that is present in the cognate sequence of the target protein. If this is not possible, choose a peptide that contains no cysteine at all. Chemical synthesis of the peptide, which proceeds from the carboxyl terminus to the amino terminus, is then initiated from an "artificial" cysteine that has no counterpart in the cognate sequence of the target protein. This cysteine residue is used as a linker to couple the synthetic peptide to the carrier protein.

COUPLING OF SYNTHETIC PEPTIDES TO KEYHOLE LIMPET HEMOCYANIN

This method, from the laboratory of M. Brown and J. Goldstein (pers. comm.), is a modification of a technique described originally by Green et al. (1982).

1. Dilute KLH with phosphate-buffered saline (pH 6.0) (PBS; see Appendix B) to a final concentration of 10 mg/ml.

   KLH (Calbiochem no. 374817) is supplied at a concentration of 100 mg/ml in 50% glycerol. KLH should be stored at 4°C.

   The pH of PBS is adjusted to pH 6.0 with 3 N HCl.

2. Transfer 1 ml of the KLH solution to a clean 15-ml tube. Put a small, Teflon-covered magnet in the tube, and stir the solution vigorously on a magnetic stirrer. Add 50 μl of dimethyl sulfoxide (DMSO) containing 1.5 mg MBS *under the surface of the KLH solution*. Continue stirring for another 30 minutes.

   MBS is *m*-maleimidobenzoic acid *N*-hydroxysuccinimide ester (Sigma M 8759). It was first used as a coupling agent by Liu et al. (1979).

3. Prepare a column (20-ml bed volume) of Sephadex G-25, equilibrated in PBS (pH 7.4). Apply the sample of KLH to the column and begin to collect 1-ml samples. Immediately after the void volume, a cloudy, grayish material elutes from the column; this is KLH-MBS.

   The purpose of the column is to remove unconjugated MBS, DMSO, and glycerol from the sample.

   If necessary, the elution profile of KLH-MBS can be monitored by measuring the OD_{280} of the column effluent. The recovery of KLH-MBS is usually >80%.

4. Dissolve 5 mg of the synthetic peptide in the smallest possible volume of PBS (pH 7.4). This step is best carried out as follows:

   a. Add 5 mg of the peptide to 200 μl of PBS in a 15-ml tube. Continue to add PBS in 100-μl aliquots until the peptide is completely dissolved or the volume of the solution is 500 μl.

   b. If the peptide is not dissolved, add 5 N NaOH in 5-μl portions until a total of 25 μl has been added.

   c. If the peptide is still not dissolved, add DMSO in 100-μl portions until a total of 500 μl has been added.

5. Mix the KLH-MBS solution from step 3 with the dissolved peptide. *Vortex vigorously to mix the two solutions.* Stir the mixture for 1 hour at room temperature.

6. Store the solution at −20°C until it is needed for injection into animals.

Notes

i. If the peptide is too hydrophobic, it will not dissolve, and alternative, less efficient ways must then be used to raise antiserum. For example, the hydrophobic peptide can be emulsified with Freund's complete adjuvant and injected without solubilization. This results in the production of specific antibodies that are often of low titer and avidity.

ii. Upon thawing, the cross-linked peptide may form a precipitate. If so, break up the aggregates by sonication or by passage through a 22-gauge hypodermic needle. For immunization, mix approximately 300 μg of cross-linked peptide with an equal volume of Freund's complete adjuvant and inject the emulsion into mice (subcutaneously) or rabbits (intramuscularly or intradermally) as described on pages 18.5–18.6. Two weeks later, boost the animals with a second injection. Mice should be boosted again, 1 week later, with an intraperitoneal injection of 200 μg of peptide mixed with 4 mg of alum. Rabbits should be boosted every 6 weeks with 200 μg of peptide mixed with 4 mg of alum. Mice should be bled 4–5 weeks after the initial injection; rabbits should be bled 1–2 weeks after the second booster injection and at regular intervals thereafter.

Collection and Storage of Antisera

Blood from animals that have been fasted for several hours should be collected in sterile centrifuge tubes or bottles (without citrate or heparin) and allowed to clot at room temperature. A glass rod or sealed pasteur pipette should then be used to "ring" the clot (i.e., to loosen the clot from the wall of the tube or bottle). During the next several hours, the clot will retract to about one half of its original volume, leaving the antiserum as a straw-colored liquid. Transfer as much as possible of the antiserum to a fresh tube, and then centrifuge the clot at 1500g for 10 minutes at room temperature. Combine the supernatant with the previously removed antiserum, discard the clot, and store the antiserum.

Antisera may be stored as a lyophilized powder at room temperature, as a solution frozen at $-20°C$ or $-70°C$, or at 4°C in the presence of 0.02% sodium azide. Although lyophilization is preferred on theoretical grounds, it is impractical for many laboratories and unnecessary for most purposes. Some guidelines for storing antisera are listed below.

Caution: Sodium azide is poisonous. It should be handled with great care wearing gloves, and solutions containing it should be clearly marked.

1. Antisera should be stored in small aliquots that are labeled with the name of the animal and the date of the bleed. Polyclonal antisera are best stored undiluted, whereas monoclonal antibodies should be stored at concentrations of >1 mg/ml in solutions containing 0.1 M NaCl.

2. Do not mix antisera taken from different animals or from different bleeds taken from the same animal. Individual sera can contain cross-reactive antibodies and, during a course of immunizations, an individual animal may respond to other environmental antigens that also are cross-reactive. To avoid the possibility of adulterating a batch of good antiserum with one that is less desirable, keep each bleed from each animal separate.

3. Antisera stored at $-20°C$ are stable for several years. However, repeated freezing and thawing of antisera causes denaturation and aggregation of immunoglobulins, with a consequent decrease in potency. Once a frozen aliquot is thawed, it should thereafter be stored at 4°C.

4. Immunoglobulins of the IgM class are far less robust than IgG and tend to denature when stored at $-20°C$, which is near the eutectic point of NaCl solutions. Antisera that contain high levels of IgM directed against the target antigen should therefore be stored lyophilized or at $-70°C$.

5. Antisera gradually become turbid during prolonged storage at 4°C. The precipitate, which is composed mostly of lipoproteins and other cryoproteins whose solubility is limited at 4°C, does not affect the ability of the immunoglobulins to react with their target antigens. Antisera drawn from animals that have not been fasted contain higher levels of lipoprotein and rapidly become turbid during storage at 4°C. Do not mistake this turbidity for bacterial growth. Before use, antibodies should be centrifuged briefly (at 12,000g for 2 minutes at 4°C in a microfuge) to sediment any particulate matter and to cause lipoproteins to float to the surface.

Purification of Antibodies

For most purposes, antisera need not be fractionated before use. However, if the antisera are to be radiolabeled or conjugated to enzymes such as horseradish peroxidase (HRP) or alkaline phosphatase, it is necessary to purify the IgG fraction or, in some cases, to purify the antibody of interest by binding it to its cognate antigen.

Although many techniques have been developed to purify IgG molecules, the method of choice is adsorption to, and elution from, beads coated with protein A, a component of the cell wall of *S. aureus* (Hjelm et al. 1972). For reasons that are not known, this protein ($M_r = 42,000$) binds strongly to sites in the second and third constant regions of the Fc portion of the immunoglobulin heavy chain (Deisenhofer 1981). Each IgG molecule therefore contains two binding sites for protein A. Because protein A itself has four potential sites for binding to IgG (Sjödahl 1977), it is possible to form multimeric complexes of the two types of protein.

Not all immunoglobulins bind to protein A with the same affinity. Antibodies from humans, rabbits, and guinea pigs bind most tightly, followed in decreasing order of affinity by those from pigs, mice, horses, and cows (Kronvall et al. 1970; Goudswaard et al. 1978). Immunoglobulins from goats, rats, chickens, and hamsters bind in a much weaker fashion, and a "bridging" antibody is usually required to purify them by adsorption to protein A. Within any one species, different classes of immunoglobulins vary in the sequences of their Fc regions and consequently bind to protein A with different affinities. Of the major classes of human IgG, for example, three (IgG1, 2, and 4) bind with high affinity and one (IgG3) binds very weakly, if at all. Similarly, mouse IgG2a binds with high affinity, IgG2b and IgG3 bind tolerably well, and IgG1 binds poorly (Ey et al. 1978). These differences are generally unimportant when dealing with polyclonal sera, where antibodies against the target antigen are distributed throughout all of the major subclasses of IgG. Consequently, purification of polyclonal immunoglobulins raised in rabbits, humans, and mice by binding to protein A may alter the distribution of subclasses of IgG, but it rarely changes the specificity or avidity of the final preparation. However, monoclonal antibodies secreted from hybridomas carry only one subclass of heavy chain. Before attempting to purify a given monoclonal antibody, it is essential to determine the subclass of its heavy chain, using commercially available immunological reagents directed against isotypes of the Fc region. If the monoclonal antibody falls into a class that binds poorly to protein A (e.g., human IgG3 or mouse IgG1), it should be purified by another method (e.g., ammonium sulfate precipitation followed by chromatography on DEAE-cellulose). Alternatively, a bridging antibody can be used to attach the monoclonal antibody to protein A.

PURIFICATION OF ANTIBODIES BY ADSORPTION TO PROTEIN A

Protein A coupled to a solid support by cyanogen bromide is supplied by several manufacturers (e.g., protein A–Sepharose CL-4B; Pharmacia). Each milliliter of swollen gel can bind approximately 10–20 mg of IgG (equivalent to 1–2 ml of antiserum). Antibodies bind to protein A chiefly by hydrophobic interactions (Deisenhofer 1981) that can be disrupted at low pH. Protein A is remarkably resilient and withstands repeated cycles of exposure to low pH extremely well; it can also be treated with high concentrations of denaturing agents such as urea, guanidine hydrochloride, or potassium isothiocyanate without permanent damage. Most antibodies can withstand transient exposure to low pH, and this is now the standard method to release them in an active form from protein A–Sepharose beads.

The following is a modification of the method of Goudswaard et al. (1978).

1. Prepare a column of protein A–Sepharose, equilibrated in 100 mM Tris · Cl (pH 8.0) according to the manufacturer's instructions. Each milliliter of swollen gel can adsorb between 10 mg and 20 mg of pure IgG.

   Protein A–Sepharose is very expensive and should be used sparingly. For the preparation of IgG for most laboratory purposes, 1-ml columns are poured and run in pasteur pipettes plugged with sterile glass wool.

2. Add 0.1 volume of 1 M Tris · Cl (pH 8.0) to the antibody preparation (serum or monoclonal antibodies derived from hybridomas grown either in tissue culture or in the peritoneal cavity of mice).

   The concentration of IgG in serum obtained from healthy animals is approximately 10 mg/ml. However, the concentration of antibody in supernatants of hybridomas or in ascitic fluid is highly variable, ranging from 10 μg/ml to 100 μg/ml in the former case and from 1 mg/ml to 20 mg/ml in the latter. To determine the concentration of antibody in serum, analyze an aliquot by SDS-polyacrylamide gel electrophoresis and compare the intensity of staining of the heavy and light chains with a set of immunoglobulin standards run on the same gel. When the antibody is the dominant protein in the solution (e.g., in ascitic fluid or in the supernatant of efficiently secreting hybridomas), its approximate concentration can be estimated from the OD_{280} of the solution (1 OD_{280} = 0.75 mg/ml of pure IgG).

3. Load the antibody preparation on the column, and then wash the column with 10 column volumes of 100 mM Tris · Cl (pH 8.0) followed by 10 volumes of 10 mM Tris · Cl (pH 8.0).

4. Add 1 column volume of 100 mM glycine (pH 3.0). Begin collecting 500-μl fractions of the column eluate in microfuge tubes containing 50 μl of 1 M Tris · Cl (pH 8.0).

5. Just as the column runs dry, add another 1 column volume of 100 mM glycine (pH 3.0) and continue collecting samples. Continue this process until 5 column volumes have eluted from the column.

6. Identify the fractions that contain IgG by measuring their absorbance at 280 nm, using elution buffer as a blank. The concentration of IgG in the column eluate should be at least 1 mg/ml (an OD_{280} of 1.33).

7. Dispense the IgG into aliquots and store them as described on page 18.10.

8. Remove residual proteins from the column by washing it sequentially with 10 volumes of 3 M urea, 1 M LiCl, and 100 mM glycine (pH 2.5). Finally, readjust the pH of the column to 8.0 by washing it with 10 volumes of 100 mM Tris · Cl (pH 8.0). Store the column at 4°C in 100 mM Tris · Cl (pH 8.0) containing 0.02% sodium azide.

Caution: Sodium azide is poisonous. It should be handled with great care wearing gloves, and solutions containing it should be clearly marked.

PURIFICATION OF ANTIBODIES BY ADSORPTION

Although hyperimmune antisera raised in experimental animals contain very high concentrations of immunoglobulin directed against the target antigen, such antisera also always contain antibodies directed against other antigens. In addition, the immunoglobulins in antisera may bind with low avidity to molecules that are not true target antigens. For these and other reasons, antisera can manifest a level of background reactivity that is unacceptably high. There are three ways to deal with this problem:

- An innocuous blocking agent (e.g., bovine serum albumin or normal serum) can be used to compete with the immunoglobulin for nonspecific binding sites.

- Antibodies that are directed against specific contaminating antigens (e.g., bacterial antigens) can be removed by adsorption.

- Antibodies directed against the target antigen can be separated from contaminating antibodies by affinity purification.

Blocking agents are routinely included in solutions used, for example, in immunological screening of expression libraries constructed in plasmid or bacteriophage λ vectors (see Chapter 12). Methods to remove antibacterial antibodies from antisera used for such screening are described on pages 12.25–12.28. In the section that follows, we describe techniques to remove antibodies that cross-react with components in eukaryotic cells and discuss methods that are available for affinity purification of antibodies.

Removal of Cross-reacting Antibodies from Antisera

To remove antibodies that react with antigens present in mammalian cells, use an acetone extract of a cell line or tissue that is known not to express the true target antigen. If such a cell line is not available or cannot be identified with certainty, use an acetone extract of commercially available dried yeast.

1. Resuspend the tissue or cells in 0.1 M NaCl, using 2 ml of ice-cold saline solution for every gram of tissue or cells. The suspension should be as homogeneous as possible, with no fibrous material or large clumps of cells. If necessary, filter the tissue suspension through three layers of cheesecloth.

2. Add 4 volumes of acetone, chilled to −20°C, and mix the suspension vigorously. Store the suspension on ice for 1 hour.

3. Centrifuge the suspension at 10,000g for 10 minutes at 4°C. Discard the supernatant, and resuspend the pellet in acetone (−20°C) using the same amount that was used in step 2. Vigorous vortexing may be necessary to achieve complete resuspension. Store the suspension on ice for 10 minutes.

4. Recentrifuge, and discard the supernatant. Transfer the pellet to a piece of Saran Wrap. Allow the acetone to evaporate in a chemical hood. Mix the pellet from time to time to speed evaporation.

5. When the powder is completely dry, transfer it to a container with a tight-fitting lid. Store the powder at −20°C until needed. Each gram of tissue or cells yields approximately 100 mg of powder.

6. To remove cross-reacting antibodies from antiserum, add the powder to a final concentration of 1% (w/v). Mix the suspension carefully, taking care to avoid producing bubbles and froth, which denature proteins. Store the suspension on ice for 15 minutes. Remove the powder by centrifugation at 10,000g for 10 minutes at 4°C. Store the supernatant, which should now be free of cross-reacting antibodies, in small aliquots at −20°C.

Antibodies may be purified by adsorption to, and elution from, their cognate antigen. In some cases (e.g., when the antigen is a protein), the antigen may be coupled to a matrix such as cyanogen-bromide-activated Sepharose. Antibodies directed against epitopes displayed by the protein will be retained by the column; all other immunoglobulins will pass through. The bound antibody is then released from the column by agents that disrupt the antigen-antibody complex (e.g., potassium thiocyanate, low-pH buffers, etc.). Details of the methods used to prepare antibodies by immunochromatography vary from antigen to antigen and from antibody to antibody. However, the general principles are well-described in a number of reviews (see, e.g., Hurn and Chantler 1980; Harlow and Lane 1988). When using these methods, it is essential to use highly purified antigen and to avoid the batch of antigen that was used to raise the antibody in the experimental animals. Furthermore, it is important to remember that antibodies with different affinities for the antigen will show different patterns of elution from the column; those that bind loosely to the antigen will elute first, and those that bind most tightly will elute last. In fact, antibodies with the highest avidity may be denatured by the elution buffer before they dissociate from the antigen. Thus, there is a tendency during immunopurification to select for antibodies that are specific for the antigen but that bind with low affinity.

Antibodies may be purified on a small scale by adsorption to, and elution from, protein antigens that are immobilized on diazotized paper (Olmsted 1981) or nitrocellulose filters (Burke et al. 1982; Smith and Fisher 1984; Earnshaw and Rothfield 1985) after electrophoresis through SDS-polyacrylamide gels. Antibodies prepared by this method are especially useful for confirming the identity of cDNA clones isolated from expression libraries constructed in bacteriophage λ or plasmid expression vectors. For example, many false-positive clones can be eliminated by purifying antibodies from crude sera by virtue of their ability to bind to a fusion protein partly encoded by the cloned cDNA and testing the ability of these purified antibodies to precipitate the target protein or to react with it on a western blot. However, this method works well only when the antibodies react with epitopes that are displayed on denatured proteins. Typically, about 50 ng of immunopurified antibody are recovered per microgram of target protein loaded on the original SDS-polyacrylamide gel. Because of the idiosyncratic nature of the interactions between antibodies and their target proteins, it is not possible to give conditions for binding and elution that are universally applicable. For example, most antibodies can be eluted from their immobilized antigens with glycine buffer (pH 2.8). However, Earnshaw and Rothfield (1985) found that antibodies to human centromeric proteins could be eluted only with a solution containing 3 M potassium thiocyanate and 0.5 M NH_4OH. Anyone who wishes to use this powerful technique should therefore be prepared to invest some effort in defining the optimal conditions for binding and release of their particular antibodies from their target proteins. However, the following protocol, which is a modification of Olmsted's (1981) method produced by J. Allan and M. Douglas (pers. comm.), has worked well with a number of rabbit polyclonal antibodies directed against a variety of mammalian and yeast proteins.

AFFINITY PURIFICATION OF MONOSPECIFIC ANTIBODIES USING ANTIGEN IMMOBILIZED ON NITROCELLULOSE FILTERS

1. Load a preparation containing the target antigen along the entire length of an SDS-polyacrylamide gel (see pages 18.47–18.54). Between 650 μg and 1000 μg of soluble protein can be loaded on a gel of standard size without overloading. After the gel has been run, transfer the proteins from the gel to a nitrocellulose filter as described on pages 18.64–18.66.

2. After transfer, incubate the filter in blocking buffer for 1 hour at room temperature. Use 0.25 ml of blocking buffer for each square centimeter of nitrocellulose filter.

> **Blocking buffer**
>
> 3% bovine serum albumin in phosphate-buffered saline (PBS; see Appendix B) containing 0.02% sodium azide.
>
> **Caution:** Sodium azide is poisonous. It should be handled with great care wearing gloves, and solutions containing it should be clearly marked.

The bovine serum albumin serves as a blocking agent by binding to sites on the nitrocellulose filter that are not occupied by proteins transferred from the gel.

3. Add the first antibody, and incubate for 5 hours at room temperature or for 16 hours at 4°C with gentle shaking.

The first antibody, which should react strongly with the denatured target antigen, can usually be diluted 500- to 1000-fold in the blocking buffer.

4. Discard the antibody solution, and rinse the nitrocellulose filter in 0.15 M NaCl for 20 minutes at room temperature.

5. Rinse the filter in PBS for 20 minutes at room temperature.

6. Using a sharp scalpel blade, cut a strip approximately 0.2 cm wide from each side of the filter. Transfer the strips to a petri dish, and stain them with horseradish-peroxidase (HRP)-coupled antibody directed against the first antibody (see page 18.75). Meanwhile, continue washing the main section of the nitrocellulose filter in PBS.

HRP-coupled second antibody should be used at the dilutions recommended by the manufacturer.

7. Remove the stained strips from the petri dish and align them with the main section of the filter. Using a soft-lead pencil, mark the position(s) of the band(s) of interest. Cut out strips of nitrocellulose filter that carry the target antigen.

8. Arrange the nitrocellulose strips on pieces of Parafilm that are pressed onto the bottoms of petri dishes. Add a small volume of elution buffer to each strip (200–500 μl, depending on the width of the strip). Incubate the strips on a rocking platform in a humidified atmosphere for 20 minutes at room temperature.

The Parafilm prevents the elution buffer from spreading over the surface of the dish.

A typical humidified tissue culture incubator set at room temperature provides an adequate moist atmosphere.

Elution buffer

0.2 M glycine (pH 2.8)
1 mM EGTA

9. Transfer the eluted antibody to a microfuge tube. Neutralize the elution buffer as soon as possible by adding 0.1 volume of 1 M Tris base. Check that the pH is near neutrality by spotting an aliquot of the solution on pH paper.

10. Add 0.1 volume of 10 × PBS. Add sodium azide to a final concentration of 0.02%, and store the antibody at 4°C.

Notes

i. The strip of nitrocellulose can be used several times in succession to purify antibody specific for the immobilized antigen. However, the strip should not be allowed to dry between successive rounds of purification and must be retreated with blocking buffer after each round of elution.

ii. Antibodies of the appropriate subclass purified in this way may be concentrated by adsorption to, and elution from, protein A–Sepharose.

Immunological Assays

Antibodies are used in a wide variety of assays, both qualitative and quantitative, to detect and measure the amount of target antigens. These assays include immunoprecipitation, western blotting, and solid-phase radioimmunoassays (RIAs), which will be described in detail in the remainder of this chapter.

SOLID-PHASE RADIOIMMUNOASSAY

The solid-phase RIA is a quantitative method that is capable of detecting as little as 1 pg of target antigen. This means that RIAs are sufficiently sensitive to measure, for example, the amount of foreign protein produced by transfected mammalian cell cultures. There are many different kinds of RIAs, which fall into four basic designs:

- *Competition RIAs:* In this method, the unlabeled target protein in the test sample competes with a constant amount of radiolabeled protein for binding sites on the antibody. The amount of radioactivity present in the unbound or bound target protein is then measured. This type of assay can be extremely sensitive but requires that target protein is available (preferably in a pure form) to serve both as a competitor and as a standard.

- *Immobilized antigen RIAs:* In this method, unlabeled antigen is attached to a solid support and exposed to radiolabeled antibody. Comparison of the amount of radioactivity that binds specifically to the samples under test with the amount that binds to a known amount of immobilized antigen allows the antigen in the test samples to be quantitated. Although used occasionally, this type of assay is not particularly useful for quantitation of small amounts of foreign protein in complex mixtures (e.g., in cell lysates); most of the binding sites on the solid support become occupied by proteins other than the target protein, so that the sensitivity of the assay is comparatively low.

- *Immobilized antibody RIAs:* In this method, a single antibody bound to a solid support is exposed to radiolabeled antigen. The amount of antigen in the test sample can be determined by the amount of radioactivity that binds to the antibody. This assay is not useful for quantitating the amount of foreign protein in many different samples, chiefly because of the practical difficulty of radiolabeling the protein either in vivo or in vitro.

- *Double-antibody RIAs:* In this method, one antibody bound to a solid support is exposed to the unlabeled target protein. After washing, the target protein bound to the immobilized antibody is quantitated with an excess of a second radiolabeled antibody. This assay is extremely sensitive and specific because the target protein is essentially purified and concentrated by immunoadsorption. Furthermore, many test samples can be processed simultaneously. However, the method requires that the first and second antibodies recognize nonoverlapping epitopes on the target protein. Ideally, the first antibody is monoclonal, whereas the second can be either a polyclonal antibody or a monoclonal antibody of different specificity. However, in some cases, it may be possible to use the same polyclonal antibody for both parts of the assay. If suitable antibodies are available, this is the method of choice for quantitation of target proteins in complex mixtures.

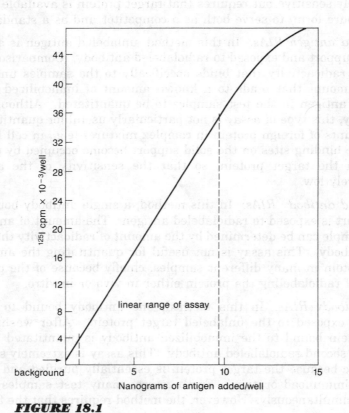

FIGURE 18.1

The main features that affect the quality of the results obtained with RIAs are the sensitivity, accuracy, and specificity of the assay.

The sensitivity of the assay is determined by two factors:

- *The amount of the first antibody that is bound to the solid support (usually a 96-well polyvinyl chloride [PVC] plate)*. There are a finite number of binding sites on the surface of the solid support and, at saturation, each well of the plate binds approximately 100 ng of pure IgG ($\sim 6 \times 10^{11}$ molecules). However, antibodies directed against the target protein must compete with immunoglobulins and other components in the serum for binding sites on the solid support. To maximize binding of relevant antibodies, immunoglobulins directed against the target protein should first be purified by immunoaffinity chromatography. Frequently, however, excellent results may be obtained using immunoglobulins purified from hyperimmune sera either by adsorption to, and elution from, protein A–Sepharose or by chromatography on DEAE-cellulose.

- *The specific activity of the second radiolabeled antibody*. Antibodies radiolabeled with ^{125}I as described later in this chapter (see pages 18.24–18.25) have an approximate specific activity of 5×10^6 cpm/μg. The amount of the radiolabeled second antibody that reacts with antigen-antibody complexes bound to a solid support varies from antigen to antigen. Typically, however, a saturating amount of protein antigen attached to an immobilized first antibody might be expected to bind approximately 2×10^4–1×10^5 cpm of radiolabeled second antibody per well.

The accuracy of the assay is determined by the concentration of antigen in solution, which should not be so high as to saturate the antibody bound to the solid support. Quantitative results can be obtained only when the amount of antigen bound to the first antibody is proportional to its concentration in the sample being tested. To compare the amount of antigen in a number of test samples, assay several different dilutions of each sample (blocking buffer should be used as a diluent). Plot the amount of radioactive second antibody bound to each well against the dilution of the sample. Within the linear range of the assay (see Figure 18.1), there is a proportional relationship between the amount of radioactivity that is specifically bound and the dilution of antigen. In some cases, it may be possible to calculate the absolute amount of antigen present in a test sample from a standard curve obtained using known amounts of pure antigen. If a standard curve with known amounts of pure antigen is not available, the relative amounts of antigen in different samples may be estimated by calculating the dilution of antigen that will give half-maximal binding of radiolabeled antibody. In this case, it is essential that the dilutions span a wide range from background to maximal binding. For accurate quantitation of antigen, it is essential that the radiolabeled second antibody be present in excess. The appropriate amount of second antibody should be determined in pilot experiments in which different amounts of radiolabeled antibody are added to saturating amounts of bound antigen-antibody complexes.

The specificity of the assay is determined chiefly by the quality of the antisera used. However, even when monoclonal antibodies or immunoglobulins purified from hyperimmune sera are used, there is always the chance of immunological cross-reactions. For this reason, it is essential to include the appropriate controls every time a solid-phase RIA is carried out. For example, when measuring the amount of a foreign protein synthesized by a transfected culture of mammalian cells, it is necessary to use extracts of mock-transfected cells as negative controls. Other essential controls include:

- *Immunoglobulins purified from preimmune serum,* which should be used in some wells in place of antibodies directed against the target antigen. If a monoclonal antibody is used as the first antibody, use a monoclonal antibody directed against an irrelevant antigen as a control. The amount of radiolabeled antibody that binds to wells treated with preimmune immunoglobulin and saturating amounts of antigen should be at least tenfold lower than the amount that binds to wells exposed to immune immunoglobulin and antigen. If the immune serum is highly specific, the signal-to-noise ratio can be as high as 100:1 in solid-phase RIAs.

- *Blank wells,* which are exposed to phosphate-buffered saline (PBS; see Appendix B) instead of first antibody. The blank wells are then exposed to blocking buffer and radiolabeled second antibody as described below. If blocking is complete, negligible amounts of radiolabeled second antibody should bind to the blank wells.

- *Positive wells,* which are exposed to known quantities of antigen.

A generalized protocol for a solid-phase RIA is given below.

1. Plan the experiment, and draw a key so that you will know what each well on the plate contains.

2. To each well of a PVC plate, add 50 μl of IgG solution (20 μg/ml in 0.2 M NaHCO$_3$). Cover the plate, and incubate it for 2 hours at room temperature in a humidified atmosphere.

   A typical humidified tissue culture incubator set at room temperature provides an adequate moist atmosphere.

   The covers supplied with the plates fit very loosely, and incubation in a humidified atmosphere is necessary to prevent drying of the samples at the edges of the plates. An alternative way to deal with this problem is to purchase plates without covers (which are less expensive) and to wrap them tightly in Saran Wrap. The plates can then be incubated in the open on the bench.

3. Aspirate the IgG solution from the wells. The solution may be saved, stored at 4°C, and reused several times.
   Important: Do not let the wells become dry at any stage during the remainder of the protocol.

4. Using a squirt bottle, wash the wells twice with blocking buffer. The IgG solution and the washes can be removed quickly and efficiently by flicking the plate over the sink or a suitable waste container.

5. Fill the wells with blocking buffer. Cover the plate and incubate it for 20 minutes at room temperature in a humidified atmosphere.

   The bovine serum albumin serves as a blocking agent by binding to sites on the PVC plate that are not occupied by immunoglobulin molecules.

   The antibody-coated plates may be stored for several days at 4°C if the wells are completely filled with blocking buffer. Before use, plates that have been stored in this fashion should be rinsed with blocking buffer.

6. Wash the wells twice more with blocking buffer (see step 4).

7. Make a series of dilutions in blocking buffer of the antigen that is to be assayed. Add 50 μl of each dilution to each of two wells. Cover the plate and incubate it for 2 hours at room temperature in a humidified atmosphere or for 8 hours at 4°C.

8. Wash the plate four times with blocking buffer (see step 4).

9. Dilute the second antibody, radiolabeled with ^{125}I as described on pages 18.24–18.25, in blocking buffer. To each well, add 50 μl of radiolabeled antibody containing 50,000–200,000 cpm (5–20 ng). Cover the plate and incubate it for at least 2 hours at room temperature in a humidified atmosphere.

10. Wash the plate several times with blocking buffer. Quickly remove as much fluid as possible from each well by aspiration, using a pasteur pipette attached to a vacuum line equipped with traps (see Appendix E, Figure E.1).

11. Using sharp scissors, cut individual wells from the plate. Place each well in a labeled counting vial, and measure the radioactivity in a gamma counter.

12. Plot the amount of radioactivity specifically bound to the test samples. Calculate the dilution of each sample that yields half-maximal binding of radiolabeled antibody. Calculate the amount of antigen present in each sample relative to the controls.

    The amount of specifically bound radioactivity varies considerably from antigen to antigen and from antiserum to antiserum. However, it would be reasonable to expect 10 ng of purified, immobilized target antigen to bind between 1000 and 10,000 cpm of radioiodinated antibody.

Iodination of Antibodies

Of the several methods that are available to radioiodinate antibodies, the most commonly used is a reaction in which $H_2O^{125}I^+$, generated by oxidation of $Na^{125}I$ with chloramine-T, attacks the side chains of tyrosine residues and, to a lesser extent, histidine residues (Greenwood et al. 1963; McConahey and Dixon 1966, 1980). This modification, if not carried out to excess, generally does not affect the ability of the antibody to interact with its target antigen. However, in some cases (presumably when the antibody has a tyrosine residue in its antigen binding site), the immunoreactivity of radioiodinated antibody may be significantly reduced. It is then necessary to use a conjugation reagent such as Bolton-Hunter reagent (Bolton and Hunter 1973). The *N*-succinimidyl group of this reagent (*N*-succinimidyl 3-[4-hydroxy 5-[^{125}I]iodophenyl]propionate) condenses with the free amino and imino in lysine and histidine residues to give a derivative in which the radioiodinated phenyl group is linked to an amide bond in the target protein. (For methods involving this reagent, see Langone 1980; Harlow and Lane 1988.)

Caution: ^{125}I is a potential health hazard. It emits low-energy γ- and X-rays, is volatile, and accumulates in the thyroid. Under United States regulations, the maximum allowable load of ^{125}I in the thyroid is 1.2 μCi. However, the procedures established by the local radiation safety committee, if strictly followed, should ensure that those working with ^{125}I stay well within this limit.

RADIOIODINATION OF ANTIBODIES USING THE CHLORAMINE-T METHOD

1. Equilibrate Sephadex G-25 in phosphate-buffered saline (PBS; see Appendix B) according to the manufacturer's instructions. Prepare a column in a pasteur pipette that is plugged with sterile glass wool. Run at least 10 column volumes of PBS containing 1% bovine serum albumin through the column. Then wash the column with 5 volumes of PBS lacking bovine serum albumin. When the column is ready, wrap a piece of Parafilm around the bottom of the pipette. This prevents the column from running dry.

   Prepared columns (PD-10 Sephadex G-25M) can be purchased from Pharmacia.

   Pretreatment with bovine serum albumin reduces nonspecific adsorption of antibody to the column and resin. It is important to remove bovine serum albumin from the column buffer before applying the radioiodinated antibody; otherwise, there is a possibility that the bovine serum albumin in the solution will become radioiodinated.

2. In a microfuge tube, add up to 70 μg of antibody (dissolved in 10 μl of PBS) to 50 μl of 50 mM sodium phosphate buffer (pH 7.2).

   Since the radioiodination of proteins involves oxidation of radioiodide by chloramine-T, it is essential that no reducing agents be present in any of the solutions used.

   Caution: The following steps must be carried out in a ventilated chemical hood behind shielding that is sufficient to protect the worker from direct radiation. Consult your radiation safety officer before beginning work with ^{125}I, and make sure you are familiar with the procedures used to monitor and dispose of the radioisotope.

3. Add 500 μCi of carrier-free Na^{125}I and mix the solution well.

Na^{125}I is usually supplied in 0.1 N NaOH, which minimizes oxidation. 50 μl of sodium phosphate buffer can accommodate up to 7 μl of 0.1 N NaOH without significant change in pH. However, the sudden addition of larger amounts of NaOH to the antibody can cause denaturation. This can be avoided by neutralizing the radioiodide solution with an equal volume of 0.1 N HCl just before use. Neutralized solutions of Na^{125}I must be used immediately to avoid loss of the volatile ^{125}I. Carrier-free Na^{125}I should be used to minimize the total amount of iodine incorporated into the antibody.

4. Add 15 μl of a freshly prepared solution of chloramine-T (2 mg/ml in 0.1 M sodium phosphate buffer [pH 7.2]).

5. Incubate the mixture for 60 seconds at room temperature.

6. Add 50 μl of stop buffer, and immediately apply the mixture to the column of Sephadex G-25.

   Stop buffer

   0.1 M sodium phosphate buffer (pH 7.2) containing
   10 mg/ml tyrosine
   2 mg/ml sodium metabisulfite
   10% glycerol
   0.1% xylene cyanol FF

7. Begin collecting 100–150-μl fractions of the eluate from the column. After the sample has entered the resin, add PBS (lacking bovine serum albumin) to the top of the column. Make sure that the column does not run dry at any stage.

8. Using a hand-held minimonitor, check the radioactivity that elutes from the column. If the reaction has gone well, approximately 90% of the radiolabel should be incorporated into the antibody. The radiolabeled antibody should elute well ahead of the xylene cyanol FF, which runs with the unincorporated radiolabel. When the antibody has eluted, seal the bottom of the pipette with modeling clay. Dispose of the entire column, which still contains unincorporated radiolabel, in the radioactive waste.

9. Combine the fractions containing the radiolabeled antibody. Add bovine serum albumin to a final concentration of 1%. Measure the radioactivity in a 5-μl aliquot, and store the antibody at 4°C until it is required for use.

If the immunoreactivity of the antibody is reduced by the above procedure, test the effects of metabisulfite on the protein. If metabisulfite damages the protein, omit it from the stop buffer. If, as is most likely, the antibody is insensitive to metabisulfite, the reduction in immunoreactivity must be caused by exposure to chloramine-T or to the incorporation of iodine into sensitive tyrosine residues. In either case, it will be necessary to try another method of radioiodination, for example, Bolton-Hunter reagent (Langone 1980; Harlow and Lane 1988).

The specific activity of the radiolabeled antibody should be ~5×10^6 cpm/μg.

IMMUNOPRECIPITATION

Immunoprecipitation is used to detect and quantitate target antigens in mixtures of proteins. The power of the technique lies in its selectivity: The specificity of the immunoglobulin for its ligand is so high that the resulting antigen-antibody complexes can be purified from contaminating proteins. Furthermore, immunoprecipitation is extremely sensitive and is capable of detecting as little as 100 pg of radiolabeled protein. When coupled with SDS-polyacrylamide gel electrophoresis, the technique is ideal for analysis of the synthesis and processing of foreign antigens expressed in prokaryotic and eukaryotic hosts or in in vitro systems.

The target protein is usually immunoprecipitated from extracts of cells that have been radiolabeled. However, immunoprecipitation can also be used to analyze unlabeled proteins as long as sufficiently sensitive methods are available to detect the target protein after it has been dissociated from the antibody. Such methods include enzymatic activity, binding of radioactive ligands, and western blotting.

Immunoprecipitation of radiolabeled target proteins and their subsequent analysis consist of the following steps, which can be completed in as little as one day or can be carried out over several successive days if desired:

- Radiolabeling of cells expressing the target protein

- Lysis of the cells

- Formation of specific immune complexes

- Collection and purification of the immune complex

- Analysis of the radiolabeled proteins in the immunoprecipitate

Radiolabeling the Target Protein

The isotope most commonly used to radiolabel proteins is ^{35}S, which emits weak β radiation and has a half-life of 87.1 days. The isotope is incorporated into proteins as ^{35}S-labeled methionine or ^{35}S-labeled methionine and cysteine. These are essential amino acids for mammalian cells and must therefore be supplied in the medium. Metabolic radiolabeling of cultured mammalian cells is therefore carried out by incubating the cells in medium containing [^{35}S]Met or a mixture of [^{35}S]Met and [^{35}S]Cys. Most strains of yeasts and prokaryotes, on the other hand, can synthesize methionine and cysteine de novo. ^{35}S-labeled sulfate is therefore generally used as the metabolic precursor to radiolabel proteins in these organisms.

Phosphorylated proteins can be radiolabeled by incubating the cells in medium that contains ^{32}P in place of the normal isotope (^{31}P). Unlike ^{35}S, ^{32}P emits strong β radiation, and slightly different methods are therefore used to detect the two isotopes. However, the basic principles of radiolabeling and immunoprecipitation remain the same and are independent of the radioisotope. The methods described below deal in detail with proteins metabolically radiolabeled with ^{35}S-labeled amino acids. However, they can easily be adapted to accommodate proteins metabolically radiolabeled with ^{32}P or radiolabeled in vitro with ^{125}I.

RADIOLABELING MAMMALIAN CELLS WITH [^{35}S]METHIONINE AND [^{35}S]CYSTEINE

For many years, ^{35}S-labeling of proteins in mammalian cells was carried out with [^{35}S]Met that was purified by high-performance liquid chromatography (HPLC) from hydrolysates of *Escherichia coli* grown on carrier-free ^{35}S-labeled sulfate. The material that elutes in the center of the peak from the HPLC column is now sold as translation-grade [^{35}S]Met and is very expensive. The side fractions are sold at a lower price and are used for radiolabeling of cultured cells. More reasonably priced still is an unpurified hydrolysate of *E. coli* grown in the presence of ^{35}S-labeled sulfate (e.g., Tran^{35}S-label; ICN Radiochemicals) that can be used as a substitute for [^{35}S]Met in metabolic labeling. About 70% of the label in the hydrolysate is in the form of [^{35}S]Met, with the remainder consisting of [^{35}S]Cys (20%) and a number of oxidized by-products. Proteins synthesized in cells incubated in the presence of the hydrolysate are radiolabeled with both [^{35}S]Met and [^{35}S]Cys. This can be a considerable advantage when analyzing proteins that are rich in cysteine and poor in methionine.

Media used to grow mammalian cells in culture contain high concentrations of both methionine and cysteine. To increase the efficiency of incorporation of radiolabeled amino acids, it is necessary to incubate the cells in medium that is depleted of either methionine or both methionine and cysteine. Rapidly growing cells incorporate more radioactivity than static cultures; however, to obtain maximum radiolabeling of the protein(s) of interest, pilot experiments should be conducted under various growth conditions and for various periods of labeling.

The number of cells and the length and intensity of radiolabeling depend on the rate of synthesis of the protein of interest, its half-life, and its amino acid composition. Typically, however, transiently transfected cultures are radiolabeled for 2–4 hours with 200 μCi of ^{35}S-labeled amino acids at 60–72 hours after transfection. If the transfection is efficient (>10%), a target protein immunoprecipitated from a lysate of a small-scale culture (10^6 cells) can be detected in a dried gel after 2–4 days of exposure of an autoradiograph.

1. Aspirate the medium from cells growing in monolayers. Wash the cells twice with medium (Med$^-$AA), prewarmed to 37°C, that lacks methionine or both methionine and cysteine and contains no serum.

2. Incubate the cells for 20 minutes at 37°C in prewarmed Med$^-$AA to deplete the intracellular pools of sulfur-containing amino acids.

3. Remove the Med$^-$AA by aspiration, and immediately replace it with fresh, prewarmed Med$^-$AA containing the appropriate quantity of ^{35}S-labeled amino acids. Incubate the cells at 37°C for the desired period of time based on the pilot experiments.

To increase the concentration of radiolabeled amino acids, keep the volume of medium to a minimum during short periods of radiolabeling (<1 hour).

Volume of medium	Size of petri dish or well (mm)
1.0 ml	90
250 µl	60
100 µl	35
100 µl	30 (well)

The plates should be rocked every 15 minutes to ensure that the cells do not dry out.

When radiolabeling cells for extended periods of time (>6 hours) with precursors of high specific activity, all of the radiolabeled amino acid(s) in the medium may be consumed, and it is therefore necessary to add the appropriate nonradioactive amino acids to keep the cells in a healthy condition. This is usually done empirically by adding 0.1 volume of complete medium to the Med⁻AA used for radiolabeling. Some types of cultured cells deteriorate rapidly when incubated in the absence of serum. In this case, it may be necessary to supplement the medium used for radiolabeling with serum that has been dialyzed against phosphate-buffered saline (PBS; see Appendix B) to reduce the concentration of free amino acids.

4. Aspirate the radioactive medium with a pasteur pipette. If only intracellular antigens are being examined, dispose of the radioactive medium in the radioactive waste. If the target antigen is a secreted protein, save the medium for immunoprecipitation (see pages 18.44–18.46).

5. Transfer the petri dish to a flat surface chilled to 0°C (e.g., an aluminum plate laid on a bed of crushed ice). Wash the cells twice with ice-cold PBS, disposing of both washes in the radioactive waste. Drain the last traces of PBS from the plate, and then lyse the cells by one of the methods described on pages 18.30–18.34.

Notes

i. In carrying out pulse-chase experiments, the radioactive medium should be removed and the cells should be washed rapidly with a large volume of complete medium, prewarmed to 37°C. The cells should then be incubated for the appropriate time at 37°C in complete, prewarmed medium containing five times the normal amount of nonradioactive methionine (or both methionine and cysteine) and then processed as described in step 5. Some workers also suppress incorporation of radioactive amino acids during the chase by adding inhibitors of protein synthesis (e.g., cycloheximide).

ii. Cells growing in suspension should be concentrated by centrifugation, washed thoroughly in prewarmed (37°C) Med⁻AA, and resuspended in Med⁻AA at a concentration of approximately 10^7 cells/ml for radiolabeling. In short-term radiolabeling experiments, there is no need to take any measures to keep the cells in suspension. However, when radiolabeling for periods of 1 hour or more, the cells should be placed in a closed tube and incubated at 37°C on a rocker platform with gentle agitation.

METABOLIC RADIOLABELING OF PROTEINS EXPRESSED IN YEASTS AND BACTERIA

Unlike most vectors used in mammalian cells, vectors used to express foreign proteins in yeasts and bacteria usually contain powerful, inducible promoters. The protocols used to radiolabel proteins expressed from different promoters vary widely, and it is not possible to produce generic methods that will work well for every combination of promoter and foreign protein. We therefore recommend following protocols established by the laboratory in which the expression vector was developed. The following method is intended only for radiolabeling of proteins that are constitutively expressed in bacteria and yeasts.

1. Grow the cells to be radiolabeled overnight with mild agitation at the appropriate temperature in defined minimal media supplemented only with those amino acids and cofactors required for cell growth.

2. Dilute the culture with fresh minimal medium, and continue growth until the cell density reaches 10^7 cells/ml (yeasts) or to mid-log phase (bacteria).

3. Collect the cells by centrifugation at 5000g for 10 minutes at room temperature, and resuspend the cells at a concentration of 10^7 cells/ml (yeasts) or 5×10^8 cells/ml (bacteria) in minimal medium lacking sulfate.

4. Add $H_2{}^{35}SO_4$ to a final concentration of 20 μCi/ml. Incubate the culture with gentle agitation at the appropriate temperature for 1 hour (yeasts) or 20 minutes (bacteria).

5. Collect the cells by centrifugation at 5000g for 10 minutes at room temperature. Remove the medium by aspiration and discard it in the radioactive waste.

6. Suspend the cells in ice-cold minimal medium, and collect them by centrifugation as before. Once again, discard the supernatant in the radioactive waste. Remove the last traces of supernatant from the walls of the centrifuge vessel using a pasteur pipette. The cells are now ready to be lysed as described in general on pages 18.30–18.33 and specifically on pages 18.35–18.39 (yeasts) or pages 18.40–18.41 (bacteria).

Lysis of Cells

This is perhaps the most crucial step in immunoprecipitation. The aim is to find a method that will solubilize all of the target antigen in a form that is immunoreactive, undegraded, and, for some purposes, biologically active. In view of the wide range of physical and biological properties of mammalian proteins, it is not surprising that no single method of lysis is sufficient for every purpose. Among the variables that have been found to influence the efficiency of solubilization and subsequent immunoprecipitation of proteins are the ionic strength and pH of the lysis buffer; the concentration and type of detergent used; and the presence of divalent cations, cofactors, and stabilizing ligands.

Although there are exceptions, many soluble nuclear and cytoplasmic proteins can be solubilized by lysis buffers that contain the nonionic detergent Nonidet P-40 (NP-40) and either no salt at all or relatively high concentrations of salt (e.g., 0.5 M NaCl). However, the efficiency of extraction is often greatly affected by the pH of the buffer and the presence or absence of chelating agents such as EDTA and EGTA. Extraction of membrane-bound and hydrophobic proteins is less affected by the ionic strength of the lysis buffer but often requires a mixture of ionic and nonionic detergents.

When attempting to solubilize a protein for the first time, there are two different strategies that can be employed. At one extreme, harsh conditions can be used in an effort to ensure that the protein is released quantitatively from the cells; however, this may result in loss of immunoreactivity. At the other extreme, gentle conditions can be used to help preserve the protein in a native state; however, this may result in inefficient extraction of the protein from the cells. Which of these approaches should be used is determined in large part by the properties of the antiserum available for immunoprecipitation. For example, as discussed earlier in this chapter, monospecific antisera raised against synthetic peptides may react only with denatured forms of the target protein, whereas monoclonal antibodies directed against native epitopes may be specific for the correctly folded form of the protein. To minimize problems, try to use polyclonal antisera or mixtures of monoclonal antibodies that react with all forms of the protein. It is usually then possible to tailor the extraction conditions to fit the characteristics of the target protein rather than the properties of the available antisera.

Many methods of solubilization, particularly those that involve mechanical disruption of cells, release intracellular proteases that can digest the target protein. The susceptibility of different proteins to attack by proteases varies widely, with cell-surface and secreted proteins generally being more resistant than intracellular proteins. Denatured proteins are much more likely to be degraded than native proteins. It is therefore advisable to take steps to minimize proteolytic activity in cell extracts, especially when harsh conditions of extraction are used. It is important to keep the extracts cold (i.e., at 0°C or below, depending on the sensitivity of the target protein to freezing and thawing). In addition, inhibitors of proteases are commonly included in lysis buffers (Table 18.1). For additional information about inhibitors of proteases, see Barrett and Salvesen (1986).

TABLE 18.1 Properties of Commonly Used Protease Inhibitors

Inhibitor	Active against	Inactive against	Effective concentration	Stock solution
Aprotinin[a] (Trasylol)	kallikrein trypsin chymotrypsin plasmin	papain	1–2 µg/ml	10 mg/ml in 0.01 M HEPES (pH 8.0)
Leupeptins (supplied as a 3:1 mixture of propionyl and acetyl derivatives)	plasmin trypsin papain cathepsin B	chymotrypsin pepsin cathepsins A and D	1–2 µg/ml	10 mg/ml in water
Pepstatin A	pepsin cathepsin D	trypsin plasmin chymotrypsin elastase thermolysin	1 µg/ml	1 mg/ml in ethanol
Antipain	cathepsins A and B papain trypsin	plasmin chymotrypsin pepsin	1–2 µg/ml	1 mg/ml in water
PMSF[b] (phenylmethyl-sulfonyl fluoride)	chymotrypsin trypsin		100 µg/ml	1.74 mg/ml (10 mM) in isopropanol
TLCK (tosyllysine chloromethyl ketone)	trypsin	chymotrypsin	50 µg/ml	1 mg/ml in 0.05 M sodium acetate (pH 5.0)
TPCK (tosylphenylalanine chloromethyl ketone)	chymotrypsin	trypsin	100 µg/ml	3 mg/ml in ethanol
EDTA	metalloproteases		1 mM	0.5 M in water

[a]Aprotinin is a basic polypeptide of 58 amino acids that aggregates if repeatedly frozen and thawed (for review, see Trautschold et al. 1967). The stock solution should be stored in small aliquots at −20°C. Each aliquot should be discarded after use.

[b]**Caution:** PMSF is extremely destructive to the mucous membranes of the respiratory tract, eyes, and skin. PMSF may be fatal if inhaled, swallowed, or absorbed through the skin. In case of contact, immediately flush eyes or skin with copious amounts of water. Discard contaminated clothing.

PMSF is inactivated in aqueous solutions. The rate of inactivation increases with pH and is faster at 25°C than at 4°C. The half-life of a 20 µM aqueous solution of PMSF is about 35 minutes at pH 8.0 (James 1978). This means that solutions of PMSF can be safely discarded after they have been rendered alkaline (pH > 8.6) and stored for several hours at room temperature.

The lysis buffers that are commonly used to prepare extracts of mammalian cells for immunoprecipitation are shown below. In the absence of any information about the target antigen, we recommend trying the triple- and single-detergent lysis buffers before turning to more specialized methods of extraction.

Triple-detergent lysis buffer

50 mM Tris · Cl (pH 8.0)
150 mM NaCl
0.02% sodium azide
0.1% SDS
100 μg/ml phenylmethylsulfonyl fluoride (PMSF)
1 μg/ml aprotinin
1% Nonidet P-40 (NP-40)
0.5% sodium deoxycholate

Single-detergent lysis buffer

50 mM Tris · Cl (pH 8.0)
150 mM NaCl
0.02% sodium azide
100 μg/ml PMSF
1 μg/ml aprotinin
1% Triton X-100 *or* 1% NP-40

High-salt lysis buffer

50 mM HEPES (pH 7.0)
500 mM NaCl
1% NP-40
1 μg/ml aprotinin
100 μg/ml PMSF

No-salt lysis buffer

50 mM HEPES (pH 7.0)
1% NP-40
1 μg/ml aprotinin
100 μg/ml PMSF

Cautions: PMSF is extremely destructive to the mucous membranes of the respiratory tract, the eyes, and skin. PMSF may be fatal if inhaled, swallowed, or absorbed through the skin. In case of contact, immediately flush eyes or skin with copious amounts of water. Discard contaminated clothing.

Sodium azide is poisonous. It should be handled with great care wearing gloves, and solutions containing it should be clearly marked.

PMSF, which is labile in aqueous solution, should be added from a stock solution just before the lysis buffer is used. The rate of inactivation in aqueous solution increases with pH and is faster at 25°C than at 4°C. The half-life of a 20 μM aqueous solution of PMSF is about 35 minutes at pH 8.0

(James 1978). This means that aqueous solutions of PMSF can be safely discarded after they have been rendered alkaline (pH > 8.6) and stored for several hours at room temperature. PMSF is usually stored as a 10 mM or 100 mM stock solution (1.74 or 17.4 mg/ml in isopropanol) at −20°C.

LYSIS OF CULTURED MAMMALIAN CELLS

1. Add the lysis buffer of choice (cooled to 0°C) to chilled, washed cell monolayers (step 5, page 18.28). Incubate for 20 minutes on a flat aluminum tray on a bed of crushed ice.

Volume of lysis buffer (ml)	Size of petri dish or well (mm)
1.0	90
0.5	60
0.25	35
0.25	30 (well)

2. Scrape the cells to one side of the dish with a policeman. Using an automatic pipetting device, transfer the cell debris and lysis buffer to a chilled microfuge tube.

The best policemen are made as follows:
a. Using a scalpel, slice silicon rubber stoppers (2 cm in diameter) into quarters.
b. Place each piece of stopper on the bench and impale it on a pasteur pipette.
c. Use the straight edge of the sliced stopper for scraping cells.
d. After use, store the policemen in detergent solution before washing. The policemen can be sterilized by autoclaving for 20 minutes at 15 lb/sq. in.

3. Centrifuge the lysate at 12,000g for 2 minutes at 4°C in a microfuge.

4. Transfer the supernatant to a fresh microfuge tube, and store it on ice or at −70°C, depending on the sensitivity of the target antigen to freezing and thawing.

Notes

i. When experimenting with conditions for solubilization of the protein of interest, it is sometimes useful to extract the cell debris (step 4) with a stronger lysis buffer.

ii. Cells growing in suspension should be concentrated by centrifugation and then resuspended in the appropriate lysis buffer (chilled to 0°C) at a concentration of 10^7 cells/ml. After 30 minutes at 0°C, transfer the cell suspension to a chilled microfuge tube and proceed as described in step 3.

iii. After thawing, samples that have been stored at −20°C should be centrifuged at 12,000g for 5 minutes at 0°C in a microfuge. This removes aggregates of cytoskeletal elements.

MECHANICAL LYSIS OF YEAST

This method can be used only when the antibody recognizes denatured forms of the target protein.

1. Collect the cells from a 1-ml culture by centrifugation at 12,000g for 1 minute at 0°C in a microfuge or at 3000 rpm for 5 minutes at 0°C in a Sorvall RT6000 (or equivalent). Discard the supernatant (unless the protein of interest is a secretory protein).

2. Resuspend the cells by vortexing in 1 ml of ice-cold 50 mM Tris·Cl (pH 8.0). Recover the cells by centrifugation as described in step 1. Pour off the supernatant, and remove as much of the residual fluid as possible with a disposable pipette tip attached to a vacuum line.

3. Resuspend the pellet in 5 volumes (~ 250 μl) of yeast lysis buffer.

> *Yeast lysis buffer*
>
> 50 mM Tris·Cl (pH 8.0)
> 0.1% Triton X-100
> 0.5% SDS

4. Add acid-washed glass beads (0.45–0.50 mm in diameter) to the level of the meniscus. The total volume of the mixture of glass beads and yeast lysis buffer should be approximately 400 μl.

   At one time, glass beads of the correct size could be purchased only from one supplier, located in Germany (B. Braun, Melsüngen). Recently, however, several companies in the United States have begun to sell beads of the correct size of acceptable quality.

5. Vortex the suspension on a heavy-duty vortexing machine for five periods of 20 seconds each. Cool the suspension on ice for 1 minute between each cycle of vortexing. Monitor cell lysis by phase-contrast microscopy.

6. Recover the cell extract by punching a small hole through the bottom of the microfuge tube. This is best done with a red-hot hypodermic needle (23 gauge). Place the microfuge tube containing the glass beads inside an empty microfuge tube. Centrifuge the piggybacked pair of tubes in a 15-ml Corex tube at 2000g for 2 minutes at 4°C.

7. Recover the bottom microfuge tube, which should contain the cell extract. Clarify the extract by centrifuging the tube at 12,000g for 5 minutes at 4°C in a microfuge. Transfer the supernatant to a fresh microfuge tube.

   Extracts of yeast cells are usually stable when stored at −20°C. However, if the protein of interest is sensitive to proteolysis, add protease inhibitors to the extracts (see Table 18.1) and store them at 0°C. Try to minimize the time between cell lysis and immunoprecipitation.

8. Immunoprecipitate the protein of interest from the cell extract as described on pages 18.44–18.46.

ENZYMATIC LYSIS OF YEAST

This method is used when the available antibody recognizes the native form of the target protein.

1. Collect the cells from a 1-ml culture by centrifugation at 12,000g for 1 minute at 0°C in a microfuge or at 3000 rpm for 5 minutes at 0°C in a Sorvall RT6000 (or equivalent). Discard the supernatant (unless the protein of interest is a secretory protein).

2. Resuspend the cells by vortexing in ice-cold phosphate-buffered saline (PBS; see Appendix B). Recover the cells by centrifugation as described in step 1. Discard the supernatant.

3. Resuspend the pellet in a volume of stabilizing buffer A (at room temperature) equivalent to the volume of the original culture (usually 1 ml as in step 1). Incubate the supension for 10 minutes at 30°C.

> *Stabilizing buffer A*
>
> 1 M sorbitol
> 10 mM $MgCl_2$
> 2 mM dithiothreitol
> 50 mM potassium phosphate (pH 7.8)
> 100 μg/ml phenylmethylsulfonyl fluoride (PMSF)

Caution: PMSF is extremely destructive to the mucous membranes of the respiratory tract, the eyes, and skin. PMSF may be fatal if inhaled, swallowed, or absorbed through the skin. In case of contact, immediately flush eyes or skin with copious amounts of water. Discard contaminated clothing.

PMSF, which is labile in aqueous solution, should be added from a stock solution just before the stabilizing buffer is used. The rate of inactivation in aqueous solution increases with pH and is faster at 25°C than at 4°C. The half-life of a 20 μM aqueous solution of PMSF is about 35 minutes at pH 8.0 (James 1978). This means that aqueous solutions of PMSF can be safely discarded after they have been rendered alkaline (pH > 8.6) and stored for several hours at room temperature. PMSF is usually stored as a 10 mM or 100 mM stock solution (1.74 or 17.4 mg/ml in isopropanol) at −20°C.

4. Recover the cells by centrifugation as described in step 1, and resuspend the cell pellet in a volume of stabilizing buffer B equal to the volume of the original culture. Incubate the suspension for 2 minutes in a water bath set at 30°C.

5. Add 0.25 volume of a stock solution of zymolase 100T (10 mg/ml).

Zymolase is an enzyme, isolated from cultures of *Arthrobacter luteus,* that catalyzes lysis of cell walls by hydrolyzing glucose polymers with $1 \rightarrow 3$ linkages. As supplied by commercial manufacturers, zymolase 100T is contaminated with other enzymes, including proteases. It should therefore be used in conjunction with protease inhibitors.

6. Incubate the suspension at 30°C. After 15 minutes, take out two small aliquots of the suspension. To one sample, add Nonidet P-40 (NP-40) to a final concentration of 1%. Examine the two samples by phase-contrast microscopy. Protoplasts formed by the action of zymolase will be visible in the untreated sample but should be lysed in the sample containing detergent. Continue incubation until the majority of the cells have been converted to protoplasts. This usually takes 20–30 minutes.

7. Collect the protoplasts by centrifugation at 500*g* for 10 minutes at 4°C. Discard the supernatant.

8. Resuspend the pellet (which consists of protoplasts and intact cells) in a volume of lysis buffer equivalent to 0.01 volume of the original culture. Store the suspension at 0°C for 30 minutes.

Either NP-40 high- or no-salt lysis buffer (see page 18.32) or RIPA buffer (see page 18.38) may be used, depending on the properties of the available antibody. It is essential to include protease inhibitors in the lysis buffer (see Table 18.1).

9. Centrifuge the suspension at 12,000*g* for 10 minutes at 4°C in a microfuge. Transfer the supernatant to a fresh microfuge tube.

Extracts of yeast cells are usually stable when stored at −20°C. However, if the protein of interest is sensitive to proteolysis, add protease inhibitors to the extracts (see Table 18.1) and store them at 0°C. Try to minimize the time between cell lysis and immunoprecipitation.

10. Immunoprecipitate the protein of interest from the cell extract as described on pages 18.44–18.46.

Note

When experimenting with conditions for solubilization of the protein of interest, it is sometimes useful to extract the cell debris (step 9) with a stronger lysis buffer.

This method can be used only when the antibody recognizes denatured forms of the target protein.

1. Collect the cells from a 1-ml culture by centrifugation at 12,000g for 1 minute at 0°C in a microfuge or at 3000 rpm for 5 minutes at 0°C in a Sorvall RT6000 (or equivalent). Discard the supernatant (unless the protein of interest is a secretory protein).

2. Resuspend the cells in 0.5 ml of 25% trichloroacetic acid (TCA).

3. Recover the cells by centrifugation as described in step 1, and resuspend them in 1 ml of 90% acetone.

4. Recover the cells by centrifugation as described in step 1, and estimate the volume of the pellet. Resuspend the pellet in an equal volume of 1% SDS.

5. Add acid-washed glass beads (0.45–0.50 mm in diameter) to the level of the meniscus.

   At one time, glass beads of the correct size could be purchased only from one supplier, located in Germany (B. Braun, Melsüngen). Recently, however, several companies in the United States have begun to sell beads of the correct size of acceptable quality.

6. Vortex the suspension on a heavy-duty vortexing machine for five periods of 20 seconds each. Cool the suspension on ice for 1 minute between each cycle of vortexing. Monitor cell lysis by phase-contrast microscopy.

7. Using a red-hot hypodermic needle (23 gauge), make a hole in the lid of the microfuge tube. This prevents the glass beads from exploding out of the tube during heating.

8. Heat the tube in a boiling-water bath for 3 minutes.

9. Transfer the glass beads into a fresh microfuge tube, and wash the original tube with 0.5 ml of RIPA buffer. Add the wash to the fresh microfuge tube containing the glass beads.

   RIPA buffer

   50 mM Tris · Cl (pH 7.5)
   150 mM NaCl
   1% Nonidet P-40 (NP-40)
   0.5% sodium deoxycholate
   0.1% SDS

10. Recover the cell extract by punching a small hole through the bottom of

the microfuge tube. This is best done with a red-hot hypodermic needle (23 gauge). Place the microfuge tube containing the glass beads inside an empty microfuge tube. Centrifuge the piggybacked pair of tubes in a 15-ml Corex tube at 2000g for 2 minutes at 4°C.

11. Recover the bottom microfuge tube, which should contain the cell extract. Clarify the extract by centrifuging the tube at 12,000g for 5 minutes at 4°C in a microfuge. Transfer the supernatant to a fresh microfuge tube.
    Extracts of yeast cells are usually stable when stored at −20°C. However, if the protein of interest is sensitive to proteolysis, add protease inhibitors to the extracts (see Table 18.1) and store them at 0°C. Try to minimize the time between cell lysis and immunoprecipitation.

12. Immunoprecipitate the protein of interest from the cell extract as described on pages 18.44–18.46.

LYSIS OF BACTERIA

Immunoprecipitation of target proteins from cell lysates is used far more rarely with bacterial expression systems than with mammalian expression systems because foreign proteins are usually expressed in *E. coli* from powerful, inducible promoters. Following induction, therefore, novel gene products frequently accumulate to such a level that they can be detected by the appearance of novel bands after SDS-polyacrylamide gel electrophoresis of total cell proteins. Immunological identification of these novel bands can then be obtained by western blotting using antisera or monoclonal antibodies that react with denatured forms of the target protein. If no novel proteins are detectable by this technique, it makes more sense to test the cloned DNA in a series of different *E. coli* expression systems rather than to attempt immunoprecipitation from cell extracts that contain unusably small quantities of target protein.

Total lysates of small-scale cultures of *E. coli* are easily made as follows:

1. After induction for the appropriate period of time, recover the bacteria from 1 ml of culture by centrifugation at 12,000g for 30 seconds in a microfuge.

2. Remove the medium by aspiration, and then resuspend the pellet by vortexing in 0.5 ml of ice-cold 50 mM Tris·Cl (pH 7.4). Recover the bacteria by centrifugation at 12,000g for 30 seconds at 0°C in a microfuge.

3. Remove the supernatant by aspiration, leaving the bacterial pellet as dry as possible. Take care to remove any beads of fluid that adhere to the sides of the microfuge tube.

4. Resuspend the pellet by vortexing in 25 μl of H$_2$O. As soon as the pellet is dispersed, add 25 μl of 2× SDS gel-loading buffer and continue vortexing for 20 seconds.

> *2 × SDS gel-loading buffer*
>
> 100 mM Tris·Cl (pH 6.8)
> 200 mM dithiothreitol
> 4% SDS (electrophoresis grade)
> 0.2% bromophenol blue
> 20% glycerol
>
> 2× SDS gel-loading buffer lacking dithiothreitol can be stored at room temperature. Dithiothreitol should then be added, just before the buffer is used, from a 1 M stock (see Appendix B).

5. Place the sample in a boiling-water bath for 5 minutes.

6. Shear the chromosomal DNA by sonication, using either a sonicator with an immersible tip or a sonicator that can process many samples simulta-

neously in a chilled cup. Depending on the power output of the sonicator and its state of tuning, between 30 seconds and 2 minutes at full power should be sufficient to reduce the viscous lysate to manageable levels.

Larger samples can be sheared by repeated passage through a 23-gauge hypodermic needle.

7. Centrifuge the sample at 12,000*g* for 10 minutes at room temperature. Transfer the supernatant to a fresh tube and discard the pellet.

8. Analyze 25 μl of the sheared or sonicated lysate by electrophoresis through an SDS-polyacrylamide gel as described on pages 18.47–18.54. Store the unused portion of the sample at −20°C.

Formation of Antigen-Antibody Complexes

In this procedure, specific antibody directed against the target protein is added to aliquots of cell lysates. Antigen-antibody complexes are then collected by one of three methods:

1. *Adsorption to staphylococcal protein A that has been covalently attached to Sepharose beads.* This method gives the cleanest results, since there is very little nonspecific adsorption of irrelevant proteins to protein A–Sepharose beads. However, the method is expensive and requires that the antibody be derived from a species, and be of a class, that is efficiently recognized by the staphylococcal protein. Protein A–Sepharose is available from several commercial manufacturers.

2. *Adsorption to* S. aureus *cells that have been killed by heat and fixed with formaldehyde.* This method gives higher backgrounds, but it is much cheaper than using protein A–Sepharose. Killed, fixed cells of *S. aureus* (Cowan strain) are available from several manufacturers or can be made in the laboratory as described by Kessler (1975).

3. *Immunoprecipitation with an anti-immunoglobulin antibody that reacts with species-specific determinants displayed on the antibody that interacts with the target antigen.* This method is used under the following conditions:

   • When the Fc portion of the primary antibody is not recognized by staphylococcal protein A. As discussed on page 18.11, antisera belonging to this class include virtually all antisera raised in chickens and goats, many polyclonal antisera raised in rats and sheep, and some polyclonal antisera raised in mice, hamsters, cows, and horses. Although some murine monoclonal antibodies bind efficiently to protein A, rat monoclonal antibodies rarely do so. The ability of staphylococcal protein A to adsorb complexes formed between these antibodies and the target antigen should therefore be determined empirically in pilot experiments. Polyclonal antibodies raised in humans, rabbits, guinea pigs, and pigs almost always interact efficiently with the staphylococcal protein A and rarely need to be precipitated by anti-immunoglobulin antibodies.

   • When quantitative recovery of the target antigen is required. Immunoglobulins in polyclonal antisera that interact with the target antigen may belong to several different classes, which may be recognized by protein A–Sepharose with different efficiencies. Depending on the relative concentrations and avidities of these antibodies, a portion of the antigen-antibody complexes may not be detected by staphylococcal protein A. In this case, recovery of the target protein will not be quantitative. In addition, not all forms of the protein will necessarily be recovered with equal efficiency. These problems can be circumvented by the use of anti-immunoglobulin antibody.

PRECLEARING THE CELL LYSATE

To reduce background caused by nonspecific adsorption of irrelevant cellular proteins to staphylococcal protein A or to immunoglobulins, many workers pretreat cell lysates with an antibody that has no activity against the target protein, for example, preimmune serum drawn from the same animal that was later immunized against the target antigen. Alternatively, if a monoclonal antibody is used for immunoprecipitation, the cell lysate should be treated with an irrelevant monoclonal antibody that binds to staphylococcal protein A.

1. To a sample of lysate, add *either* 0.05 volume of preimmune serum, *or* 1 volume of irrelevant hybridoma tissue culture supernatant, *or* 0.01 volume of irrelevant ascites fluid.

   The concentration of immunoglobulins in serum is approximately 10 mg/ml. Supernatants from cultured hybridoma cells typically contain 10–100 μg/ml immunoglobulin, and ascites fluid contains 1–20 mg/ml immunoglobulin.

2. Incubate the mixtures for 1 hour at 0°C.

3. Transfer the cell lysate/antibody mixture to a microfuge tube containing a pellet of killed, fixed *S. aureus* cells (Cowan strain) that have been washed in the lysis buffer used to make the original cell extract. The volume of the pellet should be approximately 1/20 the volume of the cell lysate.

   Heat-killed, formaldehyde-fixed *S. aureus* cells can be obtained from several commercial manufacturers. One milliliter of a 10% suspension of staphylococcal cells can adsorb approximately 1 mg of immunoglobulin.

   Protein A–Sepharose can also be used to adsorb and remove nonspecific antigen-antibody complexes. However, it is very expensive and does not work as effectively for this purpose as killed, fixed staphylococcal cells. One milliliter of packed beads of protein A–Sepharose can adsorb approximately 10–20 mg of IgG.

4. Incubate the mixture for 30 minutes at 0°C.

5. Remove the *S. aureus* cells by centrifugation at 12,000g for 5 minutes at 4°C in a microfuge. Transfer the supernatant to a fresh microfuge tube, and precipitate the target protein with specific antiserum as described on pages 18.44–18.46.

   If the background remains unacceptable, repeat steps 3 through 5.

Immunoprecipitation of the Target Protein

1. Divide the preparation of antigen into two equally sized aliquots and place in microfuge tubes. Adjust the volume of each aliquot to 0.5 ml with NET-gel buffer. To one aliquot, add antibody directed against the target protein. To another, add the same volume of a control antibody (preimmune serum or an irrelevant monoclonal antibody). Gently rock both aliquots for 1 hour at 0°C.

> ### NET-gel buffer
>
> 50 mM Tris · Cl (pH 7.5)
> 150 mM NaCl
> 0.1% Nonidet P-40 (NP-40)
> 1 mM EDTA (pH 8.0)
> 0.25% gelatin
> 0.02% sodium azide
>
> The gelatin is diluted from a 2.5% (w/v) stock that has been solubilized and sterilized by autoclaving for 20 minutes at 15 lb/sq. in. on liquid cycle. The stock solution can be melted just before use by brief heating in a microwave oven.
>
> **Caution:** Sodium azide is poisonous. It should be handled with great care wearing gloves, and solutions containing it should be clearly marked.

The aim is to use a quantity of antibody that is just sufficient to precipitate all of the target protein. This amount depends on several factors, including the concentration of the antigen and the titer and avidity of the antibody. The antibody should be titrated in pilot experiments in which increasing quantities of antibody are used to precipitate a fixed amount of antigen. Typically, complete immunoprecipitation of radiolabeled antigen from extracts of transfected mammalian cells requires between 0.5 μl and 5 μl of polyclonal antiserum, 5–100 μl of hybridoma tissue culture medium, or 0.1–1.0 μl of ascitic fluid. If more antibody is used than is necessary, the background of nonspecific immunoprecipitation will increase.

Dilution of the cell extract to 0.5 ml with NET-gel buffer lowers the level of nonspecific immunoprecipitation. Some workers prefer to use lysis buffer as a diluent. However, the relatively high concentrations of detergent and the absence of carrier protein may lead to partial denaturation or proteolysis of the target protein.

2. If the antibody directed against the target protein does not bind efficiently to staphylococcal protein A, add the appropriate anti-immunoglobulin antibody to the mixture. Continue to rock gently for 1 hour at 0°C.

The anti-immunoglobulin antibody should be titrated in pilot experiments in which increasing quantities are used to precipitate a fixed amount of the first antibody.

3. Add protein A–Sepharose to the antigen/antibody mixture, and incubate for 1 hour at 4°C on a rocking platform.

Heat-killed, formaldehyde-fixed *S. aureus* cells can be used in place of protein A–Sepharose, which is very expensive. However, the background of nonspecific adsorption is usually higher with intact staphylococcal cells than with protein A–Sepharose.

The amount of protein A–Sepharose required to bind all of the antibody should be determined in pilot experiments. 1 ml of packed, swollen protein A–Sepharose should bind at least 20 mg of IgG; 1 ml of a standard 10% suspension of killed, fixed *S. aureus* cells should bind at least 1 mg of immunoglobulin. The concentration of immunoglobulins in antiserum is approximately 10 mg/ml. Supernatants from cultured hybridoma cells typically contain 10–100 μg/ml immunoglobulin, and ascites fluid contains 1–20 mg/ml immunoglobulin.

4. Collect the tertiary protein A–antigen–antibody complexes by centrifugation at 12,000g for 20 seconds at 4°C in a microfuge. Remove the supernatant by gentle aspiration using a disposable pipette tip or a bent hypodermic needle attached to a vacuum line. Add 1 ml of wash buffer, and resuspend the Sepharose beads by vortexing.

Several different buffers are commonly used to wash protein A–antigen–antibody complexes. The aim is to displace proteins that are nonspecifically adsorbed to the protein-A matrix, while leaving specific tertiary complexes intact. The tighter the binding between antibody and antigen, the more stringent the washing conditions can be. The buffer best suited to a particular antigen-antibody combination should be determined in pilot experiments. In increasing order of stringency, these are:
- *NET-gel buffer:* The immunoprecipitate is washed twice with NET-gel buffer and then once with 10 mM Tris · Cl (pH 7.5), 0.1% NP-40.
- *Supplemented NET-gel buffer:* The first wash is carried out with NET-gel buffer supplemented with NaCl to a final concentration of 0.5 M. For the second wash, standard NET-gel buffer is supplemented with SDS (final concentration 0.1%). The third wash is carried out with 10 mM Tris · Cl (pH 7.5), 0.1% NP-40.
- *RIPA buffer* (see page 18.38): The first two washes are carried out with RIPA buffer. The third wash is carried out with 10 mM Tris · Cl (pH 7.5), 0.1% NP-40.

5. Incubate the resuspended beads for 20 minutes at 4°C on a rocking platform. This allows time for the wash buffer to equilibrate with the fluid within the beads.

If intact staphylococcal cells are used, incubation in wash buffer is unnecessary.

6. Repeat steps 4 and 5 twice (i.e., wash the Sepharose beads a total of three times), changing the wash buffer if desired (as indicated in the note to step 4). Collect the final washed tertiary protein A–antigen–antibody complex by centrifugation at 12,000g for 20 seconds at 4°C in a microfuge Take care to remove the last traces of the final wash from the Sepharose pellet and from the walls and lid of the microfuge tube.

7. Add 30 μl of 1 × SDS gel-loading buffer.

1 × SDS gel-loading buffer

50 mM Tris · Cl (pH 6.8)
100 mM dithiothreitol
2% SDS (electrophoresis grade)
0.1% bromophenol blue
10% glycerol

1 × SDS gel-loading buffer lacking dithiothreitol can be stored at room temperature. Dithiothreitol should then be added, just before the buffer is used, from a 1 M stock (see Appendix B).

8. Denature the proteins in the sample by heating to 100°C for 3 minutes. Remove the protein A–Sepharose by centrifugation at 12,000*g* for 20 seconds at room temperature in a microfuge. Transfer the supernatant to a fresh tube.

9. Analyze an aliquot of the sample by SDS-polyacrylamide gel electrophoresis (see pages 18.47–18.54). Store the remainder of the sample at −20°C.

SDS-Polyacrylamide Gel Electrophoresis of Proteins

Almost all analytical electrophoresis of proteins is carried out in poly-acrylamide gels under conditions that ensure dissociation of the proteins into their individual polypeptide subunits and that minimize aggregation. Most commonly, the strongly anionic detergent SDS is used in combination with a reducing agent and heat to dissociate the proteins before they are loaded on the gel. The denatured polypeptides bind SDS and become negatively charged. Because the amount of SDS bound is almost always proportional to the molecular weight of the polypeptide and is independent of its sequence, SDS-polypeptide complexes migrate through polyacrylamide gels in accordance with the size of the polypeptide. At saturation, approximately 1.4 g of detergent is bound per gram of polypeptide. By using markers of known molecular weight, it is therefore possible to estimate the molecular weight of the polypeptide chain(s). Modifications to the polypeptide backbone, such as N- or O-linked glycosylation, however, have a significant impact on the apparent molecular weight. Thus, the apparent molecular weight of glycosylated proteins is not a true reflection of the mass of the polypeptide chain.

In most cases, SDS-polyacrylamide gel electrophoresis is carried out with a discontinuous buffer system in which the buffer in the reservoirs is of a different pH and ionic strength from the buffer used to cast the gel. The SDS-polypeptide complexes in the sample that is applied to the gel are swept along by a moving boundary created when an electric current is passed between the electrodes. After migrating through a stacking gel of high porosity, the complexes are deposited in a very thin zone (or stack) on the surface of the resolving gel. The ability of discontinuous buffer systems to concentrate all of the complexes in the sample into a very small volume greatly increases the resolution of SDS-polyacrylamide gels.

The discontinuous buffer system that is most widely used was originally devised by Ornstein (1964) and Davis (1964). The sample and the stacking gel contain Tris · Cl (pH 6.8), the upper and lower buffer reservoirs contain Tris-glycine (pH 8.3), and the resolving gel contains Tris · Cl (pH 8.8). All components of the system contain 0.1% SDS (Laemmli 1970). The chloride ions in the sample and stacking gel form the leading edge of the moving boundary, and the trailing edge is composed of glycine molecules. Between the leading and trailing edges of the moving boundary is a zone of lower conductivity and steeper voltage gradient which sweeps the polypeptides from the sample and deposits them on the surface of the resolving gel. There, the higher pH of the resolving gel favors the ionization of glycine, and the resulting glycine ions migrate through the stacked polypeptides and travel through the resolving gel immediately behind the chloride ions. Freed from the moving boundary, the SDS-polypeptide complexes move through the resolving gel in a zone of uniform voltage and pH and are separated according to size by sieving.

Polyacrylamide gels are composed of chains of polymerized acrylamide that are cross-linked by a bifunctional agent such as N,N'-methylenebisacrylamide (see Chapter 6, page 6.36). The effective range of separation of SDS-polyacrylamide gels depends on the concentration of polyacrylamide used to cast the gel and on the amount of cross-linking. Polymerization of acrylamide in the absence of cross-linking agents generates viscous solutions

TABLE 18.2 Effective Range of Separation of SDS-Polyacrylamide Gels

Acrylamide[a] concentration (%)	Linear range of separation (kD)
15	12–43
10	16–68
7.5	36–94
5.0	57–212

[a]Molar ratio of bisacrylamide:acrylamide is 1:29.

that are of no practical use. Cross-links formed from bisacrylamide add rigidity and tensile strength to the gel and form pores through which the SDS-polypeptide complexes must pass. The size of these pores decreases as the bisacrylamide:acrylamide ratio increases, reaching a minimum when the ratio is approximately 1:20. Most SDS-polyacrylamide gels are cast with a molar ratio of bisacrylamide:acrylamide of 1:29, which has been shown empirically to be capable of resolving polypeptides that differ in size by as little as 3%.

The sieving properties of the gel are determined by the size of the pores, which is a function of the absolute concentrations of acrylamide and bisacrylamide used to cast the gel. Table 18.2 shows the linear range of separation obtained with gels cast with concentrations of acrylamide that range from 5% to 15%.

PREPARATION OF SDS-POLYACRYLAMIDE GELS

Reagents

- *Acrylamide and N,N'-methylenebisacrylamide:* Several manufacturers sell electrophoresis grade acrylamide that is free of contaminating metal ions. A stock solution containing 29% (w/v) acrylamide and 1% (w/v) *N,N'-*methylenebisacrylamide should be prepared in deionized, warm water (to assist the dissolution of the bisacrylamide). Acrylamide and bisacrylamide are slowly converted during storage to acrylic acid and bisacrylic acid. This deamination reaction is catalyzed by light and alkali. Check that the pH of the solution is 7.0 or less, and store the solution in dark bottles at room temperature. Fresh solutions should be prepared every few months.

  Caution: Acrylamide and bisacrylamide are potent neurotoxins and are absorbed through the skin. Their effects are cumulative. Wear gloves and a mask when weighing acrylamide and bisacrylamide. Polyacrylamide is considered to be nontoxic, but it should be treated with care because it may contain small quantities of unpolymerized material.

- *Sodium dodecyl sulfate (SDS):* Several manufacturers sell special grades of SDS that are sufficiently pure for electrophoresis. Although any one of these will give reproducible results, they are not interchangeable. The pattern of migration of polypeptides may change quite drastically when one manufacturer's SDS is substituted for another's. We recommend that you find a grade that you like and use it exclusively. A 10% (w/v) stock solution should be prepared in deionized water and stored at room temperature.

  If proteins are to be eluted from the gel for sequencing, electrophoresis-grade SDS should be further purified as described by Hunkapiller et al. (1983).

- *Tris buffers for the preparation of resolving and stacking gels:* It is essential that these buffers be prepared with Tris base. After the Tris base has been dissolved in deionized water, the pH of the solution should be adjusted with HCl as described in Appendix B. If Tris·Cl or Trizma is used to prepare buffers, the concentration of salt will be too high and polypeptides will migrate anomalously through the gel, yielding extremely diffuse bands.

- *TEMED (N,N,N',N'-tetramethylethylenediamine):* TEMED accelerates the polymerization of acrylamide and bisacrylamide by catalyzing the formation of free radicals from ammonium persulfate. Use the electrophoresis grade that is sold by several manufacturers. Because TEMED works only as a free base, polymerization is inhibited at low pH.

- *Ammonium persulfate:* Ammonium persulfate provides the free radicals that drive polymerization of acrylamide and bisacrylamide. A small amount of a 10% (w/v) stock solution should be prepared in deionized water and stored at 4°C. Ammonium persulfate decomposes slowly, and fresh solutions should be prepared weekly.

- *Tris-glycine electrophoresis buffer:* This buffer contains 25 mM Tris base, 250 mM glycine (electrophoresis grade) (pH 8.3), 0.1% SDS. A 5 × stock can be made by dissolving 15.1 g of Tris base and 94 g of glycine in 900 ml of deionized H_2O. Then, 50 ml of a 10% (w/v) stock solution of electrophoresis-grade SDS is added, and the volume is adjusted to 1000 ml with H_2O.

Apparatus

The use of discontinuous buffer systems requires SDS-polyacrylamide electrophoresis to be carried out in vertical gels. Although the basic design of the electrophoresis tanks and plates has changed little since Studier (1973) introduced the system, many small improvements have been incorporated into the apparatuses that are now sold by many manufacturers. Which of these systems to purchase is a matter of personal choice, but it is sensible for a laboratory to use only one type of apparatus. This type of standardization makes it easier to compare the results obtained by different workers and allows parts of broken apparatuses to be scavenged and reused.

POURING SDS-POLYACRYLAMIDE GELS

1. Assemble the glass plates according to the manufacturer's instructions.

2. Determine the volume of the gel mold (this information is usually provided by the manufacturer). In an Erlenmeyer flask, prepare the appropriate volume of solution containing the desired concentration of acrylamide for the resolving gel, using the values given in Table 18.3 on page 18.52. Mix the components in the order shown. Polymerization will begin as soon as the TEMED has been added. Without delay, swirl the mixture rapidly and proceed to the next step.

The concentration of ammonium persulfate that we recommend is higher than that used by some workers. This eliminates the need to rid the acrylamide solution of dissolved oxygen (which retards polymerization) by degassing.

3. Pour the acrylamide solution into the gap between the glass plates. Leave sufficient space for the stacking gel (the length of the teeth of the comb plus 1 cm). Using a pasteur pipette, carefully overlay the acrylamide solution with 0.1% SDS (for gels containing ≤ 8% acrylamide) or isobutanol (for gels containing ≥ 10% acrylamide). Place the gel in a vertical position at room temperature.

The overlay prevents oxygen from diffusing into the gel and inhibiting polymerization.

4. After polymerization is complete (30 minutes), pour off the overlay and wash the top of the gel several times with deionized water to remove any unpolymerized acrylamide. Drain as much fluid as possible from the top of the gel, and then remove any remaining water with the edge of a paper towel.

5. Prepare the stacking gel as follows: In a disposable plastic tube, prepare the appropriate volume of solution containing the desired concentration of acrylamide, using the values given in Table 18.4 on page 18.52. Mix the components in the order shown. Polymerization will begin as soon as the TEMED has been added. Without delay, swirl the mixture rapidly and proceed to the next step.

The concentration of ammonium persulfate is higher than that used by some workers. This eliminates the need to rid the acrylamide solution of dissolved oxygen (which retards polymerization) by degassing.

6. Pour the stacking gel solution directly onto the surface of the polymerized resolving gel. Immediately insert a clean Teflon comb into the stacking gel solution, being careful to avoid trapping air bubbles. Add more stacking gel solution to fill the spaces of the comb completely. Place the gel in a vertical position at room temperature.

Teflon combs should be cleaned with water and dried with ethanol just before use.

TABLE 18.3 Solutions for Preparing Resolving Gels for Tris-glycine SDS-Polyacrylamide Gel Electrophoresis

Solution components	Component volumes (ml) per gel mold volume of							
	5 ml	10 ml	15 ml	20 ml	25 ml	30 ml	40 ml	50 ml
6%								
H$_2$O	2.6	5.3	7.9	10.6	13.2	15.9	21.2	26.5
30% acrylamide mix	1.0	2.0	3.0	4.0	5.0	6.0	8.0	10.0
1.5 M Tris (pH 8.8)	1.3	2.5	3.8	5.0	6.3	7.5	10.0	12.5
10% SDS	0.05	0.1	0.15	0.2	0.25	0.3	0.4	0.5
10% ammonium persulfate	0.05	0.1	0.15	0.2	0.25	0.3	0.4	0.5
TEMED	0.004	0.008	0.012	0.016	0.02	0.024	0.032	0.04
8%								
H$_2$O	2.3	4.6	6.9	9.3	11.5	13.9	18.5	23.2
30% acrylamide mix	1.3	2.7	4.0	5.3	6.7	8.0	10.7	13.3
1.5 M Tris (pH 8.8)	1.3	2.5	3.8	5.0	6.3	7.5	10.0	12.5
10% SDS	0.05	0.1	0.15	0.2	0.25	0.3	0.4	0.5
10% ammonium persulfate	0.05	0.1	0.15	0.2	0.25	0.3	0.4	0.5
TEMED	0.003	0.006	0.009	0.012	0.015	0.018	0.024	0.03
10%								
H$_2$O	1.9	4.0	5.9	7.9	9.9	11.9	15.9	19.8
30% acrylamide mix	1.7	3.3	5.0	6.7	8.3	10.0	13.3	16.7
1.5 M Tris (pH 8.8)	1.3	2.5	3.8	5.0	6.3	7.5	10.0	12.5
10% SDS	0.05	0.1	0.15	0.2	0.25	0.3	0.4	0.5
10% ammonium persulfate	0.05	0.1	0.15	0.2	0.25	0.3	0.4	0.5
TEMED	0.002	0.004	0.006	0.008	0.01	0.012	0.016	0.02
12%								
H$_2$O	1.6	3.3	4.9	6.6	8.2	9.9	13.2	16.5
30% acrylamide mix	2.0	4.0	6.0	8.0	10.0	12.0	16.0	20.0
1.5 M Tris (pH 8.8)	1.3	2.5	3.8	5.0	6.3	7.5	10.0	12.5
10% SDS	0.05	0.1	0.15	0.2	0.25	0.3	0.4	0.5
10% ammonium persulfate	0.05	0.1	0.15	0.2	0.25	0.3	0.4	0.5
TEMED	0.002	0.004	0.006	0.008	0.01	0.012	0.016	0.02
15%								
H$_2$O	1.1	2.3	3.4	4.6	5.7	6.9	9.2	11.5
30% acrylamide mix	2.5	5.0	7.5	10.0	12.5	15.0	20.0	25.0
1.5 M Tris (pH 8.8)	1.3	2.5	3.8	5.0	6.3	7.5	10.0	12.5
10% SDS	0.05	0.1	0.15	0.2	0.25	0.3	0.4	0.5
10% ammonium persulfate	0.05	0.1	0.15	0.2	0.25	0.3	0.4	0.5
TEMED	0.002	0.004	0.006	0.008	0.01	0.012	0.016	0.02

TABLE 18.4 Solutions for Preparing 5% Stacking Gels for Tris-glycine SDS-Polyacrylamide Gel Electrophoresis

Solution components	Component volumes (ml) per gel mold volume of							
	1 ml	2 ml	3 ml	4 ml	5 ml	6 ml	8 ml	10 ml
H$_2$O	0.68	1.4	2.1	2.7	3.4	4.1	5.5	6.8
30% acrylamide mix	0.17	0.33	0.5	0.67	0.83	1.0	1.3	1.7
1.0 M Tris (pH 6.8)	0.13	0.25	0.38	0.5	0.63	0.75	1.0	1.25
10% SDS	0.01	0.02	0.03	0.04	0.05	0.06	0.08	0.1
10% ammonium persulfate	0.01	0.02	0.03	0.04	0.05	0.06	0.08	0.1
TEMED	0.001	0.002	0.003	0.004	0.005	0.006	0.008	0.01

Tables 18.3 and 18.4 are modified from Harlow and Lane (1988).

7. While the stacking gel is polymerizing, prepare the samples by heating them to 100°C for 3 minutes in 1 × SDS gel-loading buffer to denature the proteins.

> *1 × SDS gel-loading buffer*
>
> 50 mM Tris · Cl (pH 6.8)
> 100 mM dithiothreitol
> 2% SDS (electrophoresis grade)
> 0.1% bromophenol blue
> 10% glycerol
>
> 1 × SDS gel-loading buffer lacking dithiothreitol can be stored at room temperature. Dithiothreitol should then be added, just before the buffer is used, from a 1 M stock (see Appendix B).

Be sure to denature a sample containing marker proteins of known molecular weights. Mixtures of appropriately sized polypeptides are available from commercial sources.

8. After polymerization is complete (30 minutes), remove the Teflon comb carefully. Using a squirt bottle, wash the wells immediately with deionized water to remove any unpolymerized acrylamide. If necessary, straighten the teeth of the stacking gel with a blunt hypodermic needle attached to a syringe. Mount the gel in the electrophoresis apparatus. Add Tris-glycine electrophoresis buffer to the top and bottom reservoirs. Remove any bubbles that become trapped at the bottom of the gel between the glass plates. This is best done with a bent hypodermic needle attached to a syringe.

> *Tris-glycine electrophoresis buffer*
>
> 25 mM Tris
> 250 mM glycine (electrophoresis grade) (pH 8.3)
> 0.1% SDS
>
> A 5 × stock can be made by dissolving 15.1 g of Tris base and 94 g of glycine in 900 ml of deionized H_2O. Then, 50 ml of a 10% (w/v) stock solution of electrophoresis-grade SDS is added, and the volume is adjusted to 1000 ml with H_2O.

Do not prerun the gel before loading the samples, since this will destroy the discontinuity of the buffer systems.

9. Load up to 15 μl of each of the samples in a predetermined order into the bottom of the wells. This is best done with a Hamilton microliter syringe that is washed with buffer from the bottom reservoir after each sample is loaded. Load an equal volume of 1 × SDS gel-loading buffer into any wells that are unused.

10. Attach the electrophoresis apparatus to an electric power supply (the positive electrode should be connected to the bottom buffer reservoir). Apply a voltage of 8 V/cm to the gel. After the dye front has moved into the resolving gel, increase the voltage to 15 V/cm and run the gel until the bromophenol blue reaches the bottom of the resolving gel (about 4 hours). Then turn off the power supply.

11. Remove the glass plates from the electrophoresis apparatus and place them on a paper towel. Using a spatula, pry the plates apart. Mark the orientation of the gel by cutting a corner from the bottom of the gel that is closest to the leftmost well (slot 1).

    Important: Do not cut the corner from gels that are to be used for western blotting.

12. The gel can now be fixed (see page 18.58), stained with Coomassie Brilliant Blue (see page 18.55) or silver salts (see pages 18.56–18.57), fluorographed or autoradiographed (see Appendix E), or used to establish a western blot (see pages 18.64–18.66).

STAINING SDS-POLYACRYLAMIDE GELS WITH COOMASSIE BRILLIANT BLUE

Polypeptides separated by SDS-polyacrylamide gels can be simultaneously fixed with methanol:glacial acetic acid and stained with Coomassie Brilliant Blue R250, a triphenylmethane textile dye also known as Acid Blue 83. The gel is immersed for several hours in a concentrated methanol/acetic acid solution of the dye, and excess dye is then allowed to diffuse from the gel during a prolonged period of destaining.

1. Dissolve 0.25 g of Coomassie Brilliant Blue R250 in 90 ml of methanol:H_2O (1:1 v/v) and 10 ml of glacial acetic acid. Filter the solution through a Whatman No. 1 filter to remove any particulate matter.

2. Immerse the gel in at least 5 volumes of staining solution and place on a slowly rotating platform for a minimum of 4 hours at room temperature.

3. Remove the stain and save it for future use. Destain the gel by soaking it in the methanol/acetic acid solution (step 1) without the dye on a slowly rocking platform for 4–8 hours, changing the destaining solution three or four times.

4. The more thoroughly the gel is destained, the smaller the amount of protein that can be detected by staining with Coomassie Brilliant Blue. Destaining for 24 hours usually allows as little as 0.1 μg of protein to be detected in a single band.

   A more rapid rate of destaining can be achieved by the following methods:
   • By destaining in 30% methanol, 10% acetic acid. However, if destaining is prolonged, there will be some loss in the intensity of staining of protein bands.
   • By destaining in the normal destaining buffer at higher temperatures (45°C).
   • By including a few grams of an anion-exchange resin or a piece of sponge in the normal destaining buffer. This absorbs the stain as it leaches from the gel.
   • By destaining electrophoretically in apparatuses that are sold commercially for this purpose.

5. After destaining, gels may be stored indefinitely in water in a sealed plastic bag without any diminution in the intensity of staining. However, fixed polyacrylamide gels stored in water will swell and may distort during storage. To avoid this problem, store fixed gels in water containing 20% glycerol. Stained gels should *not* be stored in destaining buffer, which will cause the stained protein bands to fade.

6. To make a permanent record, either photograph the stained gel (see Chapter 6, pages 6.18–6.19) or dry the gel as described on pages 18.58–18.59.

STAINING SDS-POLYACRYLAMIDE GELS WITH SILVER SALTS

A number of methods have been developed to stain polypeptides with silver salts after separation by SDS-polyacrylamide gel electrophoresis. In every case, the process relies on differential reduction of silver ions that are bound to the side chains of amino acids (Switzer et al. 1979; Oakley et al. 1980; Ochs et al. 1981; Sammons et al. 1981; Merril et al. 1984). These methods fall into two major classes: those that use ammoniacal silver solutions and those that use silver nitrate. Although both types of staining are approximately 100- to 1000-fold more sensitive than staining with Coomassie Brilliant Blue R250 and are capable of detecting as little as 0.1–1.0 ng of polypeptide in a single band, silver nitrate solutions are easier to prepare and, in contrast to ammoniacal silver salts, do not generate potentially explosive by-products. The method given below is a modification of the staining procedure originally devised by Sammons et al. (1981), which has since undergone several improvements (Schoenle et al. 1984).

Note: Wear gloves and handle the gel gently because pressure and fingerprints will produce staining artifacts. In addition, it is essential to use clean glassware and deionized water because contaminants can greatly reduce the sensitivity of silver staining.

1. Separate proteins by electrophoresis through an SDS-polyacrylamide gel as described on pages 18.47–18.54. Fix the proteins by incubating the gel for 4–12 hours at room temperature with gentle shaking in at least 5 gel volumes of a solution of ethanol:glacial acetic acid:water (30:10:60).

2. Discard the fixing solution, and add at least 5 gel volumes of 30% ethanol. Incubate the gel for 30 minutes at room temperature with gentle shaking.

3. Repeat step 2.

4. Discard the ethanol and add 10 gel volumes of deionized water. Incubate the gel for 10 minutes at room temperature with gentle shaking.

5. Repeat step 4 twice.

   The gel will swell slightly during rehydration.

6. Discard the last of the water washes, and, wearing gloves, add 5 gel volumes of a 0.1% solution of $AgNO_3$ (freshly diluted from a 20% stock, stored in a tightly closed, brown glass bottle at room temperature). Incubate the gel for 30 minutes at room temperature with gentle shaking.

7. Discard the $AgNO_3$ solution, and wash both sides of the gel (20 seconds each) under a stream of deionized water.

   Do not allow the surface of the gel to dry, otherwise staining artifacts will ensue.

8. Add 5 gel volumes of a freshly made aqueous solution of 2.5% sodium carbonate, 0.02% formaldehyde. Incubate the gel at room temperature

with gentle agitation. Watch the gel carefully. Stained bands of protein should appear within a few minutes. Continue incubation until the desired contrast is obtained.

Caution: Formaldehyde vapors are toxic. Solutions containing formaldehyde should be prepared in a chemical hood.

Formaldehyde is usually obtained as a 37% solution in water. Check that the pH of the concentrated solution is greater than 4.0.

Prolonged incubation leads to a high background of silver staining within the body of the gel.

9. Quench the reaction by washing the gel in 1% acetic acid for a few minutes. Then wash the gel several times with deionized water (10 minutes per wash).

A shiny gray film of silver sometimes forms on the surface of the gel. This can be removed by washing the gel for 2–3 seconds in a 1:4 dilution of photographic reducing solution. Rinse the treated gel extensively in deionized water.
   Photographic reducing solution can be made as follows (Switzer et al. 1979):
   a. *Solution A:* Dissolve 37 g of NaCl and 37 g of $CuSO_4$ in 850 ml of deionized H_2O. Add concentrated NH_4OH until a deep blue precipitate forms and then dissolves. Adjust the volume to 1 liter with H_2O.
   b. *Solution B:* Dissolve 436 g of sodium thiosulfate in 900 ml of deionized H_2O. Adjust the volume to 1 liter with H_2O.
   c. Mix equal volumes of solution A and solution B, dilute the mixture with 3 volumes of H_2O, and use the diluted mixture immediately.

10. Preserve the gel by drying as described on pages 18.58–18.59.

SDS-polyacrylamide gels containing proteins radiolabeled with ^{35}S-labeled amino acids must be dried before autoradiographic images can be obtained. The major problems encountered when a gel is dried are (1) shrinkage and distortion and (2) cracking of the gel. The first of these problems can be minimized if the gel is attached to a piece of Whatman 3MM paper before it is dehydrated. However, there is no guaranteed solution to the second problem, which becomes more pronounced with thicker gels containing more poly-acrylamide. Cracking generally occurs when the gel is removed from the drying apparatus before it is completely dehydrated. It is therefore essential to keep the drying apparatus in good condition, to use a reliable vacuum line that has few fluctuations in pressure, and to use the thinnest gel possible to achieve the desired purpose.

Gel dryers are available from a number of commercial sources. It is best to purchase the dryer from the manufacturer of the SDS-polyacrylamide gel electrophoresis tanks: There is then a better chance that the size of the dryer will be tailored to that of the gels and will accommodate several SDS-polyacrylamide gels simultaneously.

1. Remove the gel from the electrophoresis apparatus and fix it at room temperature in 5–10 volumes of glacial acetic acid:methanol:water (10:20:70). The bromophenol blue will turn yellow as the acidic fixing solution diffuses into the gel. Continue fixation for 5 minutes after all of the blue color has disappeared. Wash the gel briefly in deionized water.

   If cracking of polyacrylamide gels during drying is a constant problem, soak the fixed gel in 20% methanol, 3% glycerol overnight before proceeding to step 2.

2. On a piece of Saran Wrap slightly larger than the gel, arrange the gel with its cut corner on the lower right-hand side.

3. Place a piece of dry Whatman 3MM paper on the damp gel. The paper should be large enough to create a border (1–2 cm) around the gel and small enough to fit on the gel dryer. Do not attempt to move the 3MM paper once contact has been made with the gel.

4. Arrange another piece of dry 3MM paper on the drying surface of the gel dryer. This piece should be large enough to accommodate all of the gels that are to be dried at the same time.

5. Place the sandwich of 3MM paper/gel/Saran Wrap on the piece of 3MM paper on the gel dryer. The Saran Wrap should be uppermost.

6. Close the lid of the gel dryer, and apply suction so that the lid makes a tight seal around the gels. If the dryer is equipped with a heater, apply low heat (50–65°C) to speed up the drying process.

7. Dry the gel for the time recommended by the manufacturer (usually 2 hours for standard 0.75-mm gels). If heat was applied, turn off the heat for a few minutes before releasing the vacuum.

8. Remove the gel, which is now attached to a piece of 3MM paper, from the dryer.

9. Remove the piece of Saran Wrap and establish an autoradiograph as described in Appendix E, or store the dehydrated gel as a record of the experiment.

TRANSFER OF PROTEINS FROM SDS-POLYACRYLAMIDE GELS TO SOLID SUPPORTS: IMMUNOLOGICAL DETECTION OF IMMOBILIZED PROTEINS (WESTERN BLOTTING)

Western blotting (Towbin et al. 1979; Burnette 1981) is to proteins what Southern blotting is to DNA. In both techniques, electrophoretically separated components are transferred from a gel to a solid support and probed with reagents that are specific for particular sequences of amino acids (western blotting) or nucleotides (Southern hybridization). In the case of proteins, the probes usually are antibodies that react specifically with antigenic epitopes displayed by the target protein attached to the solid support. Western blotting is therefore extremely useful for the identification and quantitation of specific proteins in complex mixtures of proteins that are not radiolabeled. The technique is almost as sensitive as standard solid-phase radioimmunoassays and, unlike immunoprecipitation, does not require that the target protein be radiolabeled. Furthermore, because electrophoretic separation of proteins is almost always carried out under denaturing conditions, any problems of solubilization, aggregation, and coprecipitation of the target protein with adventitious proteins are eliminated.

The critical difference between Southern and western blotting lies in the nature of the probes. Whereas nucleic acid probes hybridize with a specificity and rate that can be predicted by simple equations (see Chapter 10), antibodies behave in a much more idiosyncratic manner. As discussed earlier in this chapter, an individual immunoglobulin may preferentially recognize a particular conformation of its target epitope (e.g., denatured or native). Consequently, not all monoclonal antibodies are suitable for use as probes in western blots, where the target proteins are thoroughly denatured. Polyclonal antisera, on the other hand, are undefined mixtures of individual immunoglobulins, whose specificity, affinity, and concentration are often unknown. Consequently, it is not possible to predict the efficiency with which a given polyclonal antiserum will detect different antigenic epitopes of an immobilized, denatured target protein.

Although there is an obvious danger that comes from using undefined reagents to assay a target protein that may also be poorly characterized, most problems that arise with western blotting in practice can be solved by designing adequate controls. These include the use of (1) antibodies (i.e., preimmune sera or irrelevant monoclonal antibodies) that should not react with the target protein and (2) control preparations that either contain known amounts of target antigen or lack it altogether.

Often, there is little choice of immunological reagents for western blotting—it is simply necessary to work with whatever antibodies are at hand. However, if a choice is available, either a high-titer polyclonal antiserum or a mixture of monoclonal antibodies raised against the denatured protein should be used. Reliance on a single monoclonal antibody is hazardous because of the high frequency of spurious cross-reactions with irrelevant proteins. If, as is usually the case, monoclonal and polyclonal antibodies have been raised against native target protein, it will be necessary to verify that they react with epitopes that either (1) resist denaturation with SDS and reducing agents or (2) are created by such treatment. This can be done by using denatured target antigen in a solid-phase radioimmunoassay (see pages 18.19–18.23) or in western dot blots.

In western blotting, the samples to be assayed are solubilized with detergents and reducing agents, separated by SDS-polyacrylamide gel electrophoresis, and transferred to a solid support (usually a nitrocellulose filter), which may then be stained. The filter is subsequently exposed to unlabeled antibodies specific for the target protein. Finally, the bound antibody is detected by one of several secondary immunological reagents (^{125}I-labeled protein A or anti-immunoglobulin, or anti-immunoglobulin or protein A coupled to horseradish peroxidase or alkaline phosphatase). As little as 1–5 ng of an average-sized protein can be detected by western blotting.

Preparation and Electrophoresis of Samples

Two methods are used to extract proteins for western blotting from cells: Either the intact cells are dissolved directly in sample buffer or an extract is made as described earlier for samples to be immunoprecipitated (see pages 18.30–18.41). Which of these methods is best in any individual case depends on the type of cells and on the properties of the antigen.

- In general, bacteria expressing the target protein are lysed directly in SDS gel-loading buffer as described on pages 18.40–18.41.

- Yeasts are first lysed by vortexing in the presence of glass beads or enzymatically (see pages 18.35–18.36), and the resulting extracts are then prepared (see pages 18.35–18.39).

- Mammalian tissues are usually dispersed mechanically and then dissolved directly in SDS gel-loading buffer.

- Mammalian cells in tissue culture may be lysed gently with detergents as described on page 18.34, or, alternatively, the protocol on pages 18.62–18.63, in which the cells are lysed directly in SDS gel-loading buffer, may be used if the target antigen is resistant to this type of extraction.

In any case, the samples are analyzed by SDS-polyacrylamide gel electrophoresis as described on pages 18.47–18.54.

LYSIS OF MAMMALIAN CELLS AND TISSUE IN GEL-LOADING BUFFER

1. *For cells growing in monolayers*

   a. Wash the cells twice with phosphate-buffered saline (PBS; see Appendix B) and drain thoroughly. Remove the last traces of PBS by aspiration.

   b. For a 90-ml petri dish, add 100–200 μl of hot (85°C) 1 × SDS gel-loading buffer to lyse the cells. Scrape the viscous lysate into a microfuge tube with a policeman and proceed with step 2.

   For cells growing in suspension and fragments of tissue

   a. Wash the cells or tissue fragments thoroughly in PBS at 0°C. After centrifugation at 3000g for 5 minutes at 4°C, estimate the volume of the pellet in the bottom of the centrifuge tube.

   b. Remove all of the supernatant by aspiration. Use a disposable pipette tip attached to a vacuum line to remove any droplets of fluid that adhere to the walls of the centrifuge tube.

   c. Disperse the cells or tissue fragments in 5 volumes of ice-cold suspension buffer.

   Suspension buffer

   0.1 M NaCl
   0.01 M Tris · Cl (pH 7.6)
   0.001 M EDTA (pH 8.0)
   1 μg/ml aprotinin
   100 μg/ml phenylmethylsulfonyl fluoride (PMSF)

Caution: PMSF is extremely destructive to the mucous membranes of the respiratory tract, the eyes, and skin. PMSF may be fatal if inhaled, swallowed, or absorbed through the skin. In case of contact, immediately flush eyes or skin with copious amounts of water. Discard contaminated clothing.

PMSF, which is labile in aqueous solutions, should be added from a stock solution just before the suspension buffer is used. The rate of inactivation in aqueous solution increases with pH and is faster at 25°C than at 4°C. The half-life of a 20 μM aqueous solution of PMSF is about 35 minutes at pH 8.0 (James 1978). This means that aqueous solutions of PMSF can be safely discarded after they have been rendered alkaline (pH > 8.6) and stored for several hours at room temperature. PMSF is usually stored as a 10 mM or 100 mM stock solution (1.74 or 17.4 mg/ml in isopropanol) at −20°C.

The cells are suspended in buffer to prevent the formation of an insoluble mass when the 2 × SDS gel-loading buffer is added. This step should be carried out as quickly as possible, using ice-cold suspension buffer, to minimize proteolytic degradation. Most types of mammalian tissue can be rapidly teased apart with forceps or cut into small pieces with scissors beneath the surface of the suspension buffer.

d. As soon as possible, add an equal volume of $2 \times$ SDS gel-loading buffer.

> **$2 \times$ SDS gel-loading buffer**
>
> 100 mM Tris · Cl (pH 6.8)
> 200 mM dithiothreitol
> 4% SDS (electrophoresis grade)
> 0.2% bromophenol blue
> 20% glycerol
>
> $2 \times$ SDS gel-loading buffer lacking dithiothreitol can be stored at room temperature. Dithiothreitol should then be added, just before the buffer is used, from a 1 M stock (see Appendix B).

2. Place the sample in a boiling-water bath for 10 minutes.

   The sample should become highly viscous as a consequence of the release of high-molecular-weight chromosomal DNA.

3. Shear the chromosomal DNA by sonication, using either a sonicator with an immersible tip or a sonicator that can process several samples simultaneously in a chilled cup. Depending on the power output of the sonicator and its state of tuning, between 30 seconds and 2 minutes at full power should be sufficient to reduce the viscous lysate to manageable levels.

   Samples that are too large to be sonicated conveniently can be sheared by repeated passage through a 23-gauge hypodermic needle.

4. Centrifuge the sample at 10,000g for 10 minutes at room temperature. Transfer the supernatant to a fresh tube, and discard the pellet.

5. If possible, calculate the amount of sample that will be required to detect the target protein by western blotting.

   The lowest amount of an average-sized protein that can be detected by western blotting is approximately 1–5 ng. Approximately 100 μg of total cellular protein can be applied to a lane in a 0.75-mm-thick SDS-polyacrylamide gel without overloading. To be detected by western blotting, an average-sized target protein must therefore comprise at least 1 part in 10^5 by weight of the total protein in the extract. If it is feasible to do so, reconstruction experiments should be carried out in which known amounts of the target protein are added to total extracts of cells and analyzed by western blotting. These experiments should serve as positive controls when unknown samples are being analyzed, and they should provide an accurate estimate of the sensitivity of the method with the antibodies that are available.

6. Analyze an aliquot of the sample(s) by SDS-polyacrylamide gel electrophoresis and western blotting. The gel is loaded and run in the conventional manner as described on pages 18.47–18.54.

   Electrophoresis is usually carried out using SDS-polyacrylamide gels and the Laemmli discontinuous buffer system (Laemmli 1970) (see page 18.47). However, many other types of gels have been used successfully, including urea-polyacrylamide gels and nondenaturing polyacrylamide gels. For details of the methods used to cast and run such gels, see Hames and Rickwood (1981).

Transfer of Proteins from SDS-Polyacrylamide Gels to Solid Supports

A number of different solid supports have been used for western blotting. These include diazophenylthio (DPT) paper (Seed 1982a,b), diazobenzyloxy-methyl (DBM) paper (Renart et al. 1979), cyanogen-bromide-activated paper (Hitzeman et al. 1980), cyanuric chloride paper (Hunger et al. 1981), and activated nylon. In all of these cases, the proteins transferred from the gel become covalently bound to the support (for review, see Renart and Sandoval 1984). Although supports of this type may have the highest capacity and may retain the bound protein more securely, they are generally difficult to prepare and may require that glycine be soaked out of the gel before transfer. Furthermore, charged supports such as nylon membranes may not bind proteins of the same charge with high efficiency. Consequently, most western blotting nowadays is carried out by direct electrophoretic transfer of proteins from the gel to a nitrocellulose filter (Burnette 1981).

Two types of electrophoresis apparatuses are available for electroblotting. In the older type, one side of the gel is placed in contact with a piece of nitrocellulose filter. The gel and its attached filter are then sandwiched between Whatman 3MM paper, two porous pads, and two plastic supports. The entire construction is then immersed in an electrophoresis tank, equipped with standard platinum electrodes, that contains Tris-glycine electrophoresis buffer at pH 8.3. The nitrocellulose filter is placed toward the anode (see Figure 18.2). An electric current is then applied for about 12 hours; during this time, the proteins migrate from the gel toward the anode and become attached to the nitrocellulose filter. To prevent overheating, and the consequent formation of air bubbles in the sandwich, transfer is carried out in the cold.

In the newer type of apparatus, the gel and its attached nitrocellulose filter are sandwiched between pieces of Whatman 3MM paper that have been soaked in a transfer buffer containing Tris, glycine, SDS, and methanol. The sandwich is then placed between graphite plate electrodes, with the nitrocellulose filter on the anodic side. Transfer of the proteins from the gel can be carried out at room temperature and is complete in 1.5–2 hours.

Both of these methods work well, and the choice between them is largely a matter of personal preference. Because of its rapidity and because it does not require a large amount of electric current, we prefer the second method, which is described in detail below.

1. When the SDS-polyacrylamide gel is approaching the end of its run, rinse the graphite plates with distilled water and wipe off any beads of liquid that adhere to them with Kimwipes or other nonabsorbent tissues.

2. Wearing gloves, cut six pieces of Whatman 3MM paper and one piece of nitrocellulose filter (Millipore HAWP or equivalent) to the exact size of the SDS-polyacrylamide gel. If the paper or filter is larger than the gel, there is a good chance that the overhanging edges of the paper and the filter will touch, causing a short circuit that will prevent the transfer of protein from the gel. Mark one corner of the filter with a soft-lead pencil.

   It is important to wear gloves when handling the gel, 3MM papers, and nitrocellulose filter. Oils and secretions from the skin will prevent the transfer of proteins from the gel to the filter.

3. Float the nitrocellulose filter on the surface of a tray of deionized water, and allow it to wet from beneath by capillary action. Then, submerge the filter in the water for at least 5 minutes to displace trapped air bubbles.

4. Soak the six pieces of 3MM paper in a shallow tray containing a small amount of transfer buffer.

> *Transfer buffer*
>
> 39 mM glycine
> 48 mM Tris base
> 0.037% SDS (electrophoresis grade)
> 20% methanol
>
> To prepare 1 liter of transfer buffer (pH 8.3), mix 2.9 g of glycine, 5.8 g of Tris base, 0.37 g of SDS, and 200 ml of methanol.

5. Wearing gloves, set up the transfer apparatus as follows:

   a. Lay the bottom electrode (which will become the anode) flat on the bench, graphite side up.

   b. Place on the electrode three sheets of 3MM paper that have been soaked in transfer buffer. Stack the sheets one on top of the other so that they are exactly aligned. Using a glass pipette as a roller, squeeze out any air bubbles.

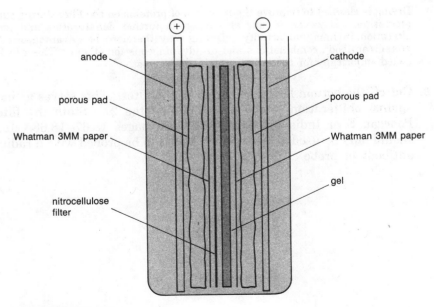

FIGURE 18.2

c. Place the nitrocellulose filter on the stack of 3MM paper. Make sure that the filter is exactly aligned and that no air bubbles are trapped between it and the 3MM paper.

d. Remove the glass plates holding the SDS-polyacrylamide gel from the electrophoresis tank. Transfer the gel briefly to a tray of deionized water, and then place it exactly on top of the nitrocellulose filter. Orient the gel so that the mark on the filter corresponds to the bottom left-hand corner of the gel. Squeeze out any trapped air bubbles with a gloved hand.

   Important: To avoid the possibility of a short circuit, do not cut off the bottom left-hand corner of the gel.

e. Place the final three sheets of 3MM paper on the gel, again making sure that they are exactly aligned and that no air bubbles are trapped.

6. Place the upper electrode (which will become the cathode) on top of the stack, graphite side down. Connect the electrical leads (positive or red lead to the bottom graphite electrode). Apply a current of 0.65 mA/sq. cm. of gel for a period of 1.5–2 hours.

7. Turn off the electric current and disconnect the leads. Disassemble the transfer apparatus from the top downward, peeling off each layer in turn. Transfer the gel to a tray containing Coomassie Brilliant Blue, and stain it as described on page 18.55. This will allow you to check whether electrophoretic transfer is complete.

8. *Optional:* Remove the nitrocellulose filter from the sandwich and transfer it to a clean piece of 3MM paper. Allow the filter to dry for 30–60 minutes at room temperature.

   Drying is claimed to improve the retention of proteins on the filter during subsequent processing. However, it may also result in further denaturation and consequent alteration in immunoreactivity. Drying may therefore be advantageous for some protein/antibody combinations and disadvantageous for others. This can be established empirically for the target proteins of interest.

9. Cut off the bottom left-hand corner of the filter. This serves as insurance against obliteration of the pencil mark (step 2). Stain the filter with Ponceau S or India ink as described on pages 18.67–18.68. India ink should only be used if the western blot is to be probed with a radiolabeled antibody or probe.

Staining Proteins Immobilized on Nitrocellulose Filters

Of the several procedures available to stain proteins immobilized on nitrocellulose filters, only one, staining with Ponceau S, is completely compatible with all methods of immunological probing because the stain is transient and is washed away during processing of the western blot. Staining with Ponceau S therefore does not interfere with the subsequent detection of antigens by chromogenic reactions catalyzed by antibody-linked enzymes such as alkaline phosphatase or lactoperoxidase. However, because the pink-purple color of Ponceau S is difficult to capture photographically, the stain does not provide a permanent record of the experiment. Instead, staining with Ponceau S is used to provide visual evidence that electrophoretic transfer of proteins has taken place and to locate molecular-weight markers, whose positions on the nitrocellulose filter are then marked with pencil or indelible ink.

If the western blot is to be probed with radiolabeled antibody or radiolabeled protein A, the proteins immobilized on the nitrocellulose filter may be stained with India ink, which is cheaper and more sensitive than Ponceau S and provides a permanent record of the location of proteins on the nitrocellulose filter. Note that proteins transferred to nylon membranes cannot be stained with India ink, since the colloidal particles become irreversibly bound.

STAINING WITH PONCEAU S

1. If the nitrocellulose filter has been dried, float it on the surface of a tray of deionized water and allow it to wet from beneath by capillary action. Then, submerge the filter in the water for at least 5 minutes to displace trapped air bubbles.

2. Transfer the filter to a tray containing a working solution of Ponceau S stain. Incubate the filter for 5–10 minutes with gentle agitation.
   For a stock solution of Ponceau S, mix:

Ponceau S	2 g
trichloroacetic acid	30 g
sulfosalicylic acid	30 g
H_2O to 100 ml	

   Dilute one part of the stock solution with nine parts of deionized water to make a working solution. Discard the working solution after use.

   Ponceau S (Sigma P 7767) is 3-hydroxy-4-(2-sulfo-4-[4-sulfophenylazo]-phenylazo)-2,7-naphthalenedisulfonic acid.

3. When the bands of protein are visible, wash the nitrocellulose filter in several changes of deionized water at room temperature.

4. Mark the positions of proteins used as molecular-weight standards with waterproof black ink.

5. Proceed with immunological probing as described on pages 18.69–18.75.

STAINING WITH INDIA INK

This method of staining should only be used when radiolabeled antibody or radiolabeled protein A is used as a probe.

1. If the nitrocellulose filter has been dried, float it on the surface of a tray of deionized water and allow it to wet from beneath by capillary action. Then, submerge the filter in the water for at least 5 minutes to displace trapped air bubbles.

2. Transfer the filter to a tray containing 0.4% Tween 20 in phosphate-buffered saline (PBS; see Appendix B). Incubate for 5 minutes at room temperature with gentle shaking, change the wash solution, and continue to incubate for a further 5 minutes.

   The Tween 20 prevents the colloidal particles of India ink from attaching to sites on the nitrocellulose filter.

3. Transfer the filter to a tray containing diluted India ink (100 μl of ink in 100 ml of 0.4% Tween 20 in PBS).

   Some brands of India ink work far better than others. Higgins India Ink and Pelikan Fount India Drawing Ink are both highly recommended.

4. Cover the tray with Saran Wrap and incubate the filter at room temperature in the solution of India ink until the protein bands reach the desired intensity. Depending on the amount of protein loaded on the gel, this may take from 15 minutes to several hours.

5. Wash the filter for several hours in PBS, changing the buffer every hour.

6. Proceed with immunological probing as described on pages 18.69–18.73.

Blocking Binding Sites for Immunoglobulins on the Nitrocellulose Filter

Just as proteins transferred from the SDS-polyacrylamide gel can bind to the nitrocellulose filter, so can proteins in the immunological reagents used for probing. The sensitivity of western blotting depends on reducing this background of nonspecific binding by blocking potential binding sites with irrelevant proteins. Of the several blocking solutions that have been devised, the best and least expensive is nonfat dried milk (Johnson et al. 1984). It is easy to use and is compatible with all of the immunological detection systems in common use. There is only one circumstance under which nonfat dried milk should not be used: when western blots are probed for proteins that may be present in milk.

1. Place the nitrocellulose filter in a heat-sealable plastic bag (Sears Seal-A-Meal or equivalent), and add 0.1 ml of blocking solution per square centimeter of filter. Seal the bag, leaving as few air bubbles as possible, and incubate the filter for 1–2 hours at room temperature with gentle agitation on a platform shaker.

> *Blocking solution*
>
> 5% (w/v) nonfat dried milk
> 0.01% antifoam A
> 0.02% sodium azide in phosphate-buffered saline (PBS; see Appendix B)
>
> **Caution:** Sodium azide is poisonous. It should be handled with great care wearing gloves, and solutions containing it should be clearly marked.

If the background of nonspecific binding of immunological probes is unacceptably high, try adding Tween 20 to a final concentration of 0.02%. In most cases, the presence of this detergent will not affect specific binding of antibodies to the target antigen.

2. Cut open the plastic bag and discard the blocking solution. Immediately incubate the filter with an antibody directed against the target protein as described on pages 18.70–18.71.

Binding of the Primary Antibody to the Target Protein

Virtually all western blots are probed in two stages: An unlabeled antibody specific to the target protein is first incubated with the nitrocellulose filter in the presence of blocking solution. The filter is then washed and incubated with a secondary reagent—anti-immunoglobulin or protein A that is either radiolabeled or coupled to an enzyme such as horseradish peroxidase or alkaline phosphatase. After further washing, the antigen-antibody-antibody or antigen-antibody-protein A complexes on the nitrocellulose filter are located by autoradiography or in situ enzymatic reactions.

Indirect or two-stage probing has the major advantage of allowing a single secondary reagent to be used to detect a wide variety of primary antibodies, thereby eliminating the tedious task of purifying and labeling each individual primary antibody. Since secondary immunological reagents can be purchased quite inexpensively from commercial sources, the resulting savings of time and money can be considerable.

INCUBATING THE NITROCELLULOSE FILTER WITH THE PRIMARY ANTIBODY DIRECTED AGAINST THE TARGET PROTEIN

1. To the plastic bag containing the nitrocellulose filter prepared as described on page 18.69, add 0.1 ml of blocking solution per square centimeter of filter and an appropriate quantity of the primary antibody.

> *Blocking solution*
>
> 5% (w/v) nonfat dried milk
> 0.01% antifoam A
> 0.02% sodium azide in phosphate-buffered saline (PBS; see Appendix B)
>
> **Caution:** Sodium azide is poisonous. It should be handled with great care wearing gloves, and solutions containing it should be clearly marked.

If the background of nonspecific binding of immunological probes is unacceptably high, try adding Tween 20 to a final concentration of 0.02%. In most cases, the presence of this detergent will not affect specific binding of antibody to the target antigen.

The appropriate amount of primary antibody should be determined empirically in pilot experiments. The recommended test dilutions are:
• Polyclonal antibodies: 1:100 to 1:5000
• Supernatants from cultured hybridoma cells: undiluted to 1:100
• Ascitic fluid from mice bearing hybrid myelomas: 1:1000 to 1:10,000

2. Seal the bag, leaving as few air bubbles as possible, and incubate the filter for 2 hours at 4°C with gentle agitation on a platform shaker.

Longer incubation times (up to 18 hours at room temperature) have been reported to increase the sensitivity of detection of target antigens. However, the background of nonspecific binding also increases as a function of the time and temperature of incubation.

---

3. Cut open the plastic bag, and discard the blocking solution and antibody. Wash the filter three times (10 minutes each time) with 250 ml of PBS.

4. Immediately incubate the filter with the secondary immunological reagent as described on pages 18.72–18.73.

INCUBATING THE NITROCELLULOSE FILTER WITH THE SECONDARY IMMUNOLOGICAL REAGENT

The secondary reagent (usually an anti-immunoglobulin or protein A) may be radiolabeled with ^{125}I as described on pages 18.24–18.25 or may be covalently coupled to an enzyme such as horseradish peroxidase or alkaline phosphatase. Covalently coupled immunoglobulin and protein A are sold commercially.

Although both radiolabeled and enzyme-coupled secondary reagents can work very well, antibodies are sometimes inactivated if the radiolabeling process is carried out too enthusiastically. Most workers therefore prefer to use enzyme-coupled reagents that have been tested by their commercial manufacturers.

Radiolabeled secondary reagents

1. Transfer the nitrocellulose filter from the final wash in phosphate-buffered saline (PBS; see Appendix B) (step 3, page 18.71) to a heat-sealable plastic bag (e.g., Sears Seal-A-Meal) containing 0.1 ml of fresh blocking solution per square centimeter of filter and add approximately 10^4 cpm of the radiolabeled secondary reagent per square centimeter of filter.

   > **Blocking solution**
   >
   > 5% (w/v) nonfat dried milk
   > 0.01% antifoam A
   > 0.02% sodium azide in PBS
   >
   > **Caution:** Sodium azide is poisonous. It should be handled with great care wearing gloves, and solutions containing it should be clearly marked.

   If the background of nonspecific binding of immunological probes is unacceptably high, try adding Tween 20 to a final concentration of 0.02%. In most cases, the presence of this detergent will not affect specific binding of antibody to the target antigen.

2. Incubate the filter for 1–2 hours at room temperature with gentle agitation on a platform shaker.

3. Cut open the bag, and quickly transfer the filter to a tray containing 250 ml of PBS. Discard the plastic bag and radioactive fluid in the radioactive waste.

4. Wash the filter in several changes of PBS (10 minutes each change). Continue washing until no radioactivity is detected by a hand-held minimonitor in regions of the filter that carry no protein.

5. Remove the filter from the last wash, and allow it to dry for 10 minutes on a stack of paper towels.

6. Mount the filter on a piece of Whatman 3MM paper, and attach radioactive markers to the paper.

Radioactive markers can be made by applying spots of radioactive ink to adhesive labels or tape, which are then attached to the surface of the 3MM paper. Radioactive ink can be made by adding a small amount of Tran[35]S-label (ICN Radiochemicals) or ^{32}P to a microfuge tube containing India ink. The radioactive ink can be applied with a fiber-tip pen kept for that specific purpose and clearly marked with radioactive-warning tape.

7. When the radioactive ink is dry, wrap the 3MM paper in Saran Wrap and expose the filter to X-ray film at $-70°C$ with an intensifying screen (see Appendix E).

Enzyme-coupled secondary reagents

1. Transfer the nitrocellulose filter from the final wash in PBS (step 3, page 18.71) to a tray containing 200 ml of 150 mM NaCl, 50 mM Tris·Cl (pH 7.5). Incubate the filter for 10 minutes at room temperature with gentle agitation.

   It is important to remove azide and phosphate from the filter before the enzyme-coupled secondary reagent is added.

2. Transfer the filter to a heat-sealable plastic bag (e.g., Sears Seal-A-Meal), or to a shallow tray, containing 0.1 ml of phosphate-free, azide-free blocking solution per square centimeter of filter.

   Phosphate-free, azide-free blocking solution

   5% (w/v) nonfat dried milk
   150 mM NaCl
   50 mM Tris·Cl (pH 7.5)

3. Add the enzyme-coupled secondary reagent according to the manufacturer's instructions and seal the bag if a bag was used. Usually, it is recommended that the secondary reagent be diluted 1:200 to 1:2000 to yield a final concentration of 0.5–5.0 $\mu g/ml$.

4. Incubate the filter with the enzyme-coupled secondary reagent for 1 hour at room temperature with gentle agitation.

5. Transfer the filter to a tray containing 200 ml of 150 mM NaCl, 50 mM Tris·Cl (pH 7.5). Incubate the filter for 10 minutes at room temperature with gentle agitation. Repeat this step three more times using fresh NaCl/Tris·Cl solution each time.

6. Add the appropriate chromogenic substrates to the filter as described on pages 18.74–18.75.

USE OF CHROMOGENIC SUBSTRATES WITH ENZYME-COUPLED ANTIBODIES

Alkaline phosphatase

The substrate 5-bromo-4-chloro-3-indolyl phosphate/nitro blue tetrazolium (BCIP/NBT) is converted in situ into a dense blue compound by immunolocalized alkaline phosphatase.

1. Prepare the following three solutions:

> *NBT*
>
> Dissolve 0.5 g of NBT, which is available from several manufacturers, in 10 ml of 70% dimethylformamide.
>
> *BCIP*
>
> Dissolve 0.5 g of BCIP disodium salt, which is available from several manufacturers, in 10 ml of 100% dimethylformamide.
>
> *Alkaline phosphatase buffer*
>
> 100 mM NaCl
> 5 mM $MgCl_2$
> 100 mM Tris · Cl (pH 9.5)
>
> These solutions are stable when stored in closed containers at room temperature.

2. Mix 66 μl of NBT stock with 10 ml of alkaline phosphatase buffer. Mix well, and add 33 μl of BCIP stock. This chromogenic substrate mixture should be used within 30 minutes.

3. Transfer the washed nitrocellulose filter (step 5, page 18.73) to a shallow tray. Add 0.1 ml of chromogenic substrate mixture per square centimeter of filter. Incubate the filter at room temperature with gentle agitation.

4. Monitor the progress of the reaction carefully. When the bands are of the desired intensity (~20 minutes), transfer the filter to a tray containing 200 μl of 0.5 M EDTA (pH 8.0) and 50 ml of phosphate-buffered saline (PBS; see Appendix B).

5. Photograph the filter to provide a permanent record of the experiment.

Horseradish peroxidase

The most sensitive substrate for immunocoupled horseradish peroxidase is 3,3′-diaminobenzidine, which is converted in situ into a brown precipitate. The color of the precipitate is enhanced and the sensitivity of the reaction is improved by carrying out the reaction in the presence of cobalt or nickel ions. However, it is impossible to suppress background staining completely with horseradish peroxidase, and it is therefore essential to monitor the course of the reaction very carefully. The reaction should be stopped as soon as the specifically stained bands are clearly visible.

1. Dissolve 6 mg of diaminobenzidine tetrahydrochloride in 9 ml of 0.01 M Tris · Cl (pH 7.6). Add 1 ml of 0.3% (w/v) $NiCl_2$ or $CoCl_2$.

   The solution of diaminobenzidine must be freshly prepared.

2. Filter the solution through a Whatman No. 1 filter to remove any precipitate that may form.

3. Add 10 μl of 30% H_2O_2. Mix well, and use the mixture immediately.

   H_2O_2 is supplied as a 30% solution. It should be stored in tightly closed brown bottles and discarded after a few weeks.

4. Transfer the washed nitrocellulose filter (step 5, page 18.73) to a shallow tray. Add 0.1 ml of the substrate solution per square centimeter of filter. Incubate the filter at room temperature with gentle agitation.

5. Monitor the progress of the reaction carefully. When the bands are of the desired intensity ($\sim$ 2–3 minutes), wash the filter briefly in water, and then transfer it to a tray containing 250 ml of phosphate-buffered saline (PBS; see Appendix B).

6. Photograph the filter to provide a permanent record of the experiment.

   The peroxidase-stained bands will fade after several hours of exposure to light.

Translation of mRNAs

TRANSLATION OF RNA IN RETICULOCYTE LYSATES

mRNA extracted from mammalian cells or generated by transcription of cloned DNA in vitro may be translated in cell-free extracts to produce proteins that can be immunoprecipitated or tested for biological activity. Excellent translation kits prepared from lysates of rabbit reticulocytes or extracts of wheat germ can be purchased from several commercial manufacturers, including New England Nuclear and Promega Biotec. However, we find that the efficiency of the commercial translation kits is much lower than that of "homemade" in vitro translation systems. We have therefore included a brief description of a procedure for preparing rabbit reticulocyte lysate.

Preparation of Rabbit Reticulocyte Lysate

The following procedure was provided by B. Roberts and M. Mathews (unpubl.).

1. To obtain anemic rabbit blood, inoculate subcutaneously six rabbits, each weighing 2–3 kg, with acetylphenylhydrazine (a 1.2% solution neutralized to pH 7.5 with 1 M HEPES [pH 7.0]) according to the following schedule:

   2.0 ml on day 1
   1.6 ml on day 2
   1.2 ml on day 3
   1.6 ml on day 4
   2.0 ml on day 5

2. On days 7 and 8, bleed the rabbits as follows: Swab one ear with alcohol-saturated cotton, and make a single incision with a new razor blade in the median ear vein midway along the length of the ear. Each rabbit should yield 50 ml of blood, which is collected in 50 ml of a chilled solution of:

   140 mM NaCl
   1.5 mM magnesium acetate
   5.0 mM KCl
   0.001% heparin

   Use discretion rather than greed when collecting blood from the rabbits on days 7 and 8. On day 9, the animal may be anesthetized with Nembutal or fentanyl and exsanguinated by heart puncture (yielding ~150 ml of blood that is collected in ~100 ml of the above salt solution).

3. Filter the blood through cheesecloth, and then centrifuge it at 2000g for 5 minutes at 4°C. Remove the buffy coat of white blood cells by aspiration. (The buffy coat is a broad band of white blood cells of heterogeneous density.) Wash the packed reticulocytes and erythrocytes three times with the salt solution (lacking heparin). The final centrifugation step should be carried out at 5000g.

4. Measure the volume of the packed cells, and lyse them at 0°C by mixing with an equal volume of cold, sterile, double-distilled water. After 1 minute, centrifuge the lysate at 20,000g for 20 minutes at 4°C.

5. Prepare the following stock solutions needed to treat the lysate with micrococcal nuclease:

   - 100 mM $CaCl_2$

   - 200 mM EGTA (pH 7.0)

   - Micrococcal nuclease (150,000 units/ml), stored in 50 mM glycine (pH 9.2), 5 mM $CaCl_2$

   - Creatine phosphokinase type I (40 mg/ml in 50% glycerol)

   - Hemin stock (5 mg/ml in ethylene glycol). Hemin stock solution is made up as follows:

a. Dissolve 30 mg of hemin (m.w. = 652) in 400 μl of 0.2 N KOH. Add the hemin slowly, a little at a time, and vortex between each addition.

Caution: Handle KOH solutions with great care. Use gloves and a face protector.

b. Add:

H_2O	600 μl
1 M Tris · Cl (pH 7.8)	100 μl

If necessary, adjust the pH to 7.8 by adding 0.1 N HCl.

c. Add 4.8 ml of ethylene glycol. Mix by vortexing.

d. Centrifuge at 5000g for 5 minutes at 4°C. Discard the insoluble pellet.

e. Estimate the hemin concentration by making a 1:100 dilution in 10 mM KCN and reading the absorbance at 540 nm. (A 1 mM solution of hemin has an absorbance of 11.1). Adjust the concentration of the hemin solution to 5 mg/ml with a 4:1 solution of ethylene glycol and water.

6. To 100 parts of reticulocyte lysate, add 1 part hemin solution and 2 parts $CaCl_2$ solution. Mix well.

7. Add 0.05 part micrococcal nuclease (150,000 units/ml) and mix. Incubate for 15 minutes at 20°C. Then chill the solution in ice water.

8. Add 1 part 200 mM EGTA (pH 7.0) and mix.

9. Add 0.4 part creatine phosphokinase (40 mg/ml). Mix well. Dispense the treated lysate in aliquots of 250–500 μl, and store them at −70°C or in liquid nitrogen.

Notes

i. This treatment destroys endogenous mRNAs so that the translational activity of the reticulocyte lysate becomes completely dependent on exogenously added mRNAs. Micrococcal nuclease is active only in the presence of calcium ions, and the digestion is terminated by the addition of EGTA. The subsequent translation reaction is not affected by the presence of EGTA or by the presence of inactive micrococcal nuclease.

ii. Hemin is included in the reticulocyte lysate because it is a powerful suppressor of an inhibitor of the initiation factor eIF-2. In the absence of hemin, protein synthesis in reticulocyte lysates ceases after a short period of incubation. For optimal translational efficiency, the hemin should be titrated over the range of 20 to 80 μM.

Translation of Reticulocyte Lysates

1. Make up the translation cocktail below, dispense in 50–100-μl aliquots, and store them at −70°C or in liquid nitrogen.

100 mM spermidine	200 μl
800 mM creatine phosphate	400 μl
5 mM amino acids (lacking methionine)	200 μl
1 M dithiothreitol	80 μl
500 mM HEPES (pH 7.4)	1600 μl
H$_2$O	720 μl

5 mM amino acids (lacking methionine) is a solution containing each amino acid, except methionine, at a concentration of 5 mM. This solution is used when proteins synthesized in vitro are to be radiolabeled with [^{35}S]Met. When proteins are to be labeled with other radiolabeled amino acids, the appropriate mixture of unlabeled amino acids should be used.

2. The standard reaction mixture for translation contains:

translation cocktail	2 μl
1 M KCl	2 μl
16.5 mM magnesium acetate	1 μl
[^{35}S]Met (80 μCi; 800–1200 Ci/mmole) (translation grade)	8 μl
microccocal-nuclease-treated reticulocyte lysate	10 μl
RNA	2 μl

Incubate for 1 hour at 30°C.

Thaw reticulocyte lysates gently (hand-heat), and return unused portions to −70°C or to liquid nitrogen as soon as possible. Repeated freezing and thawing does not seem to damage the lysates as long as care is taken to minimize the time that they are exposed to temperatures higher than −70°C.

The optimal concentration of magnesium acetate in the translation reactions varies from preparation to preparation of reticulocyte lysate and from mRNA to mRNA. The amount of magnesium acetate required to achieve the maximum efficiency of translation should be determined for every new batch of reticulocyte lysate using a standard preparation of mRNA. Titrate the magnesium acetate over the range of 0 to 1 mM. Then, using the optimum concentration of magnesium acetate, determine the effect of different concentrations of KCl (20–80 mM) on the efficiency of translation of the particular mRNA of interest. Approximately 70 mM KCl is usually optimal for translation of capped mRNAs. Approximately 40 mM KCl is usually optimal for translation of uncapped mRNAs.

Up to 10 μg of total cytoplasmic RNA or 0.2 μg of poly(A)$^+$ mRNA or RNA synthesized in vitro can be added to the reaction before saturation is reached. Usually, cleaner results are obtained with poly(A)$^+$ mRNA than with total cytoplasmic RNA.

The addition of saturating amounts of mRNA to the system should stimulate the incorporation of [^{35}S]Met into acid-precipitable material some 10- to 25-fold. Under optimal conditions, about 3×10^6 to 5×10^6 cpm (1–3 μCi) of [^{35}S]Met should be incorporated per 25 μl of reaction mixture.

Before precipitating the radiolabeled proteins with 10% trichloroacetic acid (TCA), [^{35}S]Met-tRNA should be destroyed by treating the reaction with alkali (0.3 N NaOH for 15 minutes at 37°C) or with hot TCA (5–10% for 15 minutes at 100°C). The amount of [^{35}S]Met incorporated into protein can be measured following precipitation of an aliquot with TCA (see Appendix E).

3. The products of in vitro translation reactions may be analyzed by immuno-precipitation and/or SDS-polyacrylamide gel electrophoresis. Relevant immunoprecipitation techniques are described on pages 18.44–18.46, and detailed and standardized procedures for electrophoresis in SDS-poly-acrylamide gels are described on pages 18.47–18.54.

Notes

i. For analysis by SDS-polyacrylamide gel electrophoresis, hydrolyze the [^{35}S]Met-tRNA in the sample by adding 0.2 volume of a solution of pancreatic RNAase (100 μg/ml in 50 mM EDTA [pH 8.0]). Incubate for 15 minutes at 37°C. Immediately add an equal volume of 2× SDS gel-loading buffer (see page 18.40) and boil the sample. Load and run the gel as described on pages 18.53–18.54.

ii. Unused portions of completed translation reactions should be stored at −20°C.

iii. Background bands may be a problem when reticulocyte lysates are analyzed by SDS-polyacrylamide gel electrophoresis. Frequently, a poly-peptide that migrates with an apparent molecular mass of 45 kD becomes radiolabeled. This, as well as radiolabeling of globin, is usually caused by impurities in the [^{35}S]Met. This problem can usually be avoided by using freshly bought radiolabeled methionine, preferably of the "stabilized" variety, which should be exposed minimally to air and stored at −70°C until just before it is used.

iv. Radiolabeling of many proteins (including globin) in reactions that do not contain exogenously added mRNA is a sign that the reticulocyte lysate has not been treated sufficiently with micrococcal nuclease. It is probably worthwhile treating such "dirty" lysates with micrococcal nuclease for a second time (although retreatment is hardly ever completely satisfactory). Before retreatment, adjust the concentration of CaCl$_2$ in the lysate to 2 mM. When next preparing a batch of lysate, use a higher concentration of micrococcal nuclease (rather than treating for a longer time), and make sure that the lysate is equilibrated to 20°C during treatment with the enzyme.

Synthetic mRNAs are efficiently translated in wheat germ extracts, reticulocyte extracts, and injected frog oocytes (Krieg and Melton 1984). To synthesize mRNAs in quantities sufficient for translation in vitro, it is necessary to (1) clone double-stranded DNA coding for the mRNA in the appropriate orientation in a plasmid vector containing a bacteriophage promoter, (2) linearize the recombinant plasmid with a restriction enzyme that cleaves downstream from the template sequences, and (3) transcribe the template sequences in vitro using a DNA-dependent RNA polymerase encoded by a bacteriophage that specifically recognizes the promoter cloned in the plasmid. The plasmid vectors, the properties of bacteriophage-encoded DNA-dependent RNA polymerases, and the composition of the transcription reactions are described in detail in Chapters 1, 5, and 10, and the reader should be familiar with these before proceeding further.

The structures of the 5′ and 3′ termini of synthetic mRNA can affect its stability and the efficiency of translation. All natural eukaryotic mRNAs and many viral mRNAs have a structure at their 5′ termini consisting of a 7-methylguanosine residue linked by a triphosphate bridge to the body of the mRNA. In addition, the first (and sometimes the second) nucleotide of the mRNA is methylated. The presence of such a capped structure at the 5′ terminus increases the amount of protein synthesized in extracts of wheat germ (Paterson and Rosenberg 1979) and improves the stability of natural mRNAs (e.g., Furuichi et al. 1977) and of mRNAs injected into living cells (Krieg and Melton 1984; Drummond et al. 1985). Capped mRNAs are synthesized in vitro by including an analog of the cap structure (e.g., GpppG or its methylated derivatives) in the transcription reaction (Contreras et al. 1982). If the analog is present in molar excess over rGTP, it will be used to initiate RNA synthesis at the bacteriophage promoter. However, once RNA synthesis has begun, the peculiar chemical structure of the capped structure (with two exposed 3′ hydroxyl residues) ensures that no further incorporation of the capping nucleotide can occur.

The effect of the poly(A) tail found at the 3′ terminus of the vast majority of eukaryotic and viral mRNAs is less clear. The presence of a poly(A) tract has been shown to improve the stability of certain mRNAs; in others, it seems to have little effect. In the best-studied case, polyadenylation improved the long-term stability of mRNAs injected into *Xenopus* oocytes (Drummond et al. 1985) but appeared to have little effect on the efficiency with which they were translated in short-term (1–2-day) experiments (Krieg and Melton 1984). The structure of the 3′ terminus of the mRNA is usually dictated by the restriction enzyme used to linearize the template. However, many templates used to synthesize mRNAs in vitro are derived from cDNAs and carry tracts of poly(dA:dT) near their 3′ termini. It is conceivable that the resulting tracts of poly(A) may confer stability on synthetic mRNAs even though their location within the mRNAs is perhaps not ideal. In this confusing situation, our advice is to synthesize mRNAs with a 5′ cap, but without a 3′ tract of poly(A). In the rare cases where these mRNAs prove to be unstable or are translated inefficiently, it may be worthwhile to add tracts of poly(A) posttranscriptionally with *E. coli* poly(A) polymerase (for details, see Drummond et al. 1985).

Synthesis of Synthetic mRNAs

1. Prepare 2 pmoles of template DNA by digestion with a suitable restriction enzyme. Analyze an aliquot (100 ng) of the digested DNA by agarose gel electrophoresis. If necessary, add more restriction enzyme and continue incubation until there is no trace of the undigested DNA.

   1 pmole of a plasmid 3 kb in length weighs approximately 2 μg.

2. If necessary, remove protruding 3′ termini by treating the digested DNA with bacteriophage T4 DNA polymerase or the Klenow fragment of *E. coli* DNA polymerase I in the presence of all four dNTPs (see Appendix F).

3. Purify the template DNA by extraction with phenol:chloroform and precipitation with ethanol. Redissolve the DNA in RNAase-free TE (pH 7.6) (see Chapter 7) at a concentration of 1 μg/μl.

   When using minipreparations of plasmid DNA that have been treated with RNAase (because of contamination of the template DNA with RNAase), purify the DNA at this stage as follows:
   a. Add 0.1 volume of 10× proteinase K buffer.

   10 × Proteinase K buffer

   500 mM NaCl
   50 mM EDTA (pH 8.0)
   100 mM Tris·Cl (pH 8.0)

   b. Add 0.1 volume of 5% SDS.
   c. Add proteinase K to a final concentration of 100 μg/ml. Incubate the reaction for 1 hour at 37°C. Proteinase K is stored as a stock solution at a concentration of 20 mg/ml in water (see Appendix B).
   d. Purify the DNA by extraction with phenol:chloroform and precipitate with ethanol.
   e. Redissolve the DNA in RNAase-free TE (pH 7.6) at a concentration of 1 μg/μl.

4. Mix the following components at room temperature in the order shown:

template DNA (in a volume of 5 μl or less)	1–5 μg
RNAase-free H$_2$O to 16 μl	
100 mM dithiothreitol	5 μl
bovine serum albumin (2.5 mg/ml)	2.5 μl
5 mM rATP, rUTP, rATP	2.5 μl each
500 μM rGTP	2.5 μl
cap analog (GpppG or its methylated derivatives at 5 mM)	5 μl
10 × transcription buffer	5.0 μl
placental RNAase inhibitor (10 units/ml)	2.5 μl
bacteriophage-encoded DNA-dependent RNA polymerase (10–15 units/μl)	3.0 μl

   Mix the reagents by gently tapping the outside of the tube. Try to avoid producing air bubbles. (This can sometimes be difficult because commercial preparations of bacteriophage-encoded DNA-dependent RNA polymerase contain detergents as stabilizers.) If necessary, remove air bubbles by brief centrifugation.

Incubate the reaction for 1 hour at 37°C (bacteriophages T3 and T7 DNA-dependent RNA polymerases) or 40°C (bacteriophage SP6 DNA-dependent RNA polymerase).

---

10 × Transcription buffer

400 mM Tris · Cl (pH 7.5 at 37°C)
60 mM $MgCl_2$
20 mM spermidine HCl
50 mM NaCl

The components are added at room temperature in the order shown to avoid the possibility that the template DNA may be precipitated by the high concentrations of spermidine in the transcription buffer.

If desired, the reaction can be monitored by including 10 μCi of [^{32}P]rGTP (sp. act. 400 Ci/mmole). The specific activity of the resulting [^{32}P]RNA will be approximately 1×10^6 cpm/μg.

The yield of RNA can often be increased by adding a second batch of enzyme (30–40 units) after 1 hour of incubation and continuing incubation for a further hour. Under these conditions, each mole of template yields approximately 8 moles of transcript.

5. Add 1 μl of RNAase-free DNAase (1 mg/ml) (see Appendix B). Mix, and incubate the reaction mixture for 15 minutes at 37°C.

6. Add 100 μl of RNAase-free H_2O and purify the RNA by extraction with phenol:chloroform.

7. Transfer the aqueous phase to a fresh tube, and add 20 μl of 5 M ammonium acetate solution. Mix, and then add 250 μl of ethanol. After storage for 30 minutes at −20°C, collect the RNA by centrifugation at 12,000g for 10 minutes at 4°C in a microfuge. Redissolve the RNA in 100 μl of RNAse-free H_2O and extract with phenol:chloroform.

8. Transfer the aqueous phase to a fresh tube, and add 20 μl of 5 M ammonium acetate solution. Mix, and then add 250 μl of ethanol. Remove as much of the ethanol as possible by gentle aspiration, and leave the open tube on the bench for a few minutes to allow the last visible traces of ethanol to evaporate. Store the RNA in 70% ethanol at −70°C.

9. To recover the RNA, withdraw an aliquot of the ethanolic solution, add 0.1 volume of 5 M ammonium acetate, mix, and store the solution for at least 15 minutes at −20°C. Centrifuge the solution at 12,000g for 10 minutes at 4°C in a microfuge. Remove the ethanol, and dissolve the RNA in RNAase-free H_2O at an estimated concentration of 2 mg/ml. Read the OD_{260} of a diluted aliquot of the sample and determine the concentration. The OD_{260} of a 40 μg/ml solution is 1.

Notes

i. Usually, 5–10 μg of RNA can be synthesized per microgram of template DNA.

ii. The size of the RNA should be checked by electrophoresis through agarose or polyacrylamide gels (see Chapter 6), using as markers RNAs of known size.

Translation of Synthetic RNAs

1. If necessary, dilute the synthetic RNA to a concentration of 1 mg/ml in RNAase-free H_2O.

2. Translate the synthetic RNA in the rabbit reticulocyte lysate system as described on pages 18.79–18.80.

Note

In vitro transcription of double-stranded DNA templates often generates significant quantities of double-stranded RNA that will inhibit translation in reticulocyte lysates. If translation fails or is inefficient, we recommend the following:

- Take precautions to minimize the generation of double-stranded RNA during in vitro transcription (see Chapter 10, page 10.36).

- Carry out translation experiments in which control RNAs that are known to be active in reticulocyte systems are mixed with varying quantities of RNAs generated in in vitro transcription systems.

References

Barrett, A.J. and G. Salvesen, eds. 1986. *Proteinase inhibitors. Research monographs in cell and tissue physiology*, vol. 12. Elsevier, Amsterdam.

Bolton, A.E. and W.M. Hunter. 1973. The labelling of proteins to high specific radioactivities by conjugation to a ^{125}I-containing acylating agent. Application to the radioimmunoassay. *Biochem. J.* **133**: 529.

Burke, B., G. Griffiths, H. Reggio, D. Louvard, and G. Warren. 1982. A monoclonal antibody against a 135-K Golgi membrane protein. *EMBO J.* **1**: 1621.

Burnette, W.N. 1981. "Western blotting": Electrophoretic transfer of proteins from sodium dodecyl sulfate–polyacrylamide gels to unmodified nitrocellulose and radiographic detection with antibody and radioiodinated protein A. *Anal. Biochem.* **112**: 195.

Campbell, A.M. 1984. *Monoclonal antibody technology. Laboratory techniques in biochemistry and molecular biology* (ed. R.H. Burdon and P.H. van Knippenberg), vol. 13. Elsevier, Amsterdam.

Contreras, R., H. Cheroutre, W. Degrave, and W. Fiers. 1982. Simple, efficient *in vitro* synthesis of capped RNA useful for direct expression of cloned eukaryotic genes. *Nucleic Acids Res.* **10**: 6353.

Davis, B.J. 1964. Disc electrophoresis–II. Method and application to human serum proteins. *Ann. N. Y. Acad. Sci.* **121**: 404.

Deisenhofer, J. 1981. Crystallographic refinement and atomic models of a human Fc fragment and its complex with fragment B of protein A from *Staphylococcus aureus* at 2.9- and 2.8-Å resolution. *Biochemistry* **20**: 2361.

Drummond, D.R., J. Armstrong, and A. Colman. 1985. The effect of capping and polyadenylation on the stability, movement and translation of synthetic messenger RNAs in *Xenopus* oocytes. *Nucleic Acids Res.* **13**: 7375.

Earnshaw, W.C. and N. Rothfield. 1985. Identification of a family of human centromere proteins using autoimmune sera from patients with scleroderma. *Chromosoma* **91**: 313.

Ey, P.L., S.J. Prowse, and C.R. Jenkin. 1978. Isolation of pure IgG$_1$, IgG$_{2a}$ and IgG$_{2b}$ immunoglobulins from mouse serum using protein A–Sepharose. *Immunochemistry* **15**: 429.

Furuichi, Y., A. LaFiandra, and A.J. Shatkin. 1977. 5'-Terminal structure and mRNA stability. *Nature* **266**: 235.

Goding, J.W. 1986. *Monoclonal antibodies: Principles and practice*, 2nd edition. Academic Press, London.

Goudswaard, J., J.A. van der Donk, A. Noordzij, R.H. van Dam, and J.-P. Vaerman. 1978. Protein A reactivity of various mammalian immunoglobulins. *Scand. J. Immunol.* **8**: 21.

Green, N., H. Alexander, A. Olson, S. Alexander, T.M. Shinnick, J.G. Sutcliffe, and R.A. Lerner. 1982. Immunogenic structure of the influenza virus hemagglutinin. *Cell* **28**: 477.

Greenwood, F.C., W.M. Hunter, and J.S. Glover. 1963. The preparation of ^{131}I-labelled human growth hormone of high specific radioactivity. *Biochem. J.* **89**: 114.

Hames, B.D. and D. Rickwood, eds. 1981. *Gel electrophoresis of proteins: A practical approach.* IRL Press, Oxford.

Harlow, E. and D. Lane. 1988. *Antibodies: A laboratory manual.* Cold Spring Harbor Laboratory, Cold Spring Harbor, New York.

Hitzeman, R.A., L. Clarke, and J. Carbon. 1980. Isolation and characterization of the yeast 3-phosphoglycerokinase gene (*PGK*) by an immunological screening technique. *J. Biol. Chem.* **255**: 12073.

Hjelm, H., K. Hjelm, and J. Sjöquist. 1972. Protein A from *Staphylococcus aureus*. Its isolation by affinity chromatography and its use as an immunosorbent for isolation of immunoglobulins. *FEBS Lett.* **28**: 73.

Hunger, H.-D., H. Grützmann, and C. Coutelle. 1981. Nucleic acid fixation to cyanuric chloride-activated paper. Use for nucleic acid transfer and affinity binding. *Biochim. Biophys. Acta* **653**: 344.

Hunkapiller, M.W., E. Lujan, F. Ostrander, and L.E. Hood. 1983. Isolation of microgram quantities of proteins from polyacrylamide gels for amino acid sequence analysis. *Methods Enzymol.* **91:** 227.

Hurn, B.A.L. and S.M. Chantler. 1980. Production of reagent antibodies. *Methods Enzymol.* **70:** 104.

James, G.T. 1978. Inactivation of the protease inhibitor phenylmethylsulfonyl fluoride in buffers. *Anal. Biochem.* **86:** 574.

Johnson, D.A., J.W. Gautsch, J.R. Sportsman, and J.H. Elder. 1984. Improved technique utilizing nonfat dry milk for analysis of proteins and nucleic acids transferred to nitrocellulose. *Gene Anal. Tech.* **1:** 3.

Kessler, S.W. 1975. Rapid isolation of antigens from cells with a staphylococcal protein A-antibody adsorbent: Parameters of the interaction of antibody-antigen complexes with protein A. *J. Immunol.* **115:** 1617.

Kipps, T.J. and L.A. Herzenberg. 1986. Schemata for the production of monoclonal antibody-producing hybridomas. In *Handbook of experimental immunology: Applications of immunological methods in biomedical sciences*, 4th edition (ed. D.M. Weir et al.), vol. 4, p. 108. Blackwell Scientific Publications, Oxford.

Köhler, G. and C. Milstein. 1975. Continuous cultures of fused cells secreting antibody of predefined specificity. *Nature* **256:** 495.

———. 1976. Derivation of specific antibody-producing tissue culture and tumor lines by cell fusion. *Eur. J. Immunol.* **6:** 511.

Krieg, P.A. and D.A. Melton. 1984. Formation of the 3′ end of histone mRNA by post-transcriptional processing. *Nature* **308:** 203.

Kronvall, G., U.S. Seal, J. Finstad, and R.C. Williams, Jr. 1970. Phylogenetic insight into evolution of mammalian Fc fragment of γG globulin using staphylococcal protein A. *J. Immunol.* **104:** 140.

Laemmli, U.K. 1970. Cleavage of structural proteins during the assembly of the head of bacteriophage T4. *Nature* **227:** 680.

Langone, J.J. 1980. Radioiodination by use of the Bolton–Hunter and related reagents. *Methods Enzymol.* **70:** 221.

Lerner, R.A. 1984. Antibodies of predetermined specificity in biology and medicine. *Adv. Immunol.* **36:** 1.

Liu, F.-T., M. Zinnecker, T. Hamaoka, and D.H. Katz. 1979. New procedures for preparation and isolation of conjugates of proteins and a synthetic copolymer of D-amino acids and immunochemical characterization of such conjugates. *Biochemistry* **18:** 690.

McConahey, P.J. and F.J. Dixon. 1966. A method of trace iodination of proteins for immunologic studies. *Int. Arch. Allergy Appl. Immunol.* **29:** 185.

———. 1980. Radioiodination of proteins by the use of the chloramine-T method. *Methods Enzymol.* **70:** 210.

Merril, C.R., D. Goldman, and M.L. Van Keuren. 1984. Gel protein stains: Silver stain. *Methods Enzymol.* **104:** 441.

Oakley, B.R., D.R. Kirsch, and N.R. Morris. 1980. A simplified ultrasensitive silver stain for detecting proteins in polyacrylamide gels. *Anal. Biochem.* **105:** 361.

Ochs, D.C., E.H. McConkey, and D.W. Sammons. 1981. Silver stains for proteins in polyacrylamide gels: A comparison of six methods. *Electrophoresis* **2:** 304.

Olmsted, J.B. 1981. Affinity purification of antibodies from diazotized paper blots of heterogeneous protein samples. *J. Biol. Chem.* **256:** 11955.

Ornstein, L. 1964. Disc electrophoresis—I. Background and theory. *Ann. N. Y. Acad. Sci.* **121:** 321.

Paterson, B.M. and M. Rosenberg. 1979. Efficient translation of prokaryotic mRNAs in a eukaryotic cell-free system requires addition of a cap structure. *Nature* **279:** 692.

Renart, J. and I.V. Sandoval. 1984. Western blots. *Methods Enzymol.* **104:** 455.

Renart, J., J. Reiser, and G.R. Stark. 1979. Transfer of proteins from gels to diazobenzyloxymethyl-paper and detection with antisera: A method for studying antibody specificity and antigen structure. *Proc. Natl. Acad. Sci.* **76:** 3116.

Sammons, D.W., L.D. Adams, and E.E. Nishizawa. 1981. Ultrasensitive silver-based color staining of polypeptides in polyacrylamide gels. *Electrophoresis* **2:** 135.

Schoenle, E.J., L.D. Adams, and D.W. Sammons. 1984. Insulin-induced rapid decrease of a major protein in fat cell plasma membranes. *J. Biol. Chem.* **259:** 12112.

Seed, B. 1982a. Attachment of nucleic acids to nitrocellulose and diazonium-substituted supports. In *Genetic engineering: Principles and methods* (ed. J.K. Setlow and A. Hollaender), vol. 4, p. 91. Plenum Publishing, New York.

——. 1982b. Diazotizable arylamine cellulose papers for the coupling and hybridization of nucleic acids. *Nucleic Acids Res.* **10:** 1799.

Sjödahl, J. 1977. Structural studies on the four repetitive Fc-binding regions in protein A from *Staphylococcus aureus. Eur. J. Biochem.* **78:** 471.

Smith, D.E. and P.A. Fisher. 1984. Identification, developmental regulation, and response to heat shock of two antigenically related forms of a major nuclear envelope protein in *Drosophila* embryos: Application of an improved method for affinity purification of antibodies using polypeptides immobilized on nitrocellulose blots. *J. Cell Biol.* **99:** 20.

Studier, F.W. 1973. Analysis of bacteriophage T7 early RNAs and proteins on slab gels. *J. Mol. Biol.* **79:** 237.

Switzer, R.C., III, C.R. Merril, and S. Shifrin. 1979. A highly sensitive silver stain for detecting proteins and peptides in polyacrylamide gels. *Anal. Biochem.* **98:** 231.

Towbin, H., T. Staehelin, and J. Gordon. 1979. Electrophoretic transfer of proteins from polyacrylamide gels to nitrocellulose sheets: Procedure and some applications. *Proc. Natl. Acad. Sci.* **76:** 4350.

Trautschold, I., E. Werle, and G. Zickgraf-Rüdel. 1967. Trasylol. *Biochem. Pharmacol.* **16:** 59.

Appendixes

Appendixes

A

Appendix A: Bacterial Media, Antibiotics, and Bacterial Strains

LIQUID MEDIA

LB Medium (Luria-Bertani Medium)

Per liter:

To 950 ml of deionized H_2O, add:

bacto-tryptone	10 g
bacto-yeast extract	5 g
NaCl	10 g

Shake until the solutes have dissolved. Adjust the pH to 7.0 with 5 N NaOH (~0.2 ml). Adjust the volume of the solution to 1 liter with deionized H_2O. Sterilize by autoclaving for 20 minutes at 15 lb/sq. in. on liquid cycle.

NZCYM Medium

Per liter:

To 950 ml of deionized H_2O, add:

NZ amine	10 g
NaCl	5 g
bacto-yeast extract	5 g
casamino acids	1 g
$MgSO_4 \cdot 7H_2O$	2 g

Shake until the solutes have dissolved. Adjust the pH to 7.0 with 5 N NaOH (~0.2 ml). Adjust the volume of the solution to 1 liter with deionized H_2O. Sterilize by autoclaving for 20 minutes at 15 lb/sq. in. on liquid cycle.

NZ amine: Casein hydrolysate enzymatic (ICN Biochemicals 101290).

NZYM Medium

NZYM medium is identical to NZCYM medium, except that casamino acids are omitted.

NZM Medium

NZM medium is identical to NZYM medium, except that bacto-yeast extract is omitted.

Terrific Broth
(Tartof and Hobbs 1987)

Per liter:

To 900 ml of deionized H_2O, add:

bacto-tryptone	12 g
bacto-yeast extract	24 g
glycerol	4 ml

Shake until the solutes have dissolved and sterilize by autoclaving for 20 minutes at 15 lb/sq. in. on liquid cycle.

Allow the solution to cool to 60°C or less, and then add 100 ml of a sterile solution of 0.17 M KH_2PO_4, 0.72 M K_2HPO_4. (This solution is made by dissolving 2.31 g of KH_2PO_4 and 12.54 g of K_2HPO_4 in 90 ml of deionized H_2O. After the salts have dissolved, adjust the volume of the solution to 100 ml with deionized H_2O and sterilize by autoclaving for 20 minutes at 15 lb/sq. in. on liquid cycle.)

SOB Medium

Per liter:

To 950 ml of deionized H_2O, add:

bacto-tryptone	20 g
bacto-yeast extract	5 g
NaCl	0.5 g

Shake until the solutes have dissolved. Add 10 ml of a 250 mM solution of KCl. (This solution is made by dissolving 1.86 g of KCl in 100 ml of deionized H_2O.) Adjust the pH to 7.0 with 5 N NaOH ($\sim$ 0.2 ml). Adjust the volume of the solution to 1 liter with deionized H_2O. Sterilize by autoclaving for 20 minutes at 15 lb/sq. in. on liquid cycle.

Just before use, add 5 ml of a sterile solution of 2 M $MgCl_2$. (This solution is made by dissolving 19 g of $MgCl_2$ in 90 ml of deionized H_2O. Adjust the volume of the solution to 100 ml with deionized H_2O and sterilize by autoclaving for 20 minutes at 15 lb/sq. in. on liquid cycle.)

SOC Medium

SOC medium is identical to SOB medium, except that it contains 20 mM glucose. After the SOB medium has been autoclaved, allow it to cool to 60°C or less and then add 20 ml of a sterile 1 M solution of glucose. (This solution is made by dissolving 18 g of glucose in 90 ml of deionized H_2O. After the sugar has dissolved, adjust the volume of the solution to 100 ml with deionized H_2O and sterilize by filtration through a 0.22-micron filter.)

2 × YT Medium

Per liter:

To 900 ml of deionized H_2O, add:

bacto-tryptone	16 g
bacto-yeast extract	10 g
NaCl	5 g

Shake until the solutes have dissolved. Adjust the pH to 7.0 with 5 N NaOH. Adjust the volume of the solution to 1 liter with deionized H_2O. Sterilize by autoclaving for 20 minutes at 15 lb/sq. in. on liquid cycle.

M9 Minimal Medium

Per liter:

To 750 ml of sterile deionized H_2O (cooled to 50°C or less), add:

5 × M9 salts	200 ml
sterile deionized H_2O to 1 liter	
20% solution of the appropriate carbon source (e.g., 20% glucose)	20 ml

If necessary, supplement the M9 medium with stock solutions of the appropriate amino acids.

5 × M9 salts is made by dissolving the following salts in deionized H_2O to a final volume of 1 liter:

$Na_2HPO_4 \cdot 7H_2O$	64 g
KH_2PO_4	15 g
NaCl	2.5 g
NH_4Cl	5.0 g

The salt solution is divided into 200-ml aliquots and sterilized by autoclaving for 15 minutes at 15 lb/sq. in. on liquid cycle.

MEDIA CONTAINING AGAR OR AGAROSE

Prepare liquid media according to the recipes given above. Just before autoclaving, add one of the following:

bacto-agar	(for plates)	15 g/liter
bacto-agar	(for top agar)	7 g/liter
agarose	(for plates)	15 g/liter
agarose	(for top agarose)	7 g/liter

Sterilize by autoclaving for 20 minutes at 15 lb/sq. in. on liquid cycle. When the medium is removed from the autoclave, swirl it gently to distribute the melted agar or agarose evenly throughout the solution. *Be careful! The fluid may be superheated and may boil over when swirled.* Allow the medium to cool to 50°C before adding thermolabile substances (e.g., antibiotics). To avoid producing air bubbles, mix the medium by swirling. Plates can then be poured directly from the flask; allow about 30–35 ml of medium per 90-mm plate. To remove bubbles from medium in the plate, flame the surface of the medium with a bunsen burner before the agar or agarose hardens. Set up a color code (e.g., two red stripes for LB-ampicillin plates; one black stripe for LB plates, etc.) and mark the edges of the plates with the appropriate colored markers.

When the medium has hardened completely, invert the plates and store them at 4°C until needed. The plates should be removed from storage 1–2 hours before they are used. If the plates are fresh, they will "sweat" when incubated at 37°C. This allows bacterial colonies or bacteriophage plaques to spread across the surfaces of the plates and increases the chances of cross-contamination. This problem can be avoided by wiping off any condensation from the lids of the plates and then incubating the plates for several hours at 37°C in an inverted position before they are used. Alternatively, the liquid can be removed by shaking the lid with a single, quick motion. To minimize the possibility of contamination, hold the open plate in an inverted position while removing the liquid from the lid.

STORAGE MEDIA

Bacteria can be stored for up to 2 years in stab cultures or indefinitely in cultures containing glycerol.

Stab Cultures

Use glass vials (2–3 ml) with screw-on caps fitted with rubber gaskets. Add molten LB agar until the vials are two thirds full. Autoclave the partially filled vials (with their caps loosely screwed on) for 20 minutes at 15 lb/sq. in. on liquid cycle. Remove the vials from the autoclave, let them cool to room temperature, and then tighten the caps. Store the vials at room temperature until needed.

To store bacteria, pick a single, well-isolated colony with a sterile inoculating needle and stab the needle several times through the agar to the bottom of the vial. Replace and tighten the cap, and label both the vial and the cap. Store the vial in the dark at room temperature.

Cultures Containing Glycerol

BACTERIAL CULTURES GROWING IN LIQUID MEDIA

To 0.85 ml of bacterial culture, add 0.15 ml of sterile glycerol (sterilized by autoclaving for 20 minutes at 15 lb/sq. in. on liquid cycle). Vortex the culture to ensure that the glycerol is evenly dispersed. Transfer the culture to a labeled storage tube equipped with a screw cap and an air-tight gasket. Freeze the culture in ethanol–dry ice or in liquid nitrogen, and then transfer the tube to −70°C for long-term storage.

To recover the bacteria, scrape the frozen surface of the culture with a sterile inoculating needle, and then immediately streak the bacteria that adhere to the needle onto the surface of an LB agar plate containing the appropriate antibiotics. Return the frozen culture to storage at −70°C. Incubate the plate overnight at 37°C.

BACTERIAL CULTURES GROWING ON AGAR PLATES

Scrape the bacteria growing on the surface of an agar plate into 2 ml of LB medium in a sterile tube. Add an equal volume of LB medium containing 30% sterile glycerol. Vortex the mixture to ensure that the glycerol is completely dispersed. Dispense aliquots of the glycerinated culture into sterile tubes equipped with screw caps and air-tight gaskets. Freeze the cultures as described above.

This method is useful for storing copies of cDNA libraries established in plasmid vectors (for discussion, see Hanahan 1985).

TABLE A.1 Antibiotic Solutions

	Stock solution[a]		Working concentration	
	concentration	storage	stringent plasmids	relaxed plasmids
Ampicillin	50 mg/ml in H$_2$O	−20°C	20 μg/ml	60 μg/ml
Carbenicillin	50 mg/ml in H$_2$O	−20°C	20 μg/ml	60 μg/ml
Chloramphenicol	34 mg/ml in ethanol	−20°C	25 μg/ml	170 μg/ml
Kanamycin	10 mg/ml in H$_2$O	−20°C	10 μg/ml	50 μg/ml
Streptomycin	10 mg/ml in H$_2$O	−20°C	10 μg/ml	50 μg/ml
Tetracycline[b]	5 mg/ml in ethanol	−20°C	10 μg/ml	50 μg/ml

[a]Stock solutions of antibiotics dissolved in H$_2$O should be sterilized by filtration through a 0.22-micron filter. Antibiotics dissolved in ethanol need not be sterilized. Store solutions in light-tight containers.
[b]Magnesium ions are antagonists of tetracycline. Use media without magnesium salts (e.g., LB medium) for selection of bacteria resistant to tetracycline.

SOLUTIONS FOR WORKING WITH BACTERIOPHAGE λ

Maltose

Maltose, an inducer of the gene (*lamB*) that codes for the bacteriophage λ receptor, is often added to the medium during growth of bacteria that are to be used for plating bacteriophage λ. Add 1 ml of a sterile 20% maltose solution for every 100 ml of medium.

Make up a sterile 20% stock solution of maltose as follows:

maltose   20 g
H_2O to 100 ml

Sterilize the solution by filtration through a 0.22-micron filter. Store the sterile solution at room temperature.

SM

This buffer is used for storage and dilution of bacteriophage λ stocks.

Per liter:

NaCl	5.8 g
$MgSO_4 \cdot 7H_2O$	2 g
1 M Tris · Cl (pH 7.5)	50 ml
2% gelatin solution	5 ml
H_2O to 1 liter	

Sterilize the buffer by autoclaving for 20 minutes at 15 lb/sq. in. on liquid cycle. After the solution has cooled, dispense 50-ml aliquots into sterile containers. SM may be stored indefinitely at room temperature.

A 2% gelatin solution is made by adding 2 g of gelatin to a total volume of 100 ml of H_2O and autoclaving the solution for 15 minutes at 15 lb/sq. in. on liquid cycle.

TM

Per liter:

1 M Tris · Cl (pH 7.5)	50 ml
$MgSO_4 \cdot 7H_2O$	2 g
H_2O to 1 liter	

Sterilize the buffer by autoclaving for 20 minutes at 15 lb/sq. in. on liquid cycle. After the solution has cooled, dispense 50-ml aliquots into sterile containers. TM may be stored indefinitely at room temperature.

λ *Diluent*

Per liter:

1 M Tris · Cl (pH 7.5)   10 ml
$MgSO_4 \cdot 7H_2O$ 2 g
H_2O to 1 liter

Sterilize the buffer by autoclaving for 20 minutes at 15 lb/sq. in. on liquid cycle. After the solution has cooled, dispense 50-ml aliquots into sterile containers. λ diluent may be stored indefinitely at room temperature.

For long-term storage of bacteriophage λ stocks, the addition of 50 mM NaCl and 0.01% gelatin is sometimes recommended, especially when the bacteriophage has been purified in CsCl.

BACTERIAL STRAIN LIST

TABLE A.2 Bacterial Strains

Strain	Genotype	Remarks
71/18	supE thi Δ(lac-proAB) F′ [proAB⁺ lacI^q lacZΔM15]	A strain used for growth of phagemids. It makes high levels of lac repressor and is used for inducible expression of genes that are under the control of the lac promoter. This strain can be used for detection of recombinants expressing β-galactosidase fusion proteins (Messing et al. 1977; Dente et al. 1983; Rüther and Müller-Hill 1983).
BB4	supF58 supE44 hsdR514 galK2 galT22 trpR55 metB1 tonA ΔlacU169 F′ [proAB⁺ lacI^q lacZΔM15 Tn10(tet^r)]	A recA⁺ strain used for growth of λZAP and other λ bacteriophages. The F′ in this strain carries lacZΔM15, which permits α-complementation with the amino terminus of β-galactosidase encoded in λZAP. The F′ allows superinfection with an M13 helper bacteriophage, a step required for converting a recombinant λZAP to a pBluescript plasmid (Bullock et al. 1987).
BHB2688	(N205 recA [λimm434 cIts b2 red Eam Sam/λ])	A bacteriophage λ lysogen used to prepare packaging extracts (Hohn and Murray 1977; Hohn 1979).
BHB2690	(N205 recA [λimm434 cIts b2 red Dam Sam/λ])	A bacteriophage λ lysogen used to prepare packaging extracts (Hohn and Murray 1977; Hohn 1979).
BL21(DE3)	hsdS gal (λcIts857 ind1 Sam7 nin5 lacUV5-T7 gene1)	A strain employed for high-level expression of genes cloned into expression vectors containing bacteriophage T7 promoter. Bacteriophage T7 RNA polymerase is carried on the bacteriophage λ DE3, which is integrated into the chromosome of BL21 (Studier and Moffatt 1986).
BNN102 (C600hflA)	supE44 hsdR thi-1 thr-1 leuB6 lacY1 tonA21 hflA150[chr::Tn10(tet^r)]	An hflA strain used to select λgt10 recombinants. The high frequency lysogeny mutation suppresses plaque formation by cI⁺ bacteriophages but allows plaque formation by recombinant cI⁻ bacteriophages (Young and Davis 1983a).
C-1a	A wild-type strain.	A clone of E. coli strain C wild type maintained on minimal medium for several years. E. coli C is F⁻ and lacks host restriction and modification activity. It is a nonsuppressing host strain used in complementation tests with amber mutants of bacteriophage λ (Bertani and Weigle 1953; Borck et al. 1976).
C600 (BNN93)	supE44 hsdR thi-1 thr-1 leuB6 lacY1 tonA21	A suppressing strain often used for making lysates (Appleyard 1954) and for propagation of λgt10 (Young and Davis 1983a).
CES200	sbcB15 recB21 recC22 hsdR	A strain used for growth of Spi⁻ bacteriophages (Nader et al. 1985).
CES201	recA sbcB15 recB21 recC22 hsdR	A recombination-deficient strain used for growth of Spi⁻ bacteriophages (Wyman and Wertman 1987).
CJ236	dut1 ung1 thi-1 relA1/pCJ105(cam^r F′)	A dut⁻ ung⁻ strain used to prepare uracil-containing DNA for site-directed mutagenesis experiments (Kunkel et al. 1987). pCJ105 carries an F′ and cam^r; growth of CJ236 in the presence of chloramphenicol selects for retention of the F′.
CSH18	supE thi Δ(lac-pro) F′ [proAB⁺ lacZ⁻]	A suppressing strain used to screen recombinants made in bacteriophage λ vectors carrying a lacZ gene in the stuffer fragments. These vectors give rise to blue plaques in the presence of the chromogenic substrate X-gal; recombinants in which the stuffer fragment has been replaced by foreign DNA give rise to white plaques (Miller 1972; Williams and Blattner 1979).
DH1	supE44 hsdR17 recA1 endA1 gyrA96 thi-1 relA1	A recombination-deficient suppressing strain used for plating and growth of plasmids and cosmids (Low 1968; Meselson and Yuan 1968; Hanahan 1983).

TABLE A.2 (continued)

Strain	Genotype	Remarks
DH5	supE44 hsdR17 recA1 endA1 gyrA96 thi-1 relA1	A recombination-deficient suppressing strain used for plating and growth of plasmids and cosmids (Low 1968; Meselson and Yuan 1968; Hanahan 1983). This strain has a higher transformation efficiency than DH1.
DH5α	supE44 ΔlacU169 (φ80 lacZΔM15) hsdR17 recA1 endA1 gyrA96 thi-1 relA1	A recombination-deficient suppressing strain used for plating and growth of plasmids and cosmids. The φ80 lacZΔM15 permits α-complementation with the amino terminus of β-galactosidase encoded in pUC vectors (Hanahan 1983; Bethesda Research Laboratories 1986).
DP50supF	supE44 supF58 hsdS3($r_B^- m_B^-$) dapD8 lacY1 glnV44 Δ(gal-uvrB)47 tyrT58 gyrA29 tonA53 Δ(thyA57)	A strain used for isolation and propagation of bacteriophage λ recombinants (Leder et al. 1977; B. Bachmann, pers. comm.).
ED8654	supE supF hsdR metB lacY gal trpR	A suppressing strain commonly used to propagate bacteriophage λ vectors and their recombinants (Borck et al. 1976; Murray et al. 1977).
ED8767	supE44 supF58 hsdS3($r_B^- m_B^-$) recA56 galK2 galT22 metB1	A recombination-deficient suppressing strain used for propagation of bacteriophage λ vectors (Murray et al. 1977).
HB101	supE44 hsdS20($r_B^- m_B^-$) recA13 ara-14 proA2 lacY1 galK2 rpsL20 xyl-5 mtl-1	A suppressing strain commonly used for large-scale production of plasmids. It is an E. coli K12 × E. coli B hybrid that is highly transformable (Boyer and Roulland-Dussoix 1969; Bolivar and Backman 1979).
HMS174	recA1 hsdR rif^r	A recombination-deficient nonsuppressing strain used for high-level expression of genes cloned into expression vectors containing bacteriophage T7 promoter. Bacteriophage T7 RNA polymerase is provided by infection with a bacteriophage λ that carries bacteriophage T7 gene 1 (Campbell et al. 1978; Studier and Moffatt 1986).
JM101[a]	supE thi Δ(lac-proAB) F′ [traD36 proAB⁺ lacI^q lacZΔM15]	A strain that will support growth of vectors carrying amber mutations (Messing 1979).
JM105	supE endA sbcB15 hsdR4 rpsL thi Δ(lac-proAB) F′ [traD36 proAB⁺ lacI^q lacZΔM15]	A strain that will support growth of vectors carrying amber mutations and will modify but not restrict transfected DNA (Yanisch-Perron et al. 1985).
JM107[b]	supE44 endA1 hsdR17 gyrA96 relA1 thi Δ(lac-proAB) F′ [traD36 proAB⁺ lacI^q lacZΔM15]	A strain that will support growth of vectors carrying amber mutations and will modify but not restrict transfected DNA (Yanisch-Perron et al. 1985).
JM109[b,c]	recA1 supE44 endA1 hsdR17 gyrA96 relA1 thi Δ(lac-proAB) F′ [traD36 proAB⁺ lacI^q lacZΔM15]	A recombination-deficient strain that will support growth of vectors carrying amber mutations and will modify but not restrict transfected DNA (Yanisch-Perron et al. 1985).
JM110	dam dcm supE44 hsdR17 thi leu rpsL lacY galK galT ara tonA thr tsx Δ(lac-proAB) F′ [traD36 proAB⁺ lacI^q lacZΔM15]	A strain that will not modify BclI sites and will support growth of vectors carrying amber mutations (Yanisch-Perron et al. 1985).
K802	supE hsdR gal metB	A suppressing strain used to propagate bacteriophage λ vectors and their recombinants (Wood 1966).
KK2186	supE sbcB15 hsdR4 rpsL thi Δ(lac-proAB) F′ [traD36 proAB⁺ lacI^q lacZΔM15]	A strain that will support growth of vectors carrying amber mutations and will modify but not restrict transfected DNA (Zagursky and Berman 1984).

Strain	Genotype	Description
LE392	*supE44 supF58 hsdR514 galK2 galT22 metB1 trpR55 lacY1*	A suppressing strain commonly used to propagate bacteriophage λ vectors and their recombinants. LE392 is a derivative of ED8654 (Borck et al. 1976; Murray et al. 1977).
LG90	Δ(*lac-proAB*)	A strain in which *lacZ* is deleted that is used for detection of recombinants expressing β-galactosidase fusion proteins (Guarente and Ptashne 1981).
M5219	*lacZ trpA rpsL* (λ*bio*252 *c*Its857ΔH1)	A strain used for regulated expression of genes cloned downstream from the bacteriophage λ p_L promoter. It contains a defective λ prophage that encodes the bacteriophage λcIts857 repressor and N protein, which is an antagonist of transcription termination (Remaut et al. 1981; Shimatake and Rosenberg 1981).
MBM7014.5	*hsdR2 mcrB1 zjj*202::Tn10(*tet*[r]) *araD*139 *araCU*25am Δ*lacU*169	An *mcrB* strain used for λORF8 primary libraries. Libraries are made with DNA treated with methylases to protect *Hind*III and *Bam*HI sites. M.*Alu*I methylase is used to protect *Hind*III sites since M.*Hind*III methylase is not available commercially. This strain is defective in the restriction system that recognizes *Alu*I-methylated DNA sites (Raleigh and Wilson 1986).
MC1061	*hsdR mcrB araD*139 Δ(*araABC-leu*)7679 Δ*lacX*74 *galU galK rpsL thi*	An *mcrB* strain used for λORF8 primary libraries as described for the strain MBM7014.5 (Meissner et al. 1987).
MM294	*supE44 hsdR endA1 pro thi*	A suppressing strain used for large-scale production of plasmids. It is highly transformable (Meselson and Yuan 1968).
MV1184[d]	*ara* Δ(*lac-proAB*) *rpsL thi* (φ80 *lacZ*ΔM15) Δ(*srl-recA*)306::Tn10(*tet*[r]) F′[*traD*36 *proAB*[+] *lacI*[q] *lacZ*ΔM15]	A recombination-deficient strain used to propagate phagemids pUC118/pUC119 and to obtain single-stranded copies of phagemids (Vieira and Messing 1987).
MV1193	Δ(*lac-proAB*) *rpsL thi endA spcB*15 *hsdR4* Δ(*srl-recA*)306::Tn10(*tet*[r]) F′[*traD*36 *proAB*[+] *lacI*[q] *lacZ*ΔM15]	A recombination-deficient strain used to propagate phagemids pUC118/pUC119 and to obtain single-stranded copies of phagemids (Zoller and Smith 1987).
MZ-1	*galK*Δ8*att*LΔ*Bam*N₇N₅₃*c*Its857ΔH1 *his ilv bio* N[+]	A temperature-sensitive lysogenic strain used as a host for plasmids containing the bacteriophage λ p_L promoter (Nagai and Thøgersen 1984).
NM531	*supE supF hsdR trpR lacY recA*13 *metB gal*	A recombination-deficient suppressing strain used for propagation of bacteriophage λ vectors (Arber et al. 1983).
NM538	*supF hsdR trpR lacY*	A strain used for assay and propagation of bacteriophage λ (Frischauf et al. 1983).
NM539	*supF hsdR lacY* (P2cox)	A strain used for selection of Spi⁻ bacteriophage. NM539 is a derivative of NM538 (Frischauf et al. 1983).
Q358	*supE hsdR* φ80[r]	A *supE* host used for growth of bacteriophage λ vectors (Karn et al. 1980).
Q359	*supE hsdR* φ80[r] P2	A *supE* host used to select Spi⁻ recombinants (Karn et al. 1980).
R594	*galK2 galT*22 *rpsL*179 *lac*⁻	A nonsuppressing strain used as a nonpermissive host for vectors containing amber or ochre mutations (Campbell 1965).
RB791	W3110 *lacI*[q]L8	A strain that makes high levels of *lac* repressor and is used for inducible expression of genes under the control of the *lac* and *tac* promoters (Brent and Ptashne 1981).
RR1	*supE44 hsdS*20(r$_B^-$ m$_B^-$) *ara*-14 *proA*2 *lacY*1 *galK*2 *rpsL*20 *xyl*-5 *mtl*-1	A *recA*[+] derivative of HB101 that can be transformed with high efficiency (Bolivar et al. 1977; Peacock et al. 1981; B. Bachmann, pers. comm.).
SMR10	*E. coli* C (λ*cos*2 Δ*B xis*1 *red*3 *gam* am210 *c*Its857 *nin*5 *Sam*7/λ)	A bacteriophage λ lysogen used to prepare packaging extracts (Rosenberg 1985).

Strain	Genotype	Remarks
TAP90	supE44 supF58 hsdR pro leuB thi-1 rpsL lacY1 tonA1 recD1903::mini-tet	A host strain used for production of high-titer bacteriophage λ lysates. This restriction-deficient supE supF strain has a mini-tet insertion in recD, which improves growth of Spi− λ bacteriophages (Patterson and Dean 1987).
TG1	supE hsdΔ5 thi Δ(lac-proAB) F′[traD36 proAB+ lacIq lacZΔM15]	An EcoK− derivative of JM101 that neither modifies nor restricts transfected DNA. It will support growth of vectors carrying amber mutations (Gibson 1984).
TG2	supE hsdΔ5 thi Δ(lac-proAB) Δ(srl-recA)306::Tn10(tetr) F′[traD36 proAB+ lacIq lacZΔM15]	A recombination-deficient derivative of TG1 (M. Biggin, pers. comm.).
XL1-Blue	supE44 hsdR17 recA1 endA1 gyrA46 thi relA1 lac− F′[proAB+ lacIq lacZΔM15 Tn10(tetr)]	A recombination-deficient strain that will support the growth of vectors carrying some amber mutations, but not those with the Sam100 mutation (e.g., λZAP). Transfected DNA is modified but not restricted. XL1-Blue is used to propagate λZAPII recombinants, which are unstable in BB4. The F′ in this strain allows blue/white screening on X-gal and permits bacteriophage M13 superinfection (Bullock et al. 1987).
XS101	recA1 hsdR rpoB331 F′[kan]	A recombination-deficient strain that modifies but does not restrict transfected DNA. It carries an episome conferring resistance to kanamycin and is used for growth of phagemids (Levinson et al. 1984).
XS127	gyrA thi rpoB331 Δ(lac-proAB) argE F′[traD36 proAB+ lacIq lacZΔM15]	A strain used for growth of phagemids (Levinson et al. 1984).
Y1089	araD139 ΔlacU169 proA+ Δlon rpsL hflA150[chr::Tn10(tetr)] pMC9	A strain used for protein production from λgt11 and λgt18–23 recombinants. Expression of the foreign protein is controlled by the high levels of lac repressor made by pMC9, which carries lacIq. Y1089 is deficient in the lon protease, which may allow increased stability of the foreign proteins. Lysogens are formed at a high frequency in this strain (Young and Davis 1983b).
Y1090hsdR	supF hsdR araD139 Δlon ΔlacU169 rpsL trpC22::Tn10(tetr) pMC9	A strain used for immunological screening of expression libraries and propagation of λgt11 and λgt18–23 (Young and Davis 1983b; Jendrisak et al. 1987). Expression of the foreign protein is controlled by the high levels of lac repressor made by pMC9, which carries lacIq. Detection of proteins toxic to E. coli can be achieved by adding IPTG several hours after initiation of plaque formation. Some proteins are unstable in E. coli. Y1090hsdR is deficient in the lon protease, which may allow increased stability of antigens for antibody screening. The supF marker suppresses Sam100 to allow cell lysis (Young and Davis 1983b).
YK537	supE44 hsdR hsdM recA1 phoA8 leuB6 thi lacY rpsL20 galK2 ara-14 xyl-5 mtl-1	A recombination-deficient suppressing strain used for regulated expression of genes cloned downstream from the phoA promoter (Oka et al. 1985).

[a]Strain JM103 (Messing et al. 1981) is a restrictionless derivative of JM101 that has been used to propagate bacteriophage M13 recombinants. However, some cultivars of JM103 have lost the *hsdR4* mutation (Felton 1983) and are lysogenic for bacteriophage P1 (which codes for its own restriction/modification system). JM103 is therefore no longer recommended as a host for bacteriophage M13 vectors. Strain KK2186 (Zagursky and Berman 1984) is genetically identical to JM103 except that it is nonlysogenic for bacteriophage P1.

[b]Strains JM106 and JM108 are identical to JM107 and JM109, respectively, except that they do not carry an F' episome. These strains will not support the growth of bacteriophage M13 but may be used to propagate plasmids. However, JM106 and JM108 do not carry the *lacI*q marker (normally present on the F' episome) and are therefore unable effectively to suppress the synthesis of potentially toxic products encoded by foreign DNA sequences cloned into plasmids carrying the *lacZ* promoter.

[c]Strains JM108 and JM109 are defective for synthesis of bacterial cell walls and form mucoid colonies on minimal media. This does not affect their ability to support the growth of bacteriophage M13.

[d]The original strain of MV1184, constructed by M. Volkert (pers. comm.), did not carry an F' episome. However, the strain of MV1184 distributed by the Messing laboratory clearly carries an F' episome. It is therefore advisable to check strains of MV1184 on their arrival in the laboratory for their ability to support the growth of male-specific bacteriophages.

Appendix B: Preparation of Reagents and Buffers Used in Molecular Cloning

CONCENTRATIONS OF ACIDS AND BASES

TABLE B.1 pK$_a$s of Commonly Used Buffers

Buffer	Molecular weight	pK$_a$	Buffering range
Tris[a]	121.1	8.08	7.1–8.9
HEPES[b]	238.3	7.47	7.2–8.2
MOPS[c]	209.3	7.15	6.6–7.8
PIPES[d]	304.3	6.76	6.2–7.3
MES[e]	195.2	6.09	5.4–6.8

[a]Tris(hydroxymethyl)aminomethane
[b]N-2-hydroxyethylpiperazine-N'-2-ethanesulfonic acid
[c]3-(N-morpholino)propanesulfonic acid
[d]Piperazine-N,N'-bis(2-ethanesulfonic acid)
[e]2-(N-morpholino)ethanesulfonic acid

TABLE B.2 Preparation of Tris Buffers of Various Desired pH Values

Desired pH (25°C)	Volume of 0.1 N HCl
7.10	45.7
7.20	44.7
7.30	43.4
7.40	42.0
7.50	40.3
7.60	38.5
7.70	36.6
7.80	34.5
7.90	32.0
8.00	29.2
8.10	26.2
8.20	22.9
8.30	19.9
8.40	17.2
8.50	14.7
8.60	12.4
8.70	10.3
8.80	8.5
8.90	7.0

Tris buffers (0.05 M) of the desired pH can be made by mixing 50 ml of 0.1 M Tris base with the indicated volume of 0.1 N HCl and then adjusting the volume of the mixture to 100 ml with water.

TABLE B.3 Concentrations of Acids and Bases: Common Commercial Strengths

Substance	Formula	Molecular weight	Moles/ liter	Grams/ liter	Percentage by weight	Specific gravity	Milliliters/liter to prepare 1 M solution
Acetic acid, glacial	CH_3COOH	60.05	17.4	1045	99.5	1.05	57.5
Acetic acid		60.05	6.27	376	36	1.045	159.5
Formic acid	HCOOH	46.02	23.4	1080	90	1.20	42.7
Hydrochloric acid	HCl	36.5	11.6	424	36	1.18	86.2
			2.9	105	10	1.05	344.8
Nitric acid	HNO_3	63.02	15.99	1008	71	1.42	62.5
			14.9	938	67	1.40	67.1
			13.3	837	61	1.37	75.2
Perchloric acid	$HClO_4$	100.5	11.65	1172	70	1.67	85.8
			9.2	923	60	1.54	108.7
Phosphoric acid	H_3PO_4	80.0	18.1	1445	85	1.70	55.2
Sulfuric acid	H_2SO_4	98.1	18.0	1766	96	1.84	55.6
Ammonium hydroxide	NH_4OH	35.0	14.8	251	28	0.898	67.6
Potassium hydroxide	KOH	56.1	13.5	757	50	1.52	74.1
			1.94	109	10	1.09	515.5
Sodium hydroxide	NaOH	40.0	19.1	763	50	1.53	52.4
			2.75	111	10	1.11	363.6

TABLE B.4 Approximate pH Values for Various Concentrations of Stock Solutions

Substance	1 N	0.1 N	0.01 N	0.001 N
Acetic acid	2.4	2.9	3.4	3.9
Hydrochloric acid	0.10	1.07	2.02	3.01
Sulfuric acid	0.3	1.2	2.1	
Citric acid		2.1	2.6	
Ammonium hydroxide	11.8	11.3	10.8	10.3
Sodium hydroxide	14.05	13.07	12.12	11.13
Sodium bicarbonate		8.4		
Sodium carbonate		11.5	11.0	

PREPARATION OF ORGANIC REAGENTS

Phenol

Most batches of commercial liquified phenol are clear and colorless and can be used in molecular cloning without redistillation. Occasionally, batches of liquified phenol are pink or yellow, and these should be rejected and returned to the manufacturer. Crystalline phenol is not recommended because it must be redistilled at 160°C to remove oxidation products, such as quinones, that cause the breakdown of phosphodiester bonds or cause cross-linking of RNA and DNA.

Caution: Phenol is highly corrosive and can cause severe burns. Wear gloves, protective clothing, and safety glasses when handling phenol. All manipulations should be carried out in a chemical hood. Any areas of skin that come into contact with phenol should be rinsed with a large volume of water and washed with soap and water. Do *not* use ethanol.

EQUILIBRATION OF PHENOL

Before use, phenol must be equilibrated to a pH > 7.8 because DNA will partition into the organic phase at acid pH.

1. Liquified phenol should be stored at $-20°C$. As needed, remove the phenol from the freezer, allow it to warm to room temperature, and then melt it at 68°C. Add hydroxyquinoline to a final concentration of 0.1%. This compound is an antioxidant, a partial inhibitor of RNAase, and a weak chelator of metal ions (Kirby 1956). In addition, its yellow color provides a convenient way to identify the organic phase.

2. To the melted phenol, add an equal volume of buffer (usually 0.5 M Tris · Cl [pH 8.0] at room temperature). Stir the mixture on a magnetic stirrer for 15 minutes, and then turn off the stirrer. When the two phases have separated, aspirate as much as possible of the upper (aqueous) phase using a glass pipette attached to a vacuum line equipped with traps (see Appendix E, Figure E.1).

3. Add an equal volume of 0.1 M Tris · Cl (pH 8.0) to the phenol. Stir the mixture on a magnetic stirrer for 15 minutes, and then turn off the stirrer. Remove the upper aqueous phase as described in step 2. Repeat the extractions until the pH of the phenolic phase is > 7.8 (as measured with pH paper).

4. After the phenol is equilibrated and the final aqueous phase has been removed, add 0.1 volume of 0.1 M Tris · Cl (pH 8.0) containing 0.2% β-mercaptoethanol. The phenol solution may be stored in this form under 100 mM Tris · Cl (pH 8.0) in a light-tight bottle at 4°C for periods of up to 1 month.

Phenol:Chloroform:Isoamyl Alcohol (25:24:1)

A mixture consisting of equal parts of equilibrated phenol and chloroform:isoamyl alcohol (24:1) is frequently used to remove proteins from preparations of nucleic acids. The chloroform denatures proteins and facilitates the separation of the aqueous and organic phases, and the isoamyl alcohol reduces foaming during extraction.

Neither chloroform nor isoamyl alcohol requires treatment before use. The phenol:chloroform:isoamyl alcohol mixture may be stored under 100 mM Tris·Cl (pH 8.0) in a light-tight bottle at 4°C for periods of up to 1 month.

TABLE B.5   Atomic Weights

Element	Symbol	Atomic number	Atomic weight[a]
Actinium	Ac	89	227.02
Aluminum	Al	13	26.98
Americium	Am	95	(243)
Antimony	Sb	51	121.75
Argon	Ar	18	39.94
Arsenic	As	33	74.92
Astatine	At	85	(210)
Barium	Ba	56	137.33
Berkelium	Bk	97	(247)
Beryllium	Be	4	9.01
Bismuth	Bi	83	208.98
Boron	B	5	10.81
Bromine	Br	35	79.90
Cadmium	Cd	48	112.41
Calcium	Ca	20	40.08
Californium	Cf	98	(251)
Carbon	C	6	12.01
Cerium	Ce	58	140.12
Cesium	Cs	55	132.90
Chlorine	Cl	17	35.45
Chromium	Cr	24	51.99
Cobalt	Co	27	58.93
Copper	Cu	29	63.54
Curium	Cm	96	(247)
Dysprosium	Dy	66	162.50
Einsteinium	Es	99	(252)
Erbium	Er	68	167.26
Europium	Eu	63	151.96
Fermium	Fm	100	(257)
Fluorine	F	9	18.99
Francium	Fr	87	(223)
Gadolinium	Gd	64	157.25
Gallium	Ga	31	69.72
Germanium	Ge	32	72.59
Gold	Au	79	196.96
Hafnium	Hf	72	178.49
Helium	He	2	4.00
Holmium	Ho	67	164.93
Hydrogen	H	1	1.00
Indium	In	49	114.82
Iodine	I	53	126.90
Iridium	Ir	77	192.22
Iron	Fe	26	55.84
Krypton	Kr	36	83.80
Lanthanum	La	57	138.90
Lawrencium	Lr	103	(260)
Lead	Pb	82	207.2
Lithium	Li	3	6.94
Lutetium	Lu	71	174.96
Magnesium	Mg	12	24.30
Manganese	Mn	25	54.93
Mendelevium	Md	101	(258)
Mercury	Hg	80	200.59
Molybdenum	Mo	42	95.94

Element	Symbol	Atomic number	Atomic weight[a]
Neodymium	Nd	60	144.24
Neon	Ne	10	20.17
Neptunium	Np	93	237.04
Nickel	Ni	28	58.69
Niobium	Nb	41	92.90
Nitrogen	N	7	14.00
Nobelium	No	102	(259)
Osmium	Os	76	190.2
Oxygen	O	8	15.99
Palladium	Pd	46	106.42
Phosphorus	P	15	30.97
Platinum	Pt	78	195.08
Plutonium	Pu	94	(244)
Polonium	Po	84	(209)
Potassium	K	19	39.09
Praseodymium	Pr	59	140.90
Promethium	Pm	61	(145)
Protactinium	Pa	91	231.03
Radium	Ra	88	226.02
Radon	Rn	86	(222)
Rhenium	Re	75	186.20
Rhodium	Rh	45	102.90
Rubidium	Rb	37	85.46
Ruthenium	Ru	44	101.07
Samarium	Sm	62	150.36
Scandium	Sc	21	44.95
Selenium	Se	34	78.96
Silicon	Si	14	28.08
Silver	Ag	47	107.86
Sodium	Na	11	22.98
Strontium	Sr	38	87.62
Sulfur	S	16	32.06
Tantalum	Ta	73	180.94
Technetium	Tc	43	(98)
Tellurium	Te	52	127.60
Terbium	Tb	65	158.92
Thallium	Tl	81	204.38
Thorium	Th	90	232.03
Thulium	Tm	69	168.93
Tin	Sn	50	118.69
Titanium	Ti	22	47.88
Tungsten	W	74	183.85
Unnilhexium	(Unh)	106	(263)
Unnilpentium	(Unp)	105	(262)
Unnilquadium	(Unq)	104	(261)
Unnilseptium	(Uns)	107	(262)
Uranium	U	92	238.02
Vanadium	V	23	50.94
Xenon	Xe	54	131.29
Ytterbium	Yb	70	173.04
Yttrium	Y	39	88.90
Zinc	Zn	30	65.38
Zirconium	Zr	40	91.22

[a]Numbers in parentheses are the mass numbers of the most stable isotope of that element.

TABLE B.6 Isotopic Data

³H		³⁵S		³²P		¹²⁵I		¹³¹I	
time (years)	activity remaining (%)	time (days)	activity remaining (%)	time (days)	activity remaining (%)	time (days)	activity remaining (%)	time (days)	activity remaining (%)
1	94.5	2	98.4	1	95.3	4	95.5	0.2	98.3
2	89.3	5	96.1	2	90.8	8	91.2	0.4	96.6
3	84.4	10	92.3	3	86.5	12	87.1	0.6	95.0
4	79.8	15	88.7	4	82.4	16	83.1	1.0	91.8
5	75.4	20	85.3	5	78.5	20	79.4	1.6	87.2
6	71.3	25	82.0	6	74.8	24	75.8	2.3	81.2
7	67.4	31	78.1	7	71.2	28	72.4	3.1	76.7
8	63.7	37	74.5	8	67.8	32	69.1	4.0	71.0
9	60.2	43	71.0	9	64.7	36	66.0	5.0	65.2
10	56.9	50	67.0	10	61.5	40	63.0	6.1	59.3
11	53.8	57	63.6	11	58.7	44	60.2	7.3	53.4
12	50.9	65	59.6	12	55.9	48	57.4	8.1	50.0
12.3	50.0	73	56.0	13	53.2	52	54.8		
		81	52.5	14	50.7	56	52.4		
		87.1	50.0	14.3	50.0	60	50.0		

One curie (Ci) is equivalent to the amount of an isotope undergoing 3.7×10^{10} nuclear disintegrations per second (2.22×10^{12} disintegrations/minute). 1 Ci = 3.7×10^{10} becquerels (Bq).

$1 \text{ Bq} = 2.7 \times 10^{-11}$ Ci
$1 \text{ } \mu\text{Ci} = 37 \times 10^{3} \text{ Bq} = 37 \text{ kBq} = 2.22 \times 10^{6}$ dpm
$1 \text{ mCi} = 37 \times 10^{6} \text{ Bq} = 37 \text{ MBq} = 2.22 \times 10^{9}$ dpm
$1 \text{ Ci} = 37 \times 10^{9} \text{ Bq} = 37 \text{ GBq} = 2.22 \times 10^{12}$ dpm

STOCK SOLUTIONS

TABLE B.7   Preparation of Commonly Used Stock Solutions

Solution	Method of preparation	Comments
30% Acrylamide	Dissolve 29 g of acrylamide and 1 g of *N,N'*-methylenebisacrylamide in a total volume of 60 ml of H_2O. Heat the solution to 37°C to dissolve the chemicals. Adjust the volume to 100 ml with H_2O. Sterilize the solution by filtration through a Nalgene filter (0.45-micron pore size). Check that the pH of the acrylamide solution is 7.0 or less, and store the solution in dark bottles at room temperature.	**Caution:** Acrylamide is a potent neurotoxin and is absorbed through the skin. The effects of acrylamide are cumulative. Wear gloves and a mask when weighing powdered acrylamide and methylenebisacrylamide. Wear gloves when handling solutions containing these chemicals. Although polyacrylamide is considered to be nontoxic, it should be handled with care because of the possibility that it might contain small quantities of unpolymerized acrylamide.
		Cheaper grades of acrylamide and bisacrylamide are often contaminated with metal ions. Stock solutions of acrylamide can easily be purified by stirring overnight with about 0.2 volume of monobed resin (MB-1, Mallinckrodt), followed by filtration through Whatman No. 1 paper.
		During storage, acrylamide and bisacrylamide are slowly converted to acrylic acid and bisacrylic acid.
40% Acrylamide	Dissolve 380 g of acrylamide (DNA-sequencing grade) and 20 g of *N,N'*-methylenebisacrylamide in a total volume of 600 ml of distilled H_2O. Continue to prepare the solution as described above for 30% acrylamide, except adjust the volume to 1 liter with distilled H_2O.	See comments above for 30% acrylamide. 40% Acrylamide is used for DNA sequencing.
Actinomycin D	Dissolve 20 mg of actinomycin D in 4 ml of 100% ethanol. Read the OD_{440} of a 1:10 dilution of the stock solution in ethanol, using 100% ethanol as a blank. The molar extinction coefficient of pure actinomycin D (m.w. = 1255) in aqueous solution is 21,900. The absorbance at 440 nm of a solution containing 1 mg/ml of the drug is therefore 0.182. Stock solutions of actinomycin D are stored at −20°C in foil-wrapped tubes.	**Caution:** Actinomycin D is a teratogen and a carcinogen. It should be prepared, wearing gloves, in a chemical hood, not on open bench tops. Do not breathe in dust or get in eyes or on skin.
		Preparations of actinomycin D supplied by pharmaceutical manufacturers for therapeutic uses often contain additional substances such as sugars and salts. Such preparations can be used to suppress self-priming as long as the concentration of actinomycin D is verified by measuring the absorbance of the stock solution at 440 nm.

TABLE B.7 (continued)

Solution	Method of preparation	Comments
0.1 M Adenosine triphosphate (ATP)	Dissolve 60 mg of ATP in 0.8 ml of H_2O. Adjust the pH to 7.0 with 0.1 N NaOH. Adjust the volume to 1 ml with distilled H_2O. Dispense the solution into small aliquots and store at −70°C.	
10 M Ammonium acetate	Dissolve 770 g of ammonium acetate in 800 ml of H_2O. Adjust the volume to 1 liter with H_2O. Sterilize by filtration.	
10% Ammonium persulfate	To 1 g of ammonium persulfate, add H_2O to 10 ml. The solution may be stored for several weeks at 4°C.	
BCIP	Dissolve 0.5 g of 5-bromo-4-chloro-3-indolyl phosphate disodium salt, which is available from several manufacturers, in 10 ml of 100% dimethylformamide. Store at 4°C.	
2× BES-buffered saline	Dissolve 1.07 g of BES (N,N-bis[2-hydroxyethyl]-2-aminoethanesulfonic acid), 1.6 g of NaCl, and 0.027 g of Na_2HPO_4 in a total volume of 90 ml of distilled H_2O. Adjust the pH of the solution to 6.96 with HCl at room temperature, and then adjust the volume to 100 ml with distilled H_2O. Sterilize the solution by passage through a 0.22-micron filter, and store in aliquots at −20°C.	
1 M $CaCl_2$	Dissolve 54 g of $CaCl_2 \cdot 6H_2O$ in 200 ml of pure H_2O (Milli-Q or equivalent). Sterilize the solution by passage through a 0.22-micron filter. Store in 1-ml aliquots at −20°C.	When preparing competent cells, thaw an aliquot and dilute it to 100 ml with pure H_2O. Sterilize the solution by filtration through a Nalgene filter (0.45-micron pore size), and then chill it to 0°C.
2.5 M $CaCl_2$	Dissolve 13.5 g of $CaCl_2 \cdot 6H_2O$ in 20 ml of distilled H_2O. Sterilize the solution by passage through a 0.22-micron filter. Store in 1-ml aliquots at −20°C.	
Deoxyribonucleoside triphosphates (dNTPs)	Dissolve each dNTP in H_2O at an approximate concentration of 100 mM. Using 0.05 M Tris base and a micropipette, adjust the pH of each of the solutions to 7.0 (use pH paper to check the pH). Dilute an aliquot of the neutralized dNTP appropriately, and read the optical density at the wavelengths given in the table below. Calculate the actual concentration of each dNTP. Dilute the solutions with H_2O to a final concentration of 50 mM dNTP. Store each separately at −70°C in small aliquots.	

Base	Wavelength (nm)	Extinction Coefficient (ϵ) (M^{-1} cm^{-1})
A	259	1.54×10^4
G	253	1.37×10^4
C	271	9.10×10^3
T	260	7.40×10^3

For a cuvette with a path length of 1 cm, absorbance = ϵM.

100 mM stock solutions of each dNTP are commercially available (Pharmacia) if you do not want to prepare your own.

1 M Dithiothreitol (DTT)

Dissolve 3.09 g of DTT in 20 ml of 0.01 M sodium acetate (pH 5.2). Sterilize by filtration. Dispense into 1-ml aliquots and store at −20°C.

Do not autoclave DTT or solutions containing DTT.

0.5 M EDTA (pH 8.0)

Add 186.1 g of disodium ethylenediaminetetra-acetate·$2H_2O$ to 800 ml of H_2O. Stir vigorously on a magnetic stirrer. Adjust the pH to 8.0 with NaOH (~20 g of NaOH pellets). Dispense into aliquots and sterilize by autoclaving.

The disodium salt of EDTA will not go into solution until the pH of the solution is adjusted to approximately 8.0 by the addition of NaOH.

Ethidium bromide (10 mg/ml)

Add 1 g of ethidium bromide to 100 ml of H_2O. Stir on a magnetic stirrer for several hours to ensure that the dye has dissolved. Wrap the container in aluminum foil or transfer the solution to a dark bottle and store at room temperature.

Caution: Ethidium bromide is a powerful mutagen and is moderately toxic. Gloves should be worn when working with solutions that contain this dye, and a mask should be worn when weighing it out. After use, these solutions should be decontaminated by one of the methods described in Appendix E.

2× HEPES-buffered saline

Dissolve 1.6 g of NaCl, 0.074 g of KCl, 0.027 g of $Na_2HPO_4 \cdot 2H_2O$, 0.2 g of dextrose, and 1 g of HEPES in a total volume of 90 ml of distilled H_2O. Adjust the pH to 7.05 with 0.5 N NaOH, and then adjust the volume to 100 ml with distilled H_2O. Sterilize the solution by passage through a 0.22-micron filter. Store in 5-ml aliquots at −20°C.

IPTG

Isopropylthio-β-D-galactoside   (m.w. = 238.3). Make a solution of IPTG by dissolving 2 g of IPTG in 8 ml of distilled H_2O. Adjust the volume of the solution to 10 ml with distilled H_2O and sterilize by filtration through a 0.22-micron disposable filter. Dispense the solution into 1-ml aliquots and store them at −20°C.

Solution	Method of preparation	Comments
1 M Magnesium acetate	Dissolve 214.46 g of magnesium acetate·4H$_2$O in 800 ml of H$_2$O. Adjust the volume to 1 liter with H$_2$O. Sterilize by filtration.	
1 M MgCl$_2$	Dissolve 203.3 g of MgCl$_3$·6H$_2$O in 800 ml of H$_2$O. Adjust the volume to 1 liter with H$_2$O. Dispense into aliquots and sterilize by autoclaving.	MgCl$_2$ is extremely hygroscopic. Buy small bottles (e.g., 100 g) and do not store opened bottles for long periods of time.
β-Mercaptoethanol (BME)	Usually obtained as a 14.4 M solution. Store in a dark bottle at 4°C.	Do not autoclave BME or solutions containing BME.
NBT	Dissolve 0.5 g of nitro blue tetrazolium chloride, which is available from several manufacturers, in 10 ml of 70% dimethylformamide. Store at 4°C.	
Phenol:chloroform	Mix equal amounts of phenol and chloroform. Equilibrate the mixture by extracting several times with 0.1 M Tris·Cl (pH 7.6). Store the equilibrated mixture under an equal volume of 0.01 M Tris·Cl (pH 7.6) at 4°C in dark glass bottles.	**Caution:** Phenol is highly corrosive and can cause severe burns. Wear gloves, protective clothing, and safety glasses when handling phenol. All manipulations should be carried out in a chemical hood. Any areas of skin that come into contact with phenol should be rinsed with a large volume of water and washed with soap and water. Do *not* use ethanol.
10 mM Phenylmethyl-sulfonyl fluoride (PMSF)	Dissolve PMSF in isopropanol at a concentration of 1.74 mg/ml (10 mM). Divide the solution into aliquots and store at −20°C. If necessary, stock solutions can be prepared in concentrations as high as 17.4 mg/ml (100 mM).	**Caution:** PMSF is extremely destructive to the mucous membranes of the respiratory tract, the eyes, and skin. It may be fatal if inhaled, swallowed, or absorbed through the skin. In case of contact, immediately flush eyes or skin with copious amounts of water. Discard contaminated clothing. PMSF is inactivated in aqueous solutions. The rate of inactivation increases with pH and is faster at 25°C than at 4°C. The half-life of a 20 μM aqueous solution of PMSF is about 35 minutes at pH 8.0 (James 1978). This means that aqueous solutions of PMSF can be safely discarded after they have been rendered alkaline (pH > 8.6) and stored for several hours at room temperature.
Phosphate-buffered saline (PBS)	Dissolve 8 g of NaCl, 0.2 g of KCl, 1.44 g of Na$_2$HPO$_4$, and 0.24 g of KH$_2$PO$_4$ in 800 ml of distilled H$_2$O. Adjust the pH to 7.4 with HCl. Add H$_2$O to 1 liter. Dispense the solution into aliquots and sterilize them by autoclaving for 20 minutes at 15 lb/sq. in. on liquid cycle. Store at room temperature.	

1 M Potassium acetate (pH 7.5)

Dissolve 9.82 g of potassium acetate in 90 ml of pure H_2O (Milli-Q or equivalent). Adjust the pH to 7.5 with 2 M acetic acid. Add pure H_2O to 100 ml. Divide the solution into aliquots and store them at −20°C.

Potassium acetate (for alkaline lysis)

To 60 ml of 5 M potassium acetate, add 11.5 ml of glacial acetic acid and 28.5 ml of H_2O. The resulting solution is 3 M with respect to potassium and 5 M with respect to acetate.

3 M Sodium acetate (pH 5.2 and pH 7.0)

Dissolve 408.1 g of sodium acetate · $3H_2O$ in 800 ml of H_2O. Adjust the pH to 5.2 with glacial acetic acid or adjust the pH to 7.0 with dilute acetic acid. Adjust the volume to 1 liter with H_2O. Dispense into aliquots and sterilize by autoclaving.

5 M NaCl

Dissolve 292.2 g of NaCl in 800 ml of H_2O. Adjust the volume to 1 liter with H_2O. Dispense into aliquots and sterilize by autoclaving.

10% Sodium dodecyl sulfate (SDS) (also called sodium lauryl sulfate)

Dissolve 100 g of electrophoresis-grade SDS in 900 ml of H_2O. Heat to 68°C to assist dissolution. Adjust the pH to 7.2 by adding a few drops of concentrated HCl. Adjust the volume to 1 liter with H_2O. Dispense into aliquots.

Wear a mask when weighing SDS and wipe down the weighing area and balance after use because the fine crystals of SDS disperse easily. There is no need to sterilize 10% SDS.

20× SSC

Dissolve 175.3 g of NaCl and 88.2 g of sodium citrate in 800 ml of H_2O. Adjust the pH to 7.0 with a few drops of a 10 N solution of NaOH. Adjust the volume to 1 liter with H_2O. Dispense into aliquots. Sterilize by autoclaving.

20× SSPE

Dissolve 175.3 g of NaCl, 27.6 g of NaH_2PO_4 · H_2O and 7.4 g of EDTA in 800 ml of H_2O. Adjust the pH to 7.4 with NaOH (~6.5 ml of a 10 N solution). Adjust the volume to 1 liter with H_2O. Dispense into aliquots. Sterilize by autoclaving.

Trichloroacetic acid (TCA) 100% solution

To a bottle containing 500 g of TCA, add 227 ml of H_2O. The resulting solution will contain 100% (w/v) TCA.

1 M Tris

Dissolve 121.1 g of Tris base in 800 ml of H_2O. Adjust the pH to the desired value by adding concentrated HCl.

pH	HCl
7.4	70 ml
7.6	60 ml
8.0	42 ml

If the 1 M solution has a yellow color, discard it and obtain better quality Tris.
Although many types of electrodes do not accurately measure the pH of Tris solutions, suitable electrodes can be obtained from most manufacturers.
The pH of Tris solutions is temperature-dependent and decreases approximately 0.03 pH

TABLE B.7 (continued)

Solution	Method of preparation	Comments
1 M Tris (continued)	Allow the solution to cool to room temperature before making final adjustments to the pH. Adjust the volume of the solution to 1 liter with H$_2$O. Dispense into aliquots and sterilize by autoclaving.	units for each 1°C increase in temperature. For example, a 0.05 M solution has pH values of 9.5, 8.9, and 8.6 at 5°C, 25°C, and 37°C, respectively.
Tris-buffered saline (TBS) (25 mM Tris)	Dissolve 8 g of NaCl, 0.2 g of KCl, and 3 g of Tris base in 800 ml of distilled H$_2$O. Add 0.015 g of phenol red and adjust the pH to 7.4 with HCl. Add distilled H$_2$O to 1 liter. Dispense the solution into aliquots and sterilize them by autoclaving for 20 minutes at 15 lb/sq. in. on liquid cycle. Store at room temperature.	
X-gal	5-Bromo-4-chloro-3-indolyl-β-D-galactoside. Make a stock solution by dissolving X-gal in dimethylformamide to make a 20 mg/ml solution. Use a glass or polypropylene tube. The tube containing the solution should be wrapped in aluminum foil to prevent damage by light and should be stored at −20°C. It is not necessary to sterilize X-gal solutions by filtration.	

TABLE B.8   Blocking Agents Used to Suppress Background in Hybridization Experiments

Agent	Recommended uses
Denhardt's reagent	northern hybridizations hybridizations using RNA probes single-copy Southern hybridizations hybridizations involving DNA immobilized on nylon membranes

Denhardt's reagent (Denhardt 1966) is usually made up as a 50× stock solution, which is filtered and stored at −20°C. The stock solution is diluted tenfold into prehybridization buffer (usually 6× SSC or 6× SSPE containing 0.5% SDS and 100 μg/ml denatured, fragmented salmon sperm DNA). 50× Denhardt's reagent contains 5 g of Ficoll (Type 400, Pharmacia), 5 g of polyvinylpyrrolidone, 5 g of bovine serum albumin (Fraction V; Sigma), and H₂O to 500 ml.

BLOTTO	Grunstein/Hogness hybridization Benton/Davis hybridization all Southern hybridizations other than single-copy dot blots

1× BLOTTO (Bovine Lacto Transfer Technique Optimizer; Johnson et al. 1984) is 5% nonfat dried milk dissolved in water containing 0.02% sodium azide. It should be stored at 4°C and diluted 25-fold into prehybridization buffer before use. BLOTTO should not be used in combination with high concentrations of SDS, which will cause the milk proteins to precipitate. If background hybridization is a problem, NP-40 may be added to the hybridization solution to a final concentration of 1%. BLOTTO should not be used as a blocking agent in northern hybridizations because of the possibility that it might contain unacceptably high levels of RNAase.

Caution: Sodium azide is poisonous. It should be handled with great care, wearing gloves, and solutions containing it should be clearly marked.

Heparin	Southern hybridization in situ hybridization

Heparin (Sigma H-7005 porcine grade II or equivalent) is dissolved at a concentration of 50 mg/ml in 4× SSPE or 4× SSC and stored at 4°C. It is used as a blocking agent at a concentration of 500 μg/ml in hybridization solutions containing dextran sulfate; in hybridization solutions that do not contain dextran sulfate, heparin is used at a concentration of 50 μg/ml (Singh and Jones 1984).

Denatured, fragmented salmon sperm DNA	Southern and northern hybridizations

Salmon sperm DNA (Sigma type III sodium salt) is dissolved in water at a concentration of 10 mg/ml. If necessary, the solution is stirred on a magnetic stirrer for 2–4 hours at room temperature to help the DNA to dissolve. The concentration of NaCl is adjusted to 0.1 M, and the solution is extracted once with phenol and once with phenol:chloroform. The aqueous phase is recovered, and the DNA is sheared by passing it 12 times rapidly through a 17-gauge hypodermic needle. The DNA is precipitated by adding 2 volumes of ice-cold ethanol. It is then recovered by centrifugation and redissolved at a concentration of 10 mg/ml in water. The OD_{260} of the solution is determined and the exact concentration of the DNA is calculated. The solution is then boiled for 10 minutes and stored at −20°C in small aliquots. Just before use, the solution is heated for 5 minutes in a boiling-water bath and then chilled quickly in ice water. Denatured, fragmented salmon sperm DNA should be used at a concentration of 100 μg/ml in prehybridization solutions.

B.16 *Preparation of Reagents and Buffers Used in Molecular Cloning*

TABLE B.9 *Proteolytic Enzymes*

	Stock solution	Storage temperature	Concentration in reaction	Reaction buffer	Temperature	Pretreatment[b]
Pronase[a]	20 mg/ml in H_2O	$-20°C$	1 mg/ml	0.01 M Tris (pH 7.8) 0.01 M EDTA 0.5% SDS	37°C	self-digestion[b]
Proteinase K[c]	20 mg/ml in H_2O	$-20°C$	50 μg/ml	0.01 M Tris (pH 7.8) 0.005 M EDTA 0.5% SDS	37–56°C	none required

[a]Pronase is a mixture of serine and acid proteases isolated from *Streptomyces griseus*.

[b]Self-digestion eliminates contamination with DNAase and RNAase. Self-digested pronase is prepared by dissolving powdered pronase in 10 mM Tris·Cl (pH 7.5), 10 mM NaCl to a final concentration of 20 mg/ml and incubating for 1 hour at 37°C. Store the self-digested pronase in small aliquots at $-20°C$ in tightly capped tubes.

[c]Proteinase K is a highly active protease of the subtilisin type that is purified from the mold *Tritirachium album* Limber. The enzyme has two binding sites for Ca^{++}, which lie some distance from the active site and are not directly involved in the catalytic mechanism. However, when Ca^{++} is removed from the enzyme, approximately 80% of the catalytic activity is lost because of long-range structural changes (Bajorath et al. 1989). Because the residual activity is usually sufficient to degrade proteins that commonly contaminate preparations of nucleic acids, digestion with proteinase K is usually carried out in the presence of EDTA (to inhibit the action of Mg^{++}-dependent nucleases). However, to digest highly resistant proteins such as keratin, it may be necessary to use a buffer containing 1 mM Ca^{++} and no EDTA. At the end of the digestion, the Ca^{++} should be chelated by addition of EGTA (pH 8.0) to a final concentration of 2 mM before the nucleic acids are purified.

ENZYMES

Lysozyme

Prepare a stock solution of lysozyme at a concentration of 50 mg/ml in water. Dispense into aliquots and store at −20°C. Discard each aliquot after use.

RNAase That Is Free of DNAase

Dissolve pancreatic RNAase (RNAase A) at a concentration of 10 mg/ml in 10 mM Tris · Cl (pH 7.5), 15 mM NaCl. Heat to 100°C for 15 minutes. Allow to cool slowly to room temperature. Dispense into aliquots and store at −20°C.

DNAase That Is Free of RNAase

mRNA prepared from uninfected mammalian cells contains only small amounts of DNA that generally do not compromise northern hybridization or other types of RNA analysis. However, mRNAs prepared from transfected mammalian cells or from cells infected with DNA viruses are contaminated with large amounts of DNA that must be removed by digestion with DNAase I. Unfortunately, many commercial preparations of pancreatic DNAase I, even those sold as RNAase-free, are contaminated with significant amounts of RNAase. In addition, the use of commercially prepared RNAase-free DNAase I can become expensive when many samples are prepared. Three methods, given below, are available to remove the contaminating RNAase activity. DNAase I purified by any of these procedures should always be used in the presence of a protein inhibitor of RNAase (see Chapter 7, page 7.4).

AFFINITY CHROMATOGRAPHY ON AGAROSE 5′-(4-AMINOPHENYLPHOSPHORYL) URIDINE-2′(3′)-PHOSPHATE

The following is from Maxwell et al. (1977).

1. Suspend 10 ml of agarose-5′-(4-aminophenylphosphoryl) uridine-2′(3′)-phosphate (Miles Scientific 34 203-1) in 100 ml of 0.02 M sodium acetate (pH 5.2). Allow the agarose to settle, and then decant and repeat. Prepare a column in a 25-ml disposable syringe packed with sterile, siliconized glass wool at the bottom (see Appendix E).

2. Dissolve 20 mg of pancreatic DNAase I (Sigma D 5025) in 1 ml of 0.02 M sodium acetate (pH 5.2).

3. Apply the solution of DNAase I to the column and elute with 0.02 M sodium acetate (pH 5.2) at room temperature. Collect 1-ml fractions into RNAase-free tubes (see Chapter 7, page 7.3) until all material absorbing at 280 nm has eluted from the column.

4. Pool the fractions that contain the material absorbing at 280 nm. Read the OD_{280} and calculate the concentration of protein assuming that a solution with an OD_{280} of 1 contains 1 mg/ml protein. Add 1 M $CaCl_2$ to a final concentration of 5 mM. Dispense the preparation into small aliquots and store at −20°C.

ADSORPTION TO MACALOID

Macaloid, a clay that has been known for many years to adsorb RNAase, is available from NL Chemicals. It is prepared as follows (W. Schaffner, pers. comm.):

Preparation

1. Suspend 0.5 g of Macaloid powder in 50 ml of sterile 50 mM Tris · Cl (pH 7.6). Heat to 100°C for 5 minutes with constant agitation.

2. Centrifuge at 2500g for 5 minutes at room temperature.

3. Discard the supernatant. Completely resuspend the sticky pellet in 40 ml of sterile 50 mM Tris · Cl (pH 7.6).

4. Repeat steps 2 and 3 twice more.

5. Centrifuge the suspension at 3500g for 15 minutes at room temperature.

6. Resuspend the pellet in 30 ml of sterile 50 mM Tris · Cl (pH 7.6). The final concentration of Macaloid is approximately 17 mg/ml. The suspension may be stored indefinitely at 4°C.

Use

The Macaloid suspension is used in the following steps to remove contaminating RNAase activity from DNAase I.

1. Dissolve 100 mg of pancreatic DNAase I (Sigma D 5025) in 5 ml of:

   20 mM Tris · Cl (pH 7.6)
   50 mM NaCl
   1 mM dithiothreitol
   100 μg/ml bovine serum albumin (Fraction V; Sigma) (optional)
   50% glycerol

2. Add 15 ml of ice-cold 50 mM Tris · Cl (pH 7.6). Mix gently.

3. Add 7 ml of an ice-cold, well-dispersed suspension of Macaloid and mix on a rotating wheel for 30 minutes at 4°C.

4. Centrifuge at 8000g for 10 minutes at 4°C. Decant the supernatant into a fresh tube.

5. Add another 7 ml of Macaloid suspension and mix as before.

6. Centrifuge at 12,000g for 15 minutes at 4°C.

7. Carefully remove the supernatant and mix it gently with an equal volume of ice-cold sterile 100% glycerol.

8. Dispense into small aliquots and store at −20°C.

Notes

i. The concentration of DNAase I in the final solution is approximately 1.5 mg/ml.

ii. Smaller aliquots of DNAase I can be treated with Macaloid in microfuge tubes (see Smith 1988).

HEATING IN THE PRESENCE OF IODOACETATE

The following method reduces the activity of RNAase in preparations of DNAase by approximately 98%. The resulting preparations of DNAase are acceptable for all but the most stringent applications. However, this method should not be used to remove DNA from RNA that is to be used to generate a cDNA library, for example.

1. Dissolve 10 mg of pancreatic DNAase I (Sigma D 5025) in 10 ml of 0.1 M iodoacetic acid, 0.15 M sodium acetate (pH 5.2).

2. Heat the solution to 55°C for 45 minutes. Cool the solution to 0°C, and add 1 M $CaCl_2$ to a final concentration of 5 mM.

3. Dispense the DNAase I into small aliquots and store at −20°C.

COMMONLY USED BUFFERS

TE

pH 7.4
10 mM Tris · Cl (pH 7.4)
1 mM EDTA (pH 8.0)

pH 7.6
10 mM Tris · Cl (pH 7.6)
1 mM EDTA (pH 8.0)

pH 8.0
10 mM Tris · Cl (pH 8.0)
1 mM EDTA (pH 8.0)

STE (also called TEN)

0.1 M NaCl
10 mM Tris · Cl (pH 8.0)
1 mM EDTA (pH 8.0)

STET

0.1 M NaCl
10 mM Tris · Cl (pH 8.0)
1 mM EDTA (pH 8.0)
5% Triton X-100

TNT

10 mM Tris · Cl (pH 8.0)
150 mM NaCl
0.05% Tween 20

Phosphate Buffers

**TABLE B.10   Preparation of 0.1 M
Potassium Phosphate Buffer at 25°C**

pH	Volume of 1 M K_2HPO_4 (ml)	Volume of 1 M KH_2PO_4 (ml)
5.8	8.5	91.5
6.0	13.2	86.8
6.2	19.2	80.8
6.4	27.8	72.2
6.6	38.1	61.9
6.8	49.7	50.3
7.0	61.5	38.5
7.2	71.7	28.3
7.4	80.2	19.8
7.6	86.6	13.4
7.8	90.8	9.2
8.0	94.0	6.0

Data from Green (1933).

**TABLE B.11   Preparation of 0.1 M
Sodium Phosphate Buffer at 25°C**

pH	Volume of 1 M Na_2HPO_4 (ml)	Volume of 1 M NaH_2PO_4 (ml)
5.8	7.9	92.1
6.0	12.0	88.0
6.2	17.8	82.2
6.4	25.5	74.5
6.6	35.2	64.8
6.8	46.3	53.7
7.0	57.7	42.3
7.2	68.4	31.6
7.4	77.4	22.6
7.6	84.5	15.5
7.8	89.6	10.4
8.0	93.2	6.8

Data from ISCO (1982).

Dilute the combined 1 M stock solution to 1000 ml with distilled H_2O. pH is calculated according to the Henderson-Hasselbalch equation:

$$pH = pK' + \log\left[\frac{(\text{proton acceptor})}{\text{proton donor}}\right]$$

where pK' = 6.86 at 25°C.

Alkaline Lysis Buffers for Minipreparations of Plasmid DNA

Solution I

50 mM glucose
25 mM Tris · Cl (pH 8.0)
10 mM EDTA (pH 8.0)

Solution I can be prepared in batches of approximately 100 ml, autoclaved for 15 minutes at 10 lb/sq. in. on liquid cycle, and stored at 4°C.

Solution II

0.2 N NaOH (freshly diluted from a 10 N stock)
1% SDS

Solution III

5 M potassium acetate	60 ml
glacial acetic acid	11.5 ml
H₂O	28.5 ml

The resulting solution is 3 M with respect to potassium and 5 M with respect to acetate.

TABLE B.12   *Commonly Used Electrophoresis Buffers*

Buffer	Working solution	Concentrated stock solution (per liter)
Tris-acetate (TAE)	1×: 0.04 M Tris-acetate 0.001 M EDTA	50×: 242 g Tris base 57.1 ml glacial acetic acid 100 ml 0.5 M EDTA (pH 8.0)
Tris-phosphate (TPE)	1×: 0.09 M Tris-phosphate 0.002 M EDTA	10×: 108 g Tris base 15.5 ml 85% phosphoric acid (1.679 g/ml) 40 ml 0.5 M EDTA (pH 8.0)
Tris-borate[a] (TBE)	0.5×: 0.045 M Tris-borate 0.001 M EDTA	5×: 54 g Tris base 27.5 g boric acid 20 ml 0.5 M EDTA (pH 8.0)
Alkaline[b]	1×: 50 mN NaOH 1 mM EDTA	1×: 5 ml 10 N NaOH 2 ml 0.5 M EDTA (pH 8.0)
Tris-glycine[c]	1×: 25 mM Tris 250 mM glycine 0.1% SDS	5×: 15.1 g Tris base 94 g glycine (electrophoresis grade) (pH 8.3) 50 ml 10% SDS (electrophoresis grade)

[a]A precipitate forms when concentrated solutions of TBE are stored for long periods of time. To avoid problems, store the 5× solution in glass bottles at room temperature and discard any batches that develop a precipitate.

TBE was originally used at a working strength of 1× (i.e., a 1:5 dilution of the concentrated stock) for agarose gel electrophoresis. However, a working solution of 0.5× provides more than enough buffering power, and almost all agarose gel electrophoresis is now carried out with a 1:10 dilution of the concentrated stock.

TBE is used at a working strength of 1× for polyacrylamide gel electrophoresis (see Chapter 6, page 6.39), twice the strength usually used for agarose gel electrophoresis. The buffer reservoirs of the vertical tanks used for polyacrylamide gel electrophoresis are fairly small, and the amount of electric current passed through them is often considerable. 1× TBE is required to provide adequate buffering power.

[b]Alkaline electrophoresis buffer should be freshly made.

[c]Use Tris-glycine electrophoresis buffer for SDS-polyacrylamide gels (see Chapter 18, pages 18.47–18.54).

TABLE B.13 Gel-loading Buffers

Buffer type	6× Buffer	Storage temperature
I	0.25% bromophenol blue 0.25% xylene cyanol FF 40% (w/v) sucrose in water	4°C
II	0.25% bromophenol blue 0.25% xylene cyanol FF 15% Ficoll (Type 400; Pharmacia) in water	room temp.
III	0.25% bromophenol blue 0.25% xylene cyanol FF 30% glycerol in water	4°C
IV	0.25% bromophenol blue 40% (w/v) sucrose in water	4°C
V	*Alkaline loading buffer* 300 mN NaOH 6 mM EDTA 18% Ficoll (Type 400; Pharmacia) in water 0.15% bromocresol green 0.25% xylene cyanol FF	4°C

These gel-loading buffers serve three purposes: They increase the density of the sample, ensuring that the DNA drops evenly into the well; they add color to the sample, thereby simplifying the loading process; and they contain dyes that, in an electric field, move toward the anode at predictable rates. Bromophenol blue migrates through agarose gels approximately 2.2-fold faster than xylene cyanol FF, independent of the agarose concentration. Bromophenol blue migrates through agarose gels run in 0.5× TBE at approximately the same rate as linear double-stranded DNA 300 bp in length, whereas xylene cyanol FF migrates at approximately the same rate as linear double-stranded DNA 4 kb in length. These relationships are not significantly affected by the concentration of agarose in the gel over the range of 0.5% to 1.4%.

Which type of loading dye to use is a matter of personal preference. However, bromocresol green should be used as a tracking dye in alkaline gels because it displays a more vivid color than bromophenol blue at alkaline pH.

Sequencing gel-loading buffer

    98% deionized formamide
    10 mM EDTA (pH 8.0)
    0.025% xylene cyanol FF
    0.025% bromophenol blue

Formamide: Many batches of reagent-grade formamide are sufficiently pure to be used without further treatment. However, if any yellow color is present, the formamide should be deionized by stirring on a magnetic stirrer with Dowex XG8 mixed-bed resin for 1 hour and filtering twice through Whatman No. 1 paper. Deionized formamide should be stored in small aliquots under nitrogen at −70°C. Several companies sell formamide, distilled and packaged under nitrogen, that need not be purified before use (e.g., BRL catalog no. 5515UA).

2× SDS gel-loading buffer

    100 mM Tris · Cl (pH 6.8)
    200 mM dithiothreitol
    4% SDS (electrophoresis grade)
    0.2% bromophenol blue
    20% glycerol

2× SDS gel-loading buffer lacking dithiothreitol can be stored at room temperature. Dithiothreitol should then be added, just before the buffer is used, from a 1 M stock (see Table B.7).

Enzyme Buffers

$2 \times$ KGB

200 mM potassium glutamate
50 mM Tris-acetate (pH 7.5)
20 mM magnesium acetate
100 μg/ml bovine serum albumin (Fraction V; Sigma)
1 mM β-mercaptoethanol

This buffer is used with restriction enzymes.

$10 \times$ Klenow buffer

0.5 M Tris · Cl (pH 7.6)
0.1 M MgCl$_2$

The Klenow fragment of *E. coli* DNA polymerase I also works in a wide variety of other buffers such as restriction enzyme buffers, bacteriophage T4 DNA polymerase buffer, etc. (see Chapter 5).

$10 \times$ Nick-translation buffer

0.5 M Tris · Cl (pH 7.5)
0.1 M MgSO$_4$
1 mM dithiothreitol
500 μg/ml bovine serum albumin (Fraction V; Sigma) (optional)

Divide the stock solution into small aliquots and store them at $-20°C$.

$10 \times$ Bacteriophage T4 DNA polymerase buffer

0.33 M Tris-acetate (pH 8.0)
0.66 M potassium acetate
0.1 M magnesium acetate
5 mM dithiothreitol
1 mg/ml bovine serum albumin (Fraction V; Sigma)

The $10 \times$ stock should be divided into small aliquots and stored frozen at $-20°C$.

Reverse transcriptase buffer

50 mM Tris · Cl (pH 7.6)
60 mM KCl
10 mM MgCl$_2$
1 mM of each dNTP
1 mM dithiothreitol
1 unit/μl placental RNAase inhibitor
50 μg/ml actinomycin D

10× Ligation buffer

   0.5 M Tris · Cl (pH 7.6)
   100 mM MgCl$_2$
   100 mM dithiothreitol
   500 μg/ml bovine serum albumin (Fraction V; Sigma) (optional)

10× Blunt-end ligation buffer

   0.66 M Tris · Cl (pH 7.6)
   50 mM MgCl$_2$
   50 mM dithiothreitol
   1 mg/ml bovine serum albumin (Fraction V; Sigma) (optional)
   10 mM hexamminecobalt chloride
   2 mM ATP
   5 mM spermidine HCl

Blunt-end ligation buffer should be stored in small aliquots at −20°C.

10× CIP dephosphorylation buffer

   10 mM ZnCl$_2$
   10 mM MgCl$_2$
   100 mM Tris · Cl (pH 8.3)

10× Bacteriophage T4 DNA ligase buffer

   200 mM Tris · Cl (pH 7.6)
   50 mM MgCl$_2$
   50 mM dithiothreitol
   500 μg/ml bovine serum albumin (Fraction V; Sigma) (optional)

This buffer should be stored in small aliquots at −20°C. ATP should be
added at the time of setting up the reaction. The final concentration of ATP
in the reaction should be 0.5–1 mM.

10× Bacteriophage T4 polynucleotide kinase buffer

   0.5 M Tris · Cl (pH 7.6)
   0.1 M MgCl$_2$
   50 mM dithiothreitol
   1 mM spermidine HCl
   1 mM EDTA (pH 8.0)

5× BAL 31 buffer

3 M NaCl
60 mM $CaCl_2$
60 mM $MgCl_2$
100 mM Tris · Cl (pH 8.0)
1 mM EDTA (pH 8.0)

10× Nuclease-S1 buffer

2 M NaCl
0.5 M sodium acetate (pH 4.5)
10 mM $ZnSO_4$
5% glycerol

10× Mung-bean nuclease buffer

300 mM sodium acetate (pH 4.5)
500 mM NaCl
10 mM $ZnCl_2$
50% glycerol

10× Exonuclease III buffer

0.66 M Tris · Cl (pH 8.0)
66 mM $MgCl_2$

Appendix C: Properties of Nucleic Acids

VITAL STATISTICS OF DNA

TABLE C.1 Haploid DNA Content of Various Organisms

Organism	Size of DNA (bp)	Weight of DNA (daltons)
Mammals	$\sim 3.0 \times 10^9$	$\sim 1.9 \times 10^{12}$
Drosophila	$\sim 1.2 \times 10^8$	$\sim 7.7 \times 10^{10}$
Yeast (S. cerevisiae)	$\sim 1.6 \times 10^7$	$\sim 1.0 \times 10^{10}$
E. coli	$\sim 4.0 \times 10^6$	$\sim 2.5 \times 10^9$
Bacteriophage T2	$\sim 2.0 \times 10^5$	$\sim 1.3 \times 10^8$
Bacteriophage λ	48,514[a]	3.1×10^7
pBR322	4,363[b,c]	2.8×10^6
pUC18/pUC19	2,686[d]	1.7×10^6

[a]Daniels et al. (1983).
[b]Sutcliffe (1978).
[c]Sutcliffe (1979).
[d]Yanisch-Perron et al. (1985).

TABLE C.2 Concentration of DNA in Solution

Double-stranded DNA (50 µg/ml)	Molecules/ml	Moles/ml	Molar concentration	Molar concentration of termini
Bacteriophage λ	9.78×10^{11}	1.62×10^{-12}	1.62 nM	3.24 nM
pBR322	1.09×10^{13}	1.81×10^{-11}	18.1 nM	36.2 nM
pUC18/pUC19	1.77×10^{13}	2.94×10^{-11}	29.4 nM	58.8 nM
Segment of DNA (1 kb)	4.74×10^{13}	7.87×10^{-11}	78.7 nM	157.4 nM
Octameric double-stranded linker	5.92×10^{15}	9.83×10^{-9}	9.83 µM	19.7 µM

A solution containing 50 µg/ml of *double-stranded DNA* has an absorbance of 1 at 260 nm, i.e., $A_{260} = 1 = 50$ µg/ml of double-stranded DNA. (A solution containing 40 µg/ml of *single-stranded DNA* has an absorbance of 1 at 260 nm, i.e., $A_{260} = 1 = 40$ µg/ml of single-stranded DNA.) These values are calculated assuming that the mass of a nucleotide pair in DNA is 660 daltons.

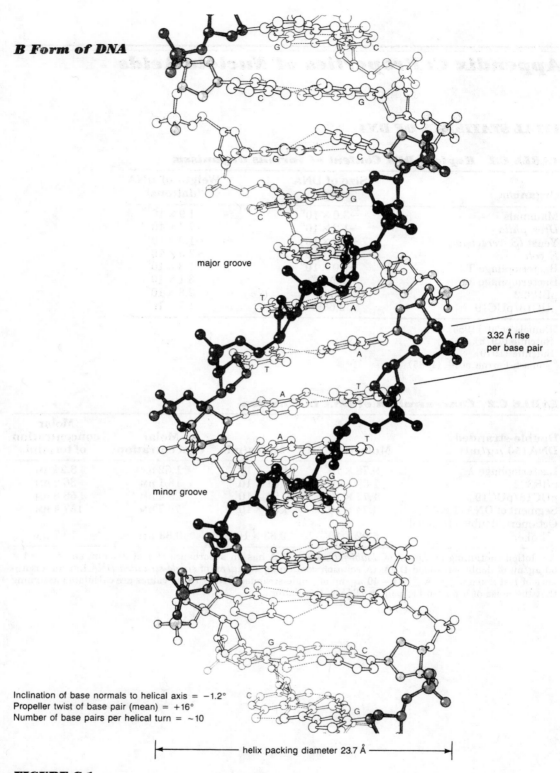

B Form of DNA

major groove

3.32 Å rise
per base pair

minor groove

Inclination of base normals to helical axis = −1.2°
Propeller twist of base pair (mean) = +16°
Number of base pairs per helical turn = ~10

|← helix packing diameter 23.7 Å →|

FIGURE C.1

The B form of DNA is a right-handed helix, whose axis passes through the base pairs. (Data from Dickerson et al. 1983; drawing from Dickerson 1983, reprinted, with permission, from I. Geis.)

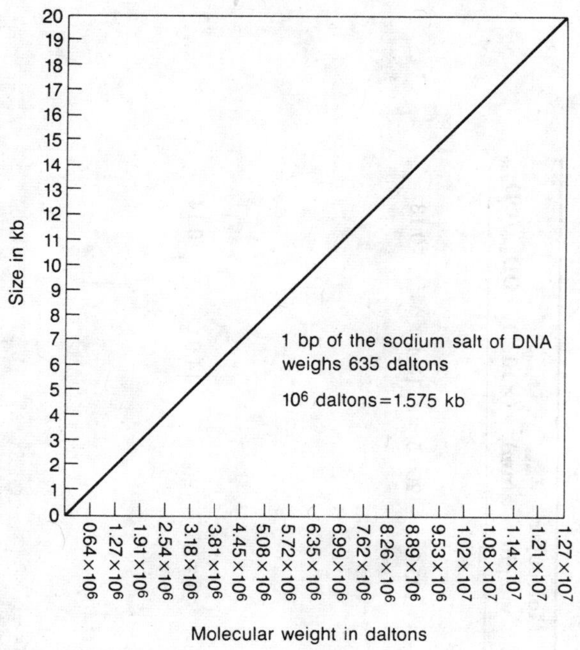
Size in kb
20
19
18
17
16
15
14
13
12
11
10
9
8
7
6
5
4
3
2
1
0
1 bp of the sodium salt of DNA
weighs 635 daltons
10^6 daltons = 1.575 kb
0.64×10^6
1.27×10^6
1.91×10^6
2.54×10^6
3.18×10^6
3.81×10^6
4.45×10^6
5.08×10^6
5.72×10^6
6.35×10^6
6.99×10^6
7.62×10^6
8.26×10^6
8.89×10^6
9.53×10^6
1.02×10^7
1.08×10^7
1.14×10^7
1.21×10^7
1.27×10^7
Molecular weight in daltons

FIGURE C.2
Relationship between the length of DNA (in kb) and its molecular weight (in daltons).

PURINES AND PYRIMIDINES

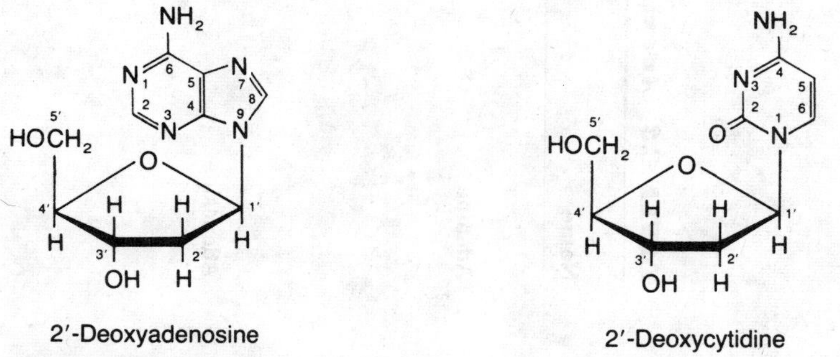
NH2
N
N
N
N
HOCH2
O
H
H
H
H
OH
H
2′-Deoxyadenosine
NH2
N
O
N
HOCH2
O
H
H
H
H
OH
H
2′-Deoxycytidine

FIGURE C.3
Numbering of atoms.

TABLE C.3 Adenine and Related Compounds

Name	Structure	Molecular weight	λ_{max} (nm)	ϵ_{max} ($\times 10^{-3}$)	OD_{280}/OD_{260}
Adenine		135.1	260.5	13.4	0.13
Adenosine		267.2	260	14.9	0.14
Adenosine 5'-phosphate (5'-AMP)		347.2	259	15.4	0.16

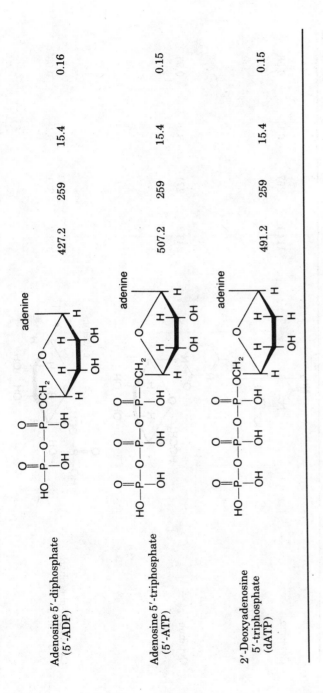

Adenosine 5'-diphosphate (5'-ADP)	427.2	259	15.4	0.16
Adenosine 5'-triphosphate (5'-ATP)	507.2	259	15.4	0.15
2'-Deoxyadenosine 5'-triphosphate (dATP)	491.2	259	15.4	0.15

TABLE C.4 Cytosine and Related Compounds

Name	Structure	Molecular weight	λ_{max} (nm)	ϵ_{max} ($\times 10^{-3}$)	OD_{280}/OD_{260}
Cytosine		111.1	267	6.1	0.58
Cytidine		243.2	271	8.3	0.93
Cytidine 5'-phosphate (5'-CMP)		323.2	271	9.1	0.98

Cytidine 5'-diphosphate (5'-CDP)   403.2   271   9.1   0.98

Cytidine 5'-triphosphate (5'-CTP)   483.2   271   9.0   0.97

2'-Deoxycytidine 5'-triphosphate (dCTP)   467.2   272   9.1   0.98

TABLE C.5 *Guanine and Related Compounds*

Name	Structure	Molecular weight	λ_{max} (nm)	ϵ_{max} ($\times 10^{-3}$)	OD_{280}/OD_{260}
Guanine		151.1	276	8.15	1.04
Guanosine		283.2	253	13.6	0.67
Guanosine 5'-phosphate (5'-GMP)		363.2	252	13.7	0.66

Compound			
Guanosine 5'-diphosphate (5'-GDP)	443.2	253	13.7	0.66
Guanosine 5'-triphosphate (5'-GTP)	523.2	253	13.7	0.66
2'-Deoxyguanosine 5'-triphosphate (dGTP)	507.2	253	13.7	0.66

TABLE C.6 Thymine and Related Compounds

Name	Structure	Molecular weight	λ_{max} (nm)	ϵ_{max} ($\times 10^{-3}$)	OD_{280}/OD_{260}
Thymine		126.1	264.5	7.9	0.53
2'-Deoxythymidine		242.2	267	9.7	0.70
2'-Deoxythymidine 5'-phosphate (TMP)		322.2	267	9.6	0.73
2'-Deoxythymidine 5'-triphosphate (dTTP)		482.2	267	9.6	0.71

TABLE C.7 Uracil and Related Compounds

Name	Structure	Molecular weight	λ_{max} (nm)	ϵ_{max} ($\times 10^{-3}$)	OD_{280}/OD_{260}
Uracil		112.1	259	8.2	0.17
Uridine		244.2	262	10.1	0.35
Uridine 5'-phosphate (5'-UMP)		324.2	260	10.0	0.38
Uridine 5'-triphosphate (UTP)		484.2	260	10.0	0.38

TABLE C.8 Unusual Bases

Name	Structure	Molecular weight	λ_{max} (nm)	ϵ_{max} ($\times 10^{-3}$)	OD_{280}/OD_{260}
Hypoxanthine		136.1	249.5	10.7	0.09
Inosine		268.2	248.5	12.3	0.25
Xanthine		152.1	267	10.3	0.61

TABLE C.9 Nucleoside Analogs Used as Chain Terminators in DNA Sequencing

2',3'-Dideoxyribonucleoside 5'-triphosphates	Structure	Molecular weight
2',3'-Dideoxyadenosine 5'-triphosphate (ddATP)	base = adenine	475.2
2',3'-Dideoxycytidine 5'-triphosphate (ddCTP)	base = cytosine	451.2
2',3'-Dideoxyguanosine 5'-triphosphate (ddGTP)	base = guanine	491.2
2',3'-Dideoxythymidine 5'-triphosphate (ddTTP)	base = thymine	466.2; Na$_4$ · H$_2$O, 608.2

Nucleotide Sequence Data Banks

Nucleotide sequence information may be submitted to the data banks by contacting the following organizations:

GenBank

GenBank Submissions
Group T-10, Mail Stop K710
Los Alamos National Laboratory
Los Alamos, NM 87545
U.S.A.

Electronic mail
  general inquiries, error reports   genbank%life@lanl.gov
  sequence submissions and forms   gb-sub%life@lanl.gov
Telephone   U.S.A. (505) 665-2177

EMBL

Data Submissions
EMBL Data Library
Postfach 10.2209
6900 Heidelberg
Federal Republic of Germany

Electronic mail
  general inquiries   datalib@embl.earn
  sequence submissions and forms   datasubs@embl.earn
Telephone   +49-6221-387-258

DDBJ

Data Submissions
Laboratory of Genetic Information Analysis
Center for Genetic Information Research
National Institute of Genetics
111 Yata
Mishima Shizuoka 411
Japan

Electronic mail
  general inquiries   ddbj@niguts.nig.junet
  submission forms   ddbjsub@niguts.nig.junet
Telephone   559 75 0771

Appendix D: Codons and Amino Acids

2nd position of codon

	U	C	A	G	
U	UUU Phe UUC Phe UUA Leu UUG Leu	UCU Ser UCC Ser UCA Ser UCG Ser	UAU Tyr UAC Tyr UAA Stop (Ochre) UAG Stop (Amber)	UGU Cys UGC Cys UGA Stop (Umber) UGG Trp	U C A G
C	CUU Leu CUC Leu CUA Leu CUG Leu	CCU Pro CCC Pro CCA Pro CCG Pro	CAU His CAC His CAA Gln CAG Gln	CGU Arg CGC Arg CGA Arg CGG Arg	U C A G
A	AUU Ile AUC Ile AUA Ile AUG Met	ACU Thr ACC Thr ACA Thr ACG Thr	AAU Asn AAC Asn AAA Lys AAG Lys	AGU Ser AGC Ser AGA Arg AGG Arg	U C A G
G	GUU Val GUC Val GUA Val GUG Val	GCU Ala GCC Ala GCA Ala GCG Ala	GAU Asp GAC Asp GAA Glu GAG Glu	GGU Gly GGC Gly GGA Gly GGG Gly	U C A G

1st position of codon (5′ terminus) — 3rd position of codon (3′ terminus)

FIGURE D.1
The genetic code (nuclear genes).

TABLE D.1 Prokaryotic Suppressors of Nonsense Mutations Used in Molecular Cloning

Suppressor	Codon recognized	Amino acid inserted
supD (suI)	amber (UAG)	serine
supE (suII)	amber (UAG)	glutamic acid
supF (suIII)	amber (UAG)	tyrosine
supB (suB)	amber (UAG) ochre (UAA)	glutamic acid
supC (suC)	amber (UAG) ochre (UAA)	tyrosine

TABLE D.2 Properties of Amino Acids

Amino acid	Three-letter symbol	One-letter symbol	Mass[a] (daltons)	pK$_a$ of ionizing side chain	Structure
Alanine	Ala	A	89.09		
Arginine	Arg	R	174.2	12.48	
Asparagine	Asn	N	132.1		
Aspartic acid	Asp	D	133.1	3.86	
Cysteine	Cys	C	121.12	8.33	
Glutamine	Gln	Q	146.15		

Name	Abbr.	Code	M.W.	pK_a
Glutamic acid	Glu	E	147.13	4.25
Glycine	Gly	G	75.07	
Histidine	His	H	155.16	6.0
Isoleucine	Ile	I	131.17	
Leucine	Leu	L	131.17	
Lysine	Lys	K	146.19	10.53

TABLE D.2 (continued)

Amino acid	Three-letter symbol	One-letter symbol	Mass[a] (daltons)	pK_a of ionizing side chain	Structure
Methionine	Met	M	149.21		$CH_3-S-CH_2-CH_2-\overset{\displaystyle H}{\underset{\displaystyle \overset{+}{N}H_3}{C}}-COO^-$
Phenylalanine	Phe	F	165.19		
Proline	Pro	P	115.13		
Serine	Ser	S	105.09		$HO-CH_2-\overset{\displaystyle H}{\underset{\displaystyle \overset{+}{N}H_3}{C}}-COO^-$

Threonine	Thr	T	119.12	
Tryptophan	Trp	W	204.22	
Tyrosine	Tyr	Y	181.19	10.07
Valine	Val	V	117.15	

Weighted mean = 126.7

[a] The polymerization of amino acids into a polypeptide chain results in a net loss of 18 daltons per peptide bond due to elimination of water during condensation.

Classification of Amino Acids

The relationship between the 20 common amino acids is shown in the form of a Venn diagram (Figure D.2). The diagram, which is based on the mutational matrix of Dayhoff (1972), takes into account a selection of physico-chemical properties important in the determination of protein structure. The amino acids are divided into three major groups: (1) those that carry a polar group, (2) those that are hydrophobic, and (3) those that are small. Each of these major groups overlaps with the other two and contains subsets of various types (aromatic, positive, etc.).

Cysteine is shown in two locations: The reduced form (C_H) contains a polarizable S—H bond and is therefore similar in some ways to serine (which carries an O—H bond). The oxidized form of cysteine (C_{S-S}) contains no polarizable bond and is therefore more hydrophobic in nature.

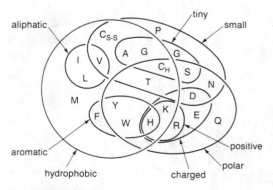

FIGURE D.2
Venn diagram showing the relationship between 20 common amino acids. (Redrawn, with permission, from Taylor 1986.)

Appendix E: Commonly Used Techniques in Molecular Cloning

GLASSWARE AND PLASTICWARE

All glassware should be sterilized by autoclaving. Much plasticware is available sterilized. Autoclaving is appropriate for some, but not all, plasticware, depending on the type of plastic. It is particularly important to have available a supply of sterilized microfuge tubes and disposable tips for automatic pipetting devices. All of the procedures commonly used in molecular cloning can be carried out in glassware or plasticware prepared in this way; there is no significant loss of material by adsorption onto the surfaces of the containers. However, for certain procedures (e.g., handling very small quantities of single-stranded DNA or sequencing by the Maxam-Gilbert technique), it is advisable to use glassware or plasticware that has been coated with a thin film of silicone. A simple procedure for siliconizing small items such as pipettes, tubes, and beakers is given below. To siliconize large items such as glass plates, refer to the note at the end of the protocol.

Siliconizing Glassware, Plasticware, and Glass Wool

The following method was supplied by B. Seed (unpubl.).

1. Place the items to be siliconized inside a large, glass desiccator.

2. Add 1 ml of dichlorodimethylsilane to a small beaker inside the desiccator.

   Caution: Dichlorodimethylsilane is toxic, volatile, and highly flammable and should be used only in a chemical hood.

3. Attach the desiccator, through a trap, to a vacuum pump. Turn on the vacuum and continue to apply suction until the dichlorodimethylsilane begins to boil. Immediately clamp the connection between the vacuum pump and the desiccator. Switch off the vacuum pump. The desiccator should maintain a vacuum.

   It is essential to turn off the vacuum pump as soon as the dichlorodimethylsilane begins to boil. Otherwise, the volatile agent will be sucked into the pump and will cause irreparable damage to the vacuum seals.

4. When the dichlorodimethylsilane has evaporated (1–2 hours), open the desiccator in a chemical hood. After the fumes of dichlorodimethylsilane have dispersed, remove the glassware or plasticware. Bake glassware and glass wool for 2 hours at 180°C before use. Plasticware should not be autoclaved but should be rinsed extensively with water before use.

Note

Large items of glassware can be siliconized by soaking or rinsing in a 5% solution of dichlorodimethylsilane in chloroform or heptane. As the organic solvent evaporates, the dichlorodimethylsilane is deposited on the glassware, which should be rinsed many times with water or baked at 180°C for 2 hours before use.

PURIFICATION OF NUCLEIC ACIDS

Perhaps the most basic of all procedures in molecular cloning is the purification of nucleic acids. The key step, the removal of proteins, can often be carried out simply by extracting aqueous solutions of nucleic acids with phenol:chloroform and chloroform. Such extractions are used whenever it is necessary to inactivate and remove enzymes that are used in one step of a cloning operation before proceeding to the next. However, additional measures are required when nucleic acids are purified from complex mixtures of molecules such as cell lysates. In these cases, it is usual to remove most of the protein by digestion with proteolytic enzymes such as pronase or proteinase K (see Appendix B, Table B.9), which are active against a broad spectrum of native proteins, before extracting with organic solvents.

Extraction with Phenol:Chloroform

The standard way to remove proteins from nucleic acid solutions is to extract first with phenol:chloroform and then with chloroform. This procedure takes advantage of the fact that deproteinization is more efficient when two different organic solvents are used instead of one. Furthermore, although phenol denatures proteins efficiently, it does not completely inhibit RNAase activity, and it is a solvent for RNA molecules that contain long tracts of poly(A) (Brawerman et al. 1972). Both of these problems can be circumvented by using a mixture of phenol:chloroform:isoamyl alcohol (25:24:1). The subsequent extraction with chloroform removes any lingering traces of phenol from the nucleic acid preparation. Extraction with ether, which was widely used for this purpose for many years, is no longer required for routine purification of DNA.

1. Add an equal volume of phenol:chloroform to the nucleic acid sample in a polypropylene tube with a plastic cap.

   The nucleic acid will tend to partition into the organic phase if the phenol has not been adequately equilibrated to a pH of 7.8–8.0.

2. Mix the contents of the tube until an emulsion forms.

3. Centrifuge the mixture at 12,000g for 15 seconds in a microfuge (or at 1600g for 3 minutes in another rotor) at room temperature. If the organic and aqueous phases are not well-separated, centrifuge again for a longer time or at a higher speed.

   Normally, the aqueous phase forms the upper phase. However, if the aqueous phase is dense because of salt (>0.5 M) or sucrose ($>10\%$), it will form the lower phase. The organic phase is easily identifiable because of the yellow color contributed by the hydroxyquinoline that is added to phenol during equilibration (see Appendix B).

4. Use a pipette to transfer the aqueous phase to a fresh tube. For small volumes (<200 μl), use an automatic pipettor fitted with a disposable tip. Discard the interface and organic phase.

   To achieve the best recovery, the organic phase and interface may be "back-extracted" as follows: After the first aqueous phase has been transferred as described above, add

---

an equal volume of TE (pH 7.8) to the organic phase and interface. Mix well. Separate the phases by centrifugation as in step 3. Combine this second aqueous phase with the first, and proceed to step 5.

5. Repeat steps 1 through 4 until no protein is visible at the interface of the organic and aqueous phases.

6. Add an equal volume of chloroform and repeat steps 2 through 4.

7. Recover the nucleic acid by precipitation with ethanol as described on page E.10.

   Occasionally, ether is used to remove traces of chloroform from preparations of high-molecular-weight DNA (see Note below).

Note

The organic and aqueous phases may be mixed by vortexing when isolating small DNA molecules (<10 kb) or by gentle shaking when isolating DNA molecules of moderate size (10–30 kb). When isolating large DNA molecules (>30 kb), the following precautions must be taken to avoid shearing (see also Chapter 9).

- The organic and aqueous phases should be mixed by rotating the tube slowly (20 rpm) on a wheel.

- Large-bore pipettes should be used to transfer the DNA from one tube to another.

- The DNA should not be precipitated with ethanol (step 7). Instead, traces of chloroform should be removed either by extensive dialysis of the DNA solution against large volumes of ice-cold TEN *or* by extraction with water-saturated ether as described below.

Caution: Ether is highly volatile and extremely flammable and should be used and stored only in an explosion-proof chemical hood.

   a. In a large glass bottle, add water to the ether and mix well. Continue to add water until additional water does not enter the ether phase and forms a layer on the bottom of the bottle.

   b. Combine the DNA sample with an equal volume of the water-saturated ether prepared in step a and mix. Let the organic and aqueous phases separate by allowing the solution to stand for 2–5 minutes.

   c. Remove and discard the upper layer (ether is less dense than water).

   d. Repeat steps b and c.

   e. Remove traces of ether by heating the DNA solution to 68°C for 5–10 minutes with gentle mixing or by blowing a stream of nitrogen gas over the surface of the solution for 10–30 minutes.

QUANTITATION OF DNA AND RNA

Two types of methods are widely used to measure the amount of nucleic acid in a preparation. If the sample is pure (i.e., without significant amounts of contaminants such as proteins, phenol, agarose, or other nucleic acids), spectrophotometric measurement of the amount of ultraviolet irradiation absorbed by the bases is simple and accurate. If the amount of DNA or RNA is very small or if the sample contains significant quantities of impurities, the amount of nucleic acid can be estimated from the intensity of fluorescence emitted by ethidium bromide.

Spectrophotometric Determination of the Amount of DNA or RNA

For quantitating the amount of DNA or RNA, readings should be taken at wavelengths of 260 nm and 280 nm. The reading at 260 nm allows calculation of the concentration of nucleic acid in the sample. An OD of 1 corresponds to approximately 50 μg/ml for double-stranded DNA, 40 μg/ml for single-stranded DNA and RNA, and ~20 μg/ml for single-stranded oligonucleotides. The ratio between the readings at 260 nm and 280 nm (OD_{260}/OD_{280}) provides an estimate of the purity of the nucleic acid. Pure preparations of DNA and RNA have OD_{260}/OD_{280} values of 1.8 and 2.0, respectively. If there is contamination with protein or phenol, the OD_{260}/OD_{280} will be significantly less than the values given above, and accurate quantitation of the amount of nucleic acid will not be possible.

Ethidium Bromide Fluorescent Quantitation of the Amount of Double-stranded DNA

Sometimes there is not sufficient DNA (<250 ng/ml) to assay spectrophotometrically, or the DNA may be heavily contaminated with other substances that absorb ultraviolet irradiation and therefore impede accurate analysis. A rapid way to estimate the amount of DNA in such samples is to utilize the ultraviolet-induced fluorescence emitted by ethidium bromide molecules intercalated into the DNA. Because the amount of fluorescence is proportional to the total mass of DNA, the quantity of DNA in the sample can be estimated by comparing the fluorescent yield of the sample with that of a series of standards. As little as 1–5 ng of DNA can be detected by this method.

Cautions: Ethidium bromide is a powerful mutagen and is moderately toxic. Gloves should be worn when working with solutions that contain this dye. After use, these solutions should be decontaminated by one of the methods described on pages E.8–E.9.

Ultraviolet radiation is dangerous, particularly to the eyes. To minimize exposure, make sure that the ultraviolet light source is adequately shielded and wear protective goggles or a full safety mask that efficiently blocks ultraviolet light.

SARAN WRAP METHOD

1. Stretch a sheet of Saran Wrap over an ultraviolet transilluminator or over a sheet of black paper.

2. Spot 1–5 μl of the DNA sample onto the Saran Wrap.

3. Spot equal volumes of a series of DNA concentration standards (0, 1, 2.5, 5, 10, and 20 μg/ml) in an ordered array on the Saran Wrap.

   The standard DNA solutions should contain a single species of DNA approximately the same size as the expected size of the unknown DNA. The DNA standards are stable for many months when stored at −20°C.

4. Add to each spot an equal volume of TE (pH 7.6) containing 2 μg/ml ethidium bromide. Mix by pipetting up and down with a micropipette.

5. Photograph the spots using short-wavelength ultraviolet illumination (see Chapter 6, page 6.19). Estimate the concentration of DNA by comparing the intensity of fluorescence in the sample with those of the standard solutions.

AGAROSE PLATE METHOD

Contaminants that may be present in the DNA sample can either contribute to or quench the fluorescence. To avoid these problems, the DNA samples and standards can be spotted onto the surface of a 1% agarose slab gel containing ethidium bromide (0.5 μg/ml). Allow the gel to stand at room temperature for a few hours so that small contaminating molecules have the chance to diffuse away. Photograph the gel as described in Chapter 6, page 6.19.

MINIGEL METHOD

Electrophoresis through minigels (see Chapter 6, page 6.14) provides a rapid and convenient way to measure the quantity of DNA and to analyze its physical state at the same time. This is the method of choice if there is a possibility that the samples may contain significant quantities of RNA.

1. Mix 2 μl of the DNA sample with 0.4 μl of gel-loading buffer IV (bromophenol blue only; see Appendix B) and load the solution into a slot in a 0.8% agarose minigel containing ethidium bromide (0.5 μg/ml).

2. Mix 2 μl of each of a series of standard DNA solutions (0, 2.5, 5, 10, 20, 30, 40, and 50 μg/ml) with 0.4 μl of gel-loading buffer IV. Load the samples into the wells of the gel.

   The standard DNA solutions should contain a single species of DNA approximately the same size as the expected size of the unknown DNA. The DNA standards are stable for many months when stored at −20°C.

3. Carry out electrophoresis until the bromophenol blue has migrated approximately 1–2 cm.

4. Destain the gel by immersing it for 5 minutes in electrophoresis buffer containing 0.01 M MgCl$_2$.

5. Photograph the gel using short-wavelength ultraviolet irradiation (see Chapter 6, page 6.19). Compare the intensity of fluorescence of the unknown DNA with that of the DNA standards and estimate the quantity of DNA in the sample.

DECONTAMINATION OF ETHIDIUM BROMIDE SOLUTIONS

Caution: Ethidium bromide is a powerful mutagen and is moderately toxic. Gloves should be worn when working with solutions that contain this dye. After use, these solutions should be decontaminated by one of the methods described below.

Decontamination of Concentrated Solutions of Ethidium Bromide
(i.e., solutions containing >0.5 mg/ml)

METHOD 1

This method (Lunn and Sansone 1987) reduces the mutagenic activity of ethidium bromide in the *Salmonella*/microsome assay by approximately 200-fold.

1. Add sufficient water to reduce the concentration of ethidium bromide to < 0.5 mg/ml.

2. To the resulting solution, add 0.2 volume of fresh 5% hypophosphorous acid and 0.12 volume of fresh 0.5 M sodium nitrite. Mix carefully.
   Important: Check that the pH of the solution is < 3.0.

   Hypophosphorous acid is usually supplied as a 50% solution, which is corrosive and should be handled with care. It should be freshly diluted immediately before use.

   Sodium nitrite solution (0.5 M) should be freshly prepared by dissolving 34.5 g of sodium nitrite in water to a final volume of 500 ml.

3. After incubation for 24 hours at room temperature, add a large excess of 1 M sodium bicarbonate. The solution may now be discarded.

METHOD 2

This method (Quillardet and Hofnung 1988) reduces the mutagenic activity of ethidium bromide in the *Salmonella*/microsome assay by approximately 3000-fold. However, there are reports (Lunn and Sansone 1987) of mutagenic activity in occasional batches of "blanks" treated with the decontaminating solutions.

1. Add sufficient water to reduce the concentration of ethidium bromide to < 0.5 mg/ml.

2. Add 1 volume of 0.5 M $KMnO_4$. Mix carefully, and then add 1 volume of 2.5 N HCl. Mix carefully, and allow the solution to stand at room temperature for several hours.

3. Add 1 volume of 2.5 N NaOH. Mix carefully, and then discard the solution.

Decontamination of Dilute Solutions of Ethidium Bromide
(e.g., electrophoresis buffer containing 0.5 μg/ml ethidium bromide)

METHOD 1

The following method is from Lunn and Sansone (1987).

1. Add 2.9 g of Amberlite XAD-16 for each 100 ml of solution. Amberlite XAD-16, a nonionic, polymeric absorbent, is available from Rohm and Haas.

2. Store the solution for 12 hours at room temperature, shaking it intermittently.

3. Filter the solution through a Whatman No. 1 filter, and discard the filtrate.

4. Seal the filter and Amberlite resin in a plastic bag, and dispose of the bag in the hazardous waste.

METHOD 2

The following method is from Bensaude (1988).

1. Add 100 mg of powdered activated charcoal for each 100 ml of solution.

2. Store the solution for 1 hour at room temperature, shaking it intermittently.

3. Filter the solution through a Whatman No. 1 filter, and discard the filtrate.

4. Seal the filter and activated charcoal in a plastic bag, and dispose of the bag in the hazardous waste.

Notes

i. Treatment of dilute solutions of ethidium bromide with hypochlorite (bleach) is not recommended as a method of decontamination. Such treatment reduces the mutagenic activity of ethidium bromide in the *Salmonella*/microsome assay by about 1000-fold, but it converts the dye into a compound that is mutagenic in the absence of microsomes (Quillardet and Hofnung 1988).

ii. Ethidium bromide decomposes at 262°C and is unlikely to be hazardous after incineration under standard conditions.

iii. Slurries of Amberlite XAD-16 or activated charcoal can be used to decontaminate surfaces that become contaminated by ethidium bromide.

CONCENTRATING NUCLEIC ACIDS

Precipitation with Ethanol or Isopropanol

The most widely used method for concentrating nucleic acids is precipitation with ethanol. The precipitate of nucleic acid, which is allowed to form in the presence of moderate concentrations of monovalent cations, is recovered by centrifugation and redissolved in an appropriate buffer at the desired concentration. The technique is rapid and is quantitative even with picogram amounts of DNA and RNA.

The three major variables are:

- *The temperature at which the precipitate is allowed to form.* Until a few years ago, ethanol precipitation was routinely carried out at low temperature (e.g., in a dry-ice/methanol bath). This is now known to be unnecessary: At 0°C in the absence of carrier, DNA at concentrations as low as 20 ng/ml will form a precipitate that can be quantitatively recovered by centrifugation in a microfuge.

- *The type and concentration of monovalent cations used in the precipitation mixture.* The most commonly used cations are shown in Table E.1. The choice among these salts is largely a matter of personal preference.

  Ammonium acetate (2.0–2.5 M) is frequently used to reduce the coprecipitation of dNTPs. For example, two sequential precipitations of DNA in the presence of 2 M ammonium acetate result in the removal of over 99% of the dNTPs from preparations of DNA (Okayama and Berg 1982). However, ammonium acetate should not be used when the precipitated nucleic acid is to be phosphorylated, since bacteriophage T4 polynucleotide kinase is inhibited by ammonium ions.

  Lithium chloride (0.8 M) is frequently used when higher concentrations of ethanol are used for precipitation, for example, when precipitating RNA. LiCl is very soluble in ethanolic solutions and is not coprecipitated with the nucleic acid. Avoid LiCl when the RNA is to be used for cell-free translation or reverse transcription. Chloride ions inhibit initiation of protein synthesis in most cell-free systems and suppress the activity of RNA-dependent DNA polymerase. Because of the differential solubility of small and large RNAs in high concentrations (0.8 M) of LiCl, LiCl (without ethanol) can be used to purify large RNAs selectively (see page E.15).

  Sodium chloride (0.2 M) should be used if the DNA sample contains SDS. The detergent then remains soluble in 70% ethanol.

  Sodium acetate (0.3 M; pH 5.2) is used for most routine precipitations of DNA and RNA.

  Notes
  i. Buffers containing >1 mM phosphate or >10 mM EDTA should not be used for ethanol precipitation, since these substances coprecipitate with the nucleic acid. High concentrations of phosphate ions and EDTA should be removed by conventional or spun-column chromatography before ethanol precipitation is attempted.
  ii. The efficiency with which very short pieces (<100 nucleotides) of nucleic acid are precipitated can be improved by adding $MgCl_2$ to a final concentration of 0.01 M.

TABLE E.1   Salt Solutions

	Stock solution (M)	Final concentration (M)
Ammonium acetate	10.0	2.0–2.5
Lithium chloride	8.0	0.8
Sodium chloride	5.0	0.2
Sodium acetate	3.0 (pH 5.2)	0.3

- *The time and speed of centrifugation.* As little as 20 ng of nucleic acid in a volume of 1 ml can be quantitatively recovered in the absence of carrier by centrifugation at 12,000*g* for 15 minutes at 0–4°C in a microfuge. However, when lower concentrations of DNA or very small fragments (<100 nucleotides) are processed, more extensive centrifugation is required to cause the pellet of nucleic acid to adhere tightly to the centrifuge tube. A number of ultracentrifuge heads are now sold that hold very small volumes of fluid. If one of these is not available, centrifugation can be carried out by floating the sealed microfuge tube containing the ethanolic solution of nucleic acid in an ultracentrifuge tube (e.g., Beckman SW28 or equivalent) that has been three-quarters filled with water. Centrifugation is then carried out at 27,000 rpm for 1–2 hours at 4°C. After ultracentrifugation, the microfuge tube is removed and the supernatant is carefully removed and discarded using an automatic micropipettor. The pellet of nucleic acid may then be washed with 70% ethanol and recovered by centrifugation at 12,000*g* for 10 minutes at 0–4°C in a microfuge. This method (Shapiro 1981) allows the recovery of picogram quantities of nucleic acid and obviates the need for carrier.

PRECIPITATION OF DNA IN MICROFUGE TUBES

1. Estimate the volume of the DNA solution.

2. Adjust the concentration of monovalent cations either by dilution with TE (pH 8.0), if the DNA solution contains a high concentration of salts, or by addition of one of the salt solutions shown in Table E.1. If the volume of the final solution is 400 μl or less, precipitation may be carried out in a single microfuge tube. Larger volumes can be divided among several microfuge tubes, or the DNA can be precipitated and centrifuged in tubes that will fit in a medium-speed centrifuge or ultracentrifuge.

3. Mix the solution well. Add exactly 2 volumes of ice-cold ethanol and again mix the solution well. Store the ethanolic solution on ice to allow the precipitate of DNA to form. Usually 15–30 minutes is sufficient, but when the size of the DNA is small (<100 nucleotides) or when it is present in small amounts (<0.1 μg/ml), the period of storage should be extended to at least 1 hour and $MgCl_2$ should be added to a final concentration of 0.01 M.

   DNA can be stored indefinitely in ethanolic solutions at 0°C or at -20°C.

4. Recover the DNA by centrifugation at 0°C. For most purposes, centrifugation at 12,000g for 10 minutes in a microfuge is sufficient. However, as discussed above, when low concentrations of DNA (<20 ng/ml) or very small fragments are being processed, more extensive centrifugation may be required.

5. Carefully remove the supernatant with an automatic micropipettor or with a disposable pipette tip attached to a vacuum line (see Figure E.1). Take care not to disturb the pellet of nucleic acid (which may be invisible). Use the pipette tip to remove any drops of fluid that adhere to the walls of the tube.

   It is best to save the supernatant from valuable DNA samples until recovery of the precipitated DNA has been verified.

6. Half fill the tube with 70% ethanol and recentrifuge at 12,000g for 2 minutes at 4°C in a microfuge.

7. Repeat step 5.

8. Store the open tube on the bench at room temperature until the last traces of fluid have evaporated.

   Until recently, it was common practice to dry pellets of nucleic acid in a lyophilizer. This is not only unnecessary, but also undesirable, since it causes denaturation of small (<400-nucleotide) fragments of DNA (Svaren et al. 1987) and greatly reduces the recovery of larger fragments of DNA.

9. Dissolve the DNA pellet (which is often invisible) in the desired volume of buffer (usually TE [pH between 7.6 and 8.0]). Rinse the walls of the tube well with the buffer.

Notes

i. After centrifugation in a microfuge, not all of the DNA is deposited on the bottom of the microfuge tube. Up to 50% of the DNA is smeared on the walls of the tube. To recover all of the DNA, it is necessary to work a bead of fluid backward and forward over the appropriate quadrant of wall. This can easily be done by pushing the bead of fluid over the surface with a disposable pipette tip attached to an automatic micropipettor.

ii. One volume of isopropanol may be used in place of 2 volumes of ethanol to precipitate DNA. Precipitation with isopropanol has the advantage that the volume of liquid to be centrifuged is smaller. However, isopropanol is

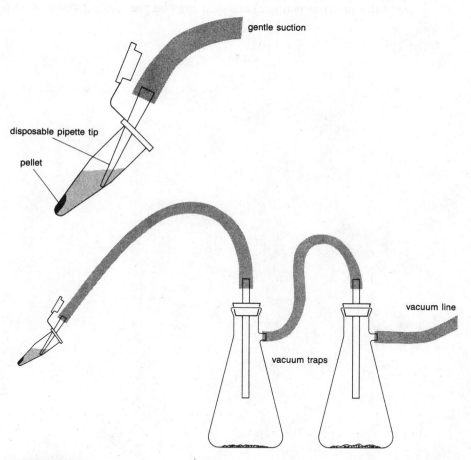

gentle suction

disposable pipette tip

pellet

vacuum line

vacuum traps

FIGURE E.1

Aspiration of supernatants. Hold the open microfuge tube at an angle, with the pellet on the upper side. Use a disposable pipette tip attached to a vacuum line to withdraw fluid from the tube. Insert the tip just beneath the meniscus on the lower side of the tube. Move the tip towards the base of the tube as the fluid is withdrawn. Use gentle suction to avoid drawing the pellet into the pipette tip. Keep the end of the tip away from the pellet. Finally, vacuum the walls of the tube to remove any adherent drops of fluid.

less volatile than ethanol and is therefore more difficult to remove; moreover, solutes such as sucrose or sodium chloride are more easily coprecipitated with DNA when isopropanol is used. In general, precipitation with ethanol is preferable, unless it is necessary to keep the volume of fluid to a minimum.

iii. In general, DNA precipitated from solution by ethanol can be easily redissolved in buffers of low ionic strength, such as TE (pH 7.6). Occasionally, difficulties arise when buffers containing $MgCl_2$ or > 0.1 M NaCl are added directly to the DNA pellet. It is therefore preferable to dissolve the DNA in a small volume of low-ionic-strength buffer and to adjust the composition of the buffer later. If the sample does not dissolve easily in a small volume, add a larger volume of buffer and repeat the precipitation with ethanol. The second precipitation may help eliminate additional salts or other components that may be preventing dissolution of the DNA.

PRECIPITATION OF RNA WITH ETHANOL

RNA is efficiently precipitated with 2.5–3.0 volumes of ethanol from solutions containing 0.8 M LiCl, 5 M ammonium acetate, or 0.3 M sodium acetate. The choice among these salts is determined by the use to which the RNA will be put later.

Since the potassium salt of dodecyl sulfate is extremely insoluble, avoid potassium acetate if the precipitated RNA is to be dissolved in buffers that contain SDS, for example, buffers that are used for chromatography on oligo(dT)-cellulose. For the same reason, avoid potassium acetate if the RNA is already dissolved in a buffer containing SDS.

Avoid LiCl when the RNA is to be used for cell-free translation or reverse transcription. Chloride ions inhibit initiation of protein synthesis in most cell-free systems and suppress the activity of RNA-dependent DNA polymerase.

Note: Solutions used for precipitation of RNA should be free of RNAase (see Chapter 7, page 7.4).

PRECIPITATION OF LARGE RNAs WITH LITHIUM CHLORIDE

Whereas small RNAs (tRNAs and 5S RNAs) are soluble in solutions of high ionic strength, large RNAs (e.g., rRNAs and mRNAs) are insoluble and can be removed by centrifugation.

1. Measure the volume of the sample and add 0.1 volume of RNAase-free 8 M LiCl. Mix the solution well and store it on ice for at least 2 hours.

2. Centrifuge the solution at 15,000g for 20 minutes at 0°C. Discard the supernatant, and redissolve the precipitated high-molecular-weight RNA in water.

3. Repeat the procedure, and finally recover the high-molecular-weight RNA from the resuspended pellet by precipitation with 2 volumes of ethanol.

CONCENTRATING NUCLEIC ACIDS BY EXTRACTION WITH BUTANOL

During extraction of aqueous solutions with solvents such as secondary butyl alcohol (2-butanol) or *n*-butyl alcohol (1-butanol), some of the water molecules are partitioned into the organic phase. By carrying out several cycles of extraction, the volume of a nucleic acid solution can be reduced significantly. This method of concentration is used to reduce the volume of dilute solutions to the point where the nucleic acid can easily be recovered by precipitation with ethanol.

1. Measure the volume of the nucleic acid solution and add an equal volume of 2-butanol. Mix the solution well by vortexing.

   Addition of too much 2-butanol can result in removal of all the water and precipitation of the nucleic acid. If this happens, add water to the organic phase until an aqueous phase (which should contain the nucleic acid) reappears.

2. Centrifuge the solution at room temperature at 12,000g for 20 seconds in a microfuge or at 1600g for 1 minute in a bench-top centrifuge. Use an automatic micropipettor to remove and discard the upper (2-butanol) phase.

3. Repeat steps 1 and 2 until the desired volume of aqueous phase is achieved.

Note

Because 2-butanol extraction does not remove salt, the salt concentration increases in proportion to the reduction in the volume of the solution. The nucleic acid can be transferred to the desired buffer by spun-column chromatography (see pages E.37–E.38) or by precipitation with ethanol as described on pages E.12–E.14.

DRYING DOWN ^{32}P-LABELED NUCLEOTIDES FROM MIXTURES OF ETHANOL AND WATER

Most commercial suppliers sell [^{32}P]dNTPs as concentrated, stabilized aqueous solutions that can be added directly to the appropriate reaction mixtures. However, some manufacturers still supply [^{32}P]dNTPs dissolved in 50% ethanol, which must be removed by evaporation before the [^{32}P]dNTP can be used.

1. Using an automatic micropipettor, carefully dispense the desired quantity of [^{32}P]dNTPs into a microfuge tube.

2. Plug the top of the tube with a small piece of cotton and cover with two or three layers of Parafilm.

3. Poke several holes in the Parafilm with a needle.

4. Place the tube securely in a beaker or rack, and evaporate the [^{32}P]dNTPs to dryness under vacuum at room temperature or in a lyophilizer.

5. Discard the Parafilm and cotton in the radioactive waste. Add a small volume (5 μl) of H_2O to the tube. Vortex for 15 seconds.

6. Add the remaining ingredients of the reaction mixture to the tube. Mix by vortexing and incubate as indicated in the relevant protocol.

Notes

i. Steps 1–3 can be eliminated by using a SpeedVac concentrator, which prevents bumping of the contents of the tube under vacuum.

ii. Wherever possible, manipulations involving ^{32}P should be carried out behind Lucite screens to shield personnel from exposure to radioactivity.

MEASUREMENT OF RADIOACTIVITY IN NUCLEIC ACIDS

Radioactive isotopes are used as tracers to monitor the progress of many reactions used to synthesize DNA and RNA. To calculate the efficiency of such reactions, it is necessary to measure accurately the proportion of the radioactive precursor that has been incorporated into the desired product. This can be achieved by two methods: (1) differential precipitation of the nucleic acid products with trichloroacetic acid (TCA) and (2) differential adsorption of the products onto positively charged surfaces (e.g., DE-81 paper).

Precipitation of Nucleic Acids with Trichloroacetic Acid

1. Using a soft-lead pencil, label the appropriate number of Whatman GF/C glass-fiber filters (2.4-cm diameter). Impale each of the filters on a pin stuck into a polystyrene support.

2. Spot an accurately known volume (up to 5 μl) of the sample to be assayed on the center of each of two labeled filters.

   One of the filters is used to measure the total amount of radioactivity in the reaction (i.e., acid-soluble and acid-precipitable radioactivity). The other filter is used to measure only the acid-precipitable radioactivity. Under the conditions described, DNA and RNA molecules more than 50 nucleotides long will be precipitated on the surface of the filter.

3. Store the filters at room temperature until all of the fluid has evaporated. This process can be accelerated by using a heat lamp, although this is not usually necessary.

4. Using blunt-end forceps (e.g., Millipore forceps), transfer one of each pair of filters to a beaker containing 200–300 ml of ice-cold 5% TCA and 20 mM sodium pyrophosphate. Swirl the filters in the acid solution for 2 minutes, and then transfer them to a fresh beaker containing the same volume of the ice-cold 5% TCA/sodium pyrophosphate mixture. Repeat the washing two more times.

   During washing, the unincorporated nucleotide precursors are eluted from the filters and the radioactive nucleic acids are fixed to them.

   Commercially available, vacuum-driven filtration manifolds that hold up to 24 filters may also be used to wash the filters.

5. Transfer the washed filters to a beaker containing 70% ethanol and allow them to remain there briefly. Then dry them either at room temperature or under a heat lamp.

6. Insert each of the filters (washed and unwashed) into a scintillation vial. Measure the amount of radioactivity on each filter.

   ^{32}P can be detected on dry filters by Cerenkov counting (in the ^{3}H channel of a liquid scintillation counter). The efficiency with which Cerenkov radiation can be measured varies from instrument to instrument and also depends on the geometry of the scintillation vials and the amount of water remaining in the filters. With dry filters, the efficiency of Cerenkov counting is approximately 25% (one radioactive decay in four can be detected). Alternatively, ^{32}P can be measured with 100% efficiency by adding a few milliliters of toluene-based scintillation fluid to the dried filters and counting in the ^{32}P channel of the liquid scintillation counter.

To measure other isotopes (^{3}H, ^{14}C, ^{35}S, etc.), it is essential to use toluene-based scintillation fluid and the appropriate channel of a liquid scintillation counter. The efficiency of counting of these isotopes varies from counter to counter and should be determined for each instrument.

7. Compare the amount of radioactivity on the unwashed filter with the amount on the washed filter, and then calculate the proportion of the precursor that has been incorporated:

$$\frac{\text{cpm in washed filter}}{\text{cpm in unwashed filter}} = \text{proportion incorporated}$$

$$\text{proportion incorporated} \times \text{total wt.} = \text{total amt. of product}$$

$$\frac{\text{cpm incorporated}}{\text{total amt. of product}} = \text{specific activity}$$

Adsorption to DE-81 Filters

DE-81 filters are positively charged and will strongly adsorb and retain nucleic acids, including oligonucleotides that are too small to be precipitated efficiently with TCA. Unincorporated nucleotides stick less tightly to the filters and are removed by washing the filter extensively in sodium phosphate. The procedure is essentially identical to that described for precipitation of nucleic acids by TCA, except that the DE-81 filters are washed in 0.5 M Na_2HPO_4 (pH 7.0) instead of TCA/sodium pyrophosphate.

STANDARD MARKERS FOR GEL ELECTROPHORESIS

TABLE E.2 Sizes of Marker Fragments in Base Pairs

λ HindIII	λ HindIII-EcoRI	λ EcoRI	pUC18 Sau3AI	φX174 HaeIII
23,130	21,226	21,226	955	1,353
9,416	5,148	7,421	585	1,078
6,557	4,973	5,804	341	872
4,361	4,268	5,643	258	603
2,322	3,530	4,878	141	310
2,027	2,027	3,530	105	281
564	1,904		78	271
125	1,584		75	234
	1,375		46	194
	947		36	118
	831			72
	564			
	125			

AUTORADIOGRAPHY

Autoradiography produces permanent images on photographic film of the distribution of radioactive atoms on a two-dimensional surface. In molecular cloning, autoradiography is used for a variety of purposes, including the visualization of bands of radioactive nucleic acids in Southern and northern hybridizations, the identification of bacterial colonies or bacteriophage plaques that hybridize to radioactive probes, and the localization of bands of DNA in gels (e.g., in DNA sequencing). A diagram of the events that occur during exposure of photographic emulsion to radioactivity is shown in Figure E.2.

The two isotopes most commonly used for autoradiography are ^{35}S and ^{32}P, both of which emit β particles. However, the energies of these particles are very different: ^{35}S emits a particle with a maximum energy of 0.167 MeV that can penetrate film emulsion only to a depth of 0.25 mm (see Figure E.3). Although this is sufficient to allow the emitted β particles to interact productively with silver halide crystals in the emulsion, it is not enough to allow the particles to pass through barriers (e.g., Saran Wrap) that might be placed between the film and the source of the radiation. Thus, when establishing autoradiographs of ^{35}S-labeled material, it is essential that the film and the source of the radiation be directly apposed to one another. Gels should be as thin as possible and should be fixed and dried before autoradiographs are taken (see Chapter 6, page 6.45). Nitrocellulose filters and nylon membranes should be thoroughly dried, and care should be taken to ensure that the surface carrying the radioactivity is placed in contact with the film. (*Note*: Damp gels and membranes stick tightly to the film and usually cannot be removed).

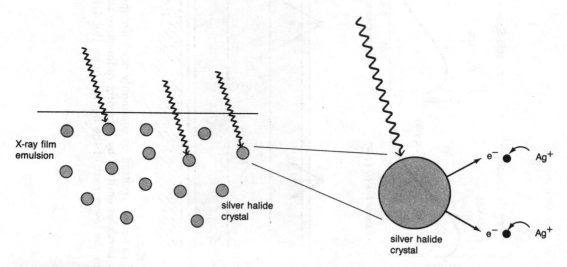

FIGURE E.2
Events leading to the formation of an autoradiographic image. The diagram shows that particles entering autoradiographic film cause ejection of electrons from silver halide crystals. These electrons attract positively charged silver ions, generating precipitates of silver atoms.

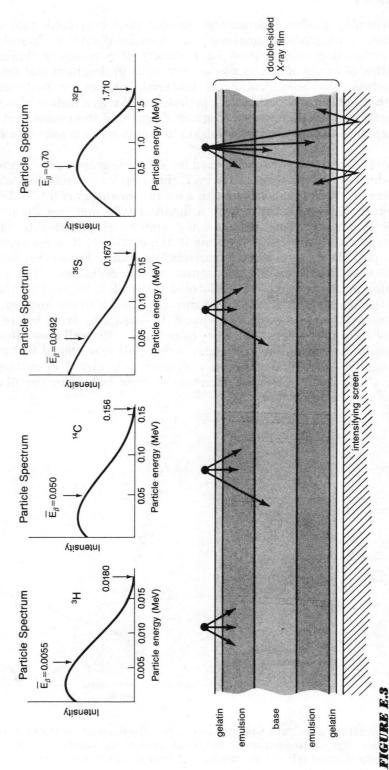

FIGURE E.3

Energy of radiation emitted by commonly used isotopes. The graphs in the upper part of the figure show the spectra of energies carried by particles emitted by decaying radioactive isotopes. In each case, the arrow marks the average energy per particle. The diagram in the lower part of the figure shows the depth to which commonly used isotopes penetrate autoradiographic film.

^{32}P, on the other hand, emits a particle with sufficient energy (1.71 MeV) to penetrate water or plastic to a depth of 6 mm and to pass completely through an X-ray film. Gels and filters therefore need not be completely dried (although the sharpness of the autoradiographic image is much improved if they are) and can be covered with Saran Wrap before they are exposed to the film. To increase the efficiency with which these strong β particles are detected, an intensifying screen may be placed behind the X-ray film. Radioactive particles that pass through the film hit the intensifying screen and cause it to emit photons that are captured by silver halide crystals in the emulsion. This leads to an approximately fivefold enhancement in the intensity of an autoradiographic image when the film is exposed at low temperature (−70°C). The best intensifying screens are du Pont Cronex Lightening Plus and Fuji Mach 2, both of which are made from calcium tungstate and emit blue light, to which X-ray films are most sensitive (Swantrom and Shank 1978). A further twofold enhancement can be obtained by preexposing the film to a short (≤ 1-msec) flash of light emitted by a stroboscope or a photographic flash unit. This activates the silver halide crystals and increases the probability that any single crystal exposed to light emitted from the intensifying screen will be reduced to silver metal during the developing process. The distance of the light source from the film during preexposure should be determined empirically as follows (Laskey and Mills 1975, 1977):

1. Cover the stroboscope or flash unit with an orange filter (Kodak, Wratten 22A). This reduces the amount of incident blue light.

2. Working in total darkness, place the film perpendicular to the light source and at least 50 cm away from it. This prevents uneven illumination. The film should be covered by a diffusing screen. If no suitable screen is available, use a piece of Whatman No. 1 filter paper.

3. Expose a series of test films to the light source for different lengths of time and then develop the films. Cut the films into pieces that will fit neatly into the cuvette holder of a spectrophotometer. Measure the absorbance at 545 nm of the exposed films against a blank consisting of film that was not preexposed. Choose an exposure time that causes the absorbance to increase by 0.15.

Preflashed film has another advantage: The intensity of the image on the film becomes proportional to the amount of radioactivity in the sample (Laskey 1980). The intensity of autoradiographic images on preexposed film can therefore be quantitated by microdensitometry and used to measure the amount of radioactivity in the original sample. In contrast, the silver halide crystals in film that is not preexposed to light are not fully activated and therefore respond in a sigmoidal fashion to increasing amounts of radioactivity. This can complicate quantitation of autoradiographic images. The best types of films for all types of autoradiography except fluorography are Kodak XAR-2 and Fuji RX. When preexposed, these films yield images whose absorbances are proportional to the intensity of the source of radioactivity over the range 0.1 to 1.0.

Fluorography

The intensity of autoradiographic images of weak β emitters such as ^{3}H, ^{14}C, and ^{35}S can be enhanced by chemicals that are fluorescent and emit many photons when they encounter a single quantum of radiation. Fluorography increases the sensitivity of detection of ^{14}C and ^{35}S about tenfold and permits detection of ^{3}H, which is otherwise invisible to conventional autoradiography. Fluorography is therefore particularly useful for the detection of radiolabeled proteins and nucleic acids in polyacrylamide gels. In the original methods (Bonner and Laskey 1974; Laskey and Mills 1975), aqueous gels containing the radioactive samples were equilibrated with dimethyl sulfoxide (DMSO), impregnated with the scintillant PPO (2,5-diphenyloxazole), soaked in water to remove the DMSO, dried, and exposed to X-ray film at $-70°C$. These procedures were costly, tedious (requiring at least 5 hours of work), and irreproducible in inexperienced hands. The most frequent cause of difficulty was the failure to remove DMSO: Complete removal is essential to avoid sticky gels after drying.

PPO has now been largely replaced as a scintillant by sodium salicylate (Chamberlain 1979). In this method, the gel is soaked in 10 volumes of 1 M sodium salicylate (pH 6.0) for 30 minutes and then dried and exposed to X-ray film. The level of enhancement is approximately equal to that obtained with organic scintillants, although the bands are slightly more diffuse.

Caution: Salicylate can elicit allergic reactions and is readily absorbed through the skin. Gloves should be worn while handling gels in salicylate solutions.

Many commercial preparations of enhancing fluors are also available in both liquid and spray-on form (e.g., En^{3}Hance, New England Nuclear). The formulations of these fluors differ from manufacturer to manufacturer, and it is important to follow carefully the instructions that accompany each scintillant.

The best type of X-ray film for fluorography is Kodak SB, which has a high sensitivity for photons of the wavelength emitted by the fluor.

Sensitivity of Different Autoradiographic Methods

Table E.3 shows the sensitivity of different autoradiographic methods for the detection of radioisotopes. The amounts of radioactivity shown in the table are those required to obtain a detectable image ($A_{545} = 0.02$) on preflashed film that is exposed to the sample for 24 hours. Much longer exposure times are necessary to obtain publishable images!

TABLE E.3 Sensitivity of Autoradiographic Methods for Detection of Radioisotopes

Isotope	Method	Sensitivity (dpm/mm^2)
^{35}S	no enhancement	15–25
^{35}S	fluorography	2
^{32}P	direct	2–5
^{32}P	intensifying screen	0.5
^{14}C	fluorography	2
^{125}I	intensifying screen	1–2
^{3}H	fluorography	10–20

Setting up Autoradiographs

1. Prepare gels for autoradiography in one of the following ways:

   a. SDS-polyacrylamide gels containing ^{35}S, ^{14}C, or ^{3}H should be fixed as described in Chapter 6, page 6.45, and dried onto Whatman 3MM paper using a commercial gel dryer.

   b. Sequencing gels containing ^{35}S should be fixed as described in Chapter 13, page 13.56, and dried onto Whatman 3MM paper using a commercial gel dryer.

   c. For maximal sensitivity and resolution, polyacrylamide gels containing ^{32}P should be fixed, dried, and mounted on backing paper. However, satisfactory images of wet, unfixed gels can also be obtained as long as the gels are sealed in a plastic bag or wrapped in Saran Wrap before they are exposed to the film (see Chapter 6, page 6.21).

   d. ^{32}P-labeled nucleic acids in agarose gels can be detected by exposing the wet gel (wrapped in Saran Wrap) to X-ray film. However, for maximal sensitivity and resolution, the radiolabeled nucleic acids should be transferred to a solid support (nitrocellulose filter or nylon membrane) as described in Chapters 7 and 9. The solid supports are then dried and covered with Saran Wrap (to prevent contamination of intensifying screens and film holders).

2. Place pieces of tape marked with radioactive ink around the edge of the sample on the backing paper or Saran Wrap. Cover the pieces of tape with Scotch Tape. This prevents contamination of the film holder or intensifying screen with the radioactive ink.

   Radioactive ink is made by mixing a small amount of ^{32}P with waterproof black drawing ink. We find it convenient to make the ink in three grades: very hot (> 2000 cps on a hand-held minimonitor), hot (> 500 cps on a hand-held minimonitor), and cool (> 50 cps on a hand-held minimonitor). Use a fiber-tip pen to apply ink of the desired hotness to the pieces of tape. Attach radioactive-warning tape to the pen, and store it in an appropriate place.

3. In a darkroom, place the sample in a light-tight X-ray film holder and cover it with a sheet of X-ray film. If preflashed film is used, the preexposed side should face the sample, unless an intensifying screen is used, in which case the preexposed side should face the intensifying screen. Acceptable orientations of sample, film, and intensifying screens are shown in Figure E.4.

4. Expose the film for an appropriate length of time (see Table E.3). When intensifying screens are used or when fluorography is employed, the film must be exposed at $-70°C$. The low temperature stabilizes the silver atoms and ions that form the latent image of the radioactive source.

5. Remove the film holder from storage (use gloves to handle holders stored at −70°C). In a darkroom, remove the film as quickly as possible and develop it immediately. This prevents the formation of excessive condensation on films exposed at −70°C.

If it is necessary to obtain another autoradiograph, apply another film immediately and return the film holder and screens to the freezer as rapidly as possible. If condensation forms before you have time to apply a new film, allow the sample and screens to reach room temperature and wipe away all condensation before applying a new film.

6. X-ray film may be developed either in an automatic X-ray film processor or by hand as follows:

X-ray developer	5 minutes
3% acetic acid stop bath or water bath	1 minute
rapid fixer	5 minutes
running water	15 minutes

The temperatures of all solutions should be 18–20°C.

7. Use the images of the radioactive markers to align the autoradiograph with the sample.

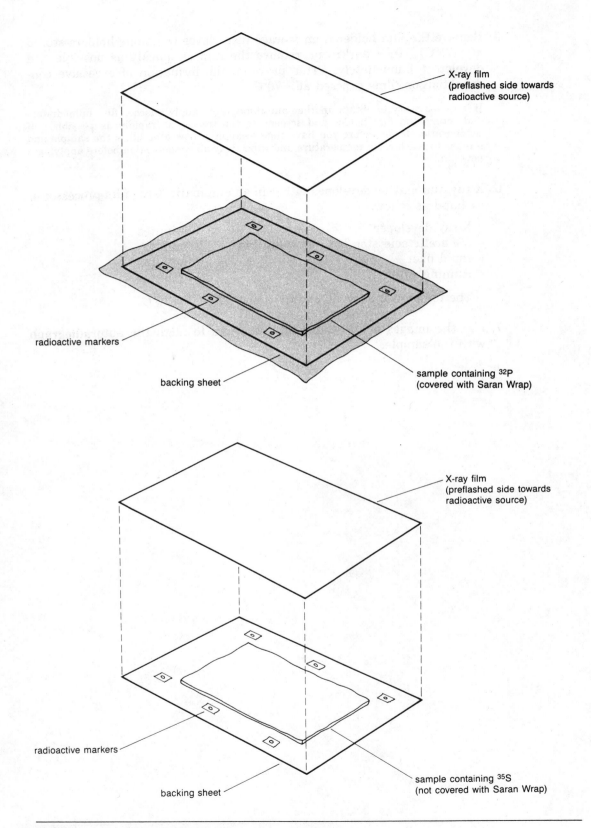

X-ray film
(preflashed side towards
radioactive source)

radioactive markers

backing sheet

sample containing ^{32}P
(covered with Saran Wrap)

X-ray film
(preflashed side towards
radioactive source)

radioactive markers

backing sheet

sample containing ^{35}S
(not covered with Saran Wrap)

E.28   *Commonly Used Techniques in Molecular Cloning*

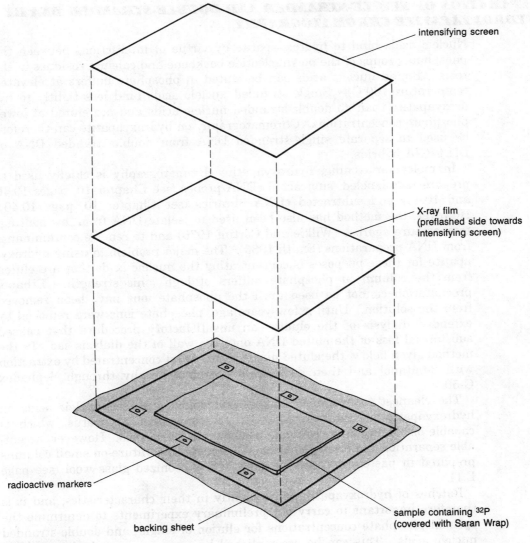

intensifying screen

X-ray film
(preflashed side towards
intensifying screen)

radioactive markers

backing sheet

sample containing 32P
(covered with Saran Wrap)

FIGURE E.4
(*Top left*) Autoradiography with ^{32}P using preflashed film and no intensifying screen. (*Bottom left*) Autoradiography with ^{35}S or ^{14}C and preflashed film. (*Above*) Autoradiography with ^{32}P using preflashed film and an intensifying screen.

SEPARATION OF SINGLE-STRANDED AND DOUBLE-STRANDED DNA BY HYDROXYAPATITE CHROMATOGRAPHY

Nucleic acids bind to hydroxyapatite by virtue of interactions between the phosphate groups of the polynucleotide backbone and calcium residues in the resin. Bound nucleic acids can be eluted in phosphate buffers at elevated temperature (60°C). Single-stranded nucleic acids bind less tightly to hydroxyapatite than do double-stranded nucleic acids and are eluted at lower phosphate concentrations. Chromatography on hydroxyapatite can therefore be used to separate single-stranded DNA from double-stranded DNA or DNA:RNA hybrids.

In molecular cloning, hydroxyapatite chromatography is chiefly used to prepare radiolabeled, subtracted cDNA probes (see Chapter 10, pages 10.46 and 10.48) and subtracted cDNA libraries (see Chapter 10, page 10.40). However, the method has also been used to isolate DNA from low-melting-temperature agarose (Wilkie and Cortini 1976) and to remove contaminants from DNA preparations (Smith 1980). The major problem in using hydroxyapatite for these purposes is concentrating the nucleic acids that are eluted from the column in phosphate buffers of high ionic strength. Ethanol precipitation cannot be used until the phosphate ions have been removed from the solution. Until a few years ago, phosphate ions were removed by extended dialysis of the eluate—an unsatisfactory procedure that caused substantial loss of the eluted DNA onto the wall of the dialysis sac. In the method given below, the eluted nucleic acid is first concentrated by extraction with 2-butanol and then desalted by chromatography through Sephadex G-50.

The cleanest separation of single- and double-stranded nucleic acids on hydroxyapatite is obtained at 60°C. A convenient apparatus, which is capable of holding many columns, is shown in Figure E.5. However, acceptable separation can often be obtained at room temperature on small columns prepared in pasteur pipettes plugged with siliconized glass wool (see page E.1).

Batches of hydroxyapatite vary slightly in their characteristics, and it is therefore important to carry out preliminary experiments to determine the optimal phosphate concentrations for elution of single- and double-stranded nucleic acids. This can be accomplished by setting up two hydroxyapatite columns as described below. One of the columns is loaded with a small amount ($\sim 10^5$ cpm) of ^{32}P-labeled DNA that has been denatured by boiling for 10 minutes in TE (pH 7.6). The other column receives an equal amount ($\sim 10^5$ cpm) of ^{32}P-labeled native DNA. Each of the columns is washed with a series of buffers containing increasing concentrations of sodium phosphate (0.01, 0.12, 0.16, 0.20, 0.24, 0.28, 0.32, 0.36, and 0.40 M). The amount of radioactivity eluting at each phosphate concentration is then measured in a liquid scintillation counter (either by Cerenkov counting or in a water-miscible fluor). Usually, single-stranded DNA elutes in 0.14–0.16 M sodium phosphate (pH 6.8), whereas double-stranded DNA is not removed from the column until the phosphate concentration exceeds 0.36 M. In the protocol that follows, SS buffer contains the phosphate concentration that is optimal for elution of single-stranded DNA; DS buffer contains the concentration that is optimal for elution of double-stranded DNA.

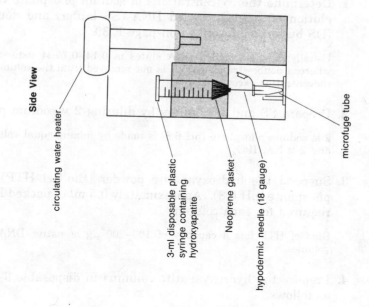

Side View

circulating water heater

3-ml disposable plastic
syringe containing
hydroxyapatite

Neoprene gasket

hypodermic needle (18 gauge)

microfuge tube

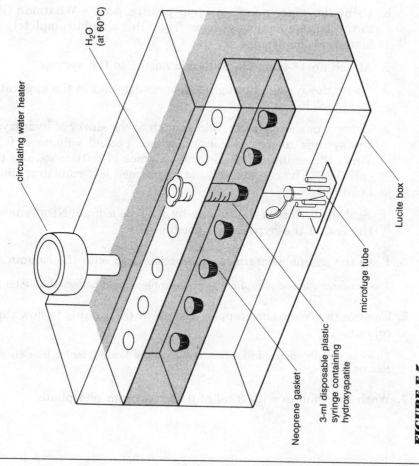

circulating water heater

H₂O
(at 60°C)

3-ml disposable plastic
syringe containing
hydroxyapatite

Neoprene gasket

microfuge tube

Lucite box

FIGURE E.5
Apparatus for hydroxyapatite chromatography.

1. Determine the concentrations of sodium phosphate that are optimal for elution of single-stranded DNA (SS buffer) and double-stranded DNA (DS buffer) as described on page E.30.

   Usually, single-stranded DNA elutes in 0.14–0.16 M sodium phosphate (pH 6.8), whereas doube-stranded DNA is not removed from the column until the phosphate concentration exceeds 0.36 M.

2. Prepare SS and DS buffers by diluting 2 M sodium phosphate (pH 6.8).

   2 M sodium phosphate (pH 6.8) is made by mixing equal volumes of 2 M NaH_2PO_4 and 2 M Na_2HPO_4.

3. Suspend the hydroxyapatite powder (Bio-Gel HTP) in 0.01 M sodium phosphate (pH 6.8). Approximately 0.5 ml of packed Bio-Gel HTP will be required for each column.

   Bio-Gel HTP has a capacity of 100–200 μg of native DNA per milliliter of bed volume.

4. Prepare the hydroxyapatite columns in disposable 3-ml plastic syringes as follows:

   a. Remove the barrel from the syringe.

   b. Using the wide end of a pasteur pipette, push a Whatman GF/C filter to the bottom of the syringe. The filter should completely cover the bottom of the syringe.

   c. Attach an 18-gauge hypodermic needle to the syringe.

   d. Insert the syringe through a Neoprene gasket in the apparatus shown in Figure E.5.

   e. Using a pasteur pipette, add enough of the slurry of hydroxyapatite to the syringe to form a column whose packed volume is 0.5–1.0 ml. Wash the column with several volumes of 0.01 M sodium phosphate (pH 6.8). The column will not be harmed if it runs dry; simply rewet before use.

   f. Seal the bottom of the column by sticking a small Neoprene stopper on the end of the hypodermic needle.

5. Load the sample containing the nucleic acid onto the column.

   The concentration of phosphate in the sample should be less than 0.08 M.

6. Remove the Neoprene stopper, and allow the sample to flow through the column.

   There is usually no need to collect and save the loading buffer that elutes from the column.

7. Wash the column with 3 ml of 0.01 M sodium phosphate.

8. Seal the bottom of the column with a Neoprene stopper, and then add 1 column volume of SS buffer preheated to 60°C.

9. After 5 minutes, remove the Neoprene stopper and collect the eluate in microfuge tubes. No more than 0.5 ml should be collected in any one microfuge tube. Repeat steps 8 and 9 two more times.

10. Seal the bottom of the column with a Neoprene stopper, and then add 1 column volume of DS buffer preheated to 60°C.

11. After 5 minutes, remove the Neoprene stopper and collect the eluate in microfuge tubes. No more than 0.5 ml should be collected in any one microfuge tube. Repeat steps 10 and 11 two more times.

12. Allow the eluates to cool to room temperature. DNA can then be extracted as follows:

   a. Add an equal volume of 2-butanol to each of the tubes containing the desired nucleic acids.

   b. Mix the two phases by vortexing, and centrifuge the mixture at 12,000g for 20 seconds at room temperature in a microfuge.

   c. Discard the upper (organic) phase.

   d. Repeat the extraction with 2-butanol until the volume of the aqueous phase is 100–125 μl.

   e. Remove salts from the DNA by chromatography on, or centrifugation through, a small column of Sephadex G-50 equilibrated in TE (pH 8.0).

   f. Recover the DNA by precipitation with 2 volumes of ethanol at 0°C.

Note

In molecular cloning, nucleic acids fractionated by hydroxyapatite are usually radiolabeled, and the tubes containing the desired fractions can be easily identified by Cerenkov counting in a liquid scintillation counter.

GEL-FILTRATION CHROMATOGRAPHY

This technique, which employs gel filtration to separate high-molecular-weight DNA from smaller molecules, is used most often to separate unincorporated labeled dNTPs from DNA that has been labeled by nick translation or by filling of recessed 3' termini. However, it is also used at several stages during the synthesis of double-stranded cDNA, during addition of linkers to blunt-ended DNA, and, in general, whenever it is necessary to change the composition of the buffer in which DNA is dissolved.

Two methods are available: conventional column chromatography, which is used when it is necessary to collect fractions that contain components of different sizes; and centrifugation through gel matrices packed in disposable syringes, which is a rapid method used to free DNA from smaller molecules. The two most commonly used gel matrices are Sephadex and Bio-Gel, both of which are available in several porosities. Sephadex G-50 and Bio-Gel P-60 are ideal for purifying DNA larger than 80 nucleotides in length. Smaller molecules are retained in the pores of gel while the DNA is excluded and passes directly through the column. Bio-Gel P-2 can be used to separate oligonucleotides from phosphate ions or dNTPs. Bio-Gel is supplied in the form of a gel and needs only to be equilibrated in running buffer before use. Sephadex is supplied as a powder that must be hydrated before use.

Preparation of Sephadex

1. Slowly add Sephadex of the desired grade to distilled, sterile water in a 500-ml beaker or bottle (10 g of Sephadex G-50 [medium] yields 160 ml of slurry). Wash the swollen resin with distilled, sterile water several times to remove soluble dextran, which can create problems by precipitating during ethanol precipitation.

2. Equilibrate the resin in TE (pH 7.6), autoclave at 10 lb/sq. in. for 15 minutes, and store at room temperature.

Column Chromatography

1. Sephadex and Bio-Gel columns can be conveniently prepared in disposable 5-ml borosilicate glass pipettes or pasteur pipettes plugged with a small amount of sterile glass wool. Use a long, narrow pipette (e.g., a disposable 1-ml plastic pipette) to push the wool to the bottom of the glass or pasteur pipette.

2. Using a pasteur pipette, fill the column with a slurry of the Sephadex or Bio-Gel, taking care to avoid producing air bubbles. There is no need to close the bottom of the column. Keep adding gel until it packs to a level 1 cm below the top of the column. Wash the gel with several volumes of 1 × TEN buffer (pH 8.0).

---

1 × TEN buffer

10 mM Tris·Cl (pH 8.0)
1 mM EDTA (pH 8.0)
100 mM NaCl

---

3. Apply the DNA sample (in a volume of 200 μl or less) to the top of the gel. Wash out the sample tube with approximately 100 μl of 1 × TEN buffer, and load the washing on the column as soon as the DNA sample has entered the gel. When the washing has entered the gel, immediately fill the column with 1 × TEN buffer.

   Caution: Columns used to separate radiolabeled DNA from radioactive precursors should be run behind Lucite screens to shield personnel from exposure to radioactivity.

4. Immediately start to collect fractions ($\sim$ 200 μl) in microfuge tubes. If the DNA is labeled with ^{32}P, measure the radioactivity in each of the tubes by using either a hand-held minimonitor or by Cerenkov counting in a liquid scintillation counter. Add more 1 × TEN buffer to the top of the gel as required from time to time.

   The DNA will be excluded from the gel and will be found in the void volume (usually $\sim$ 30% of the total column volume). The leading peak of radioactivity therefore consists of nucleotides incorporated into DNA, and the trailing peak consists of unincorporated [^{32}P]dNTPs.

5. Pool the radioactive fractions in the leading peak and store at $-20°$C.

Note

Instead of collecting individual fractions, it is possible with practice to follow the progress of the incorporated and unincorporated [^{32}P]dNTPs down the column using a hand-held minimonitor. The leading peak should be collected into a sterile polypropylene tube as it elutes from the column. The bottom of the column should then be clamped off and the buffer reservoir disconnected. The column should be discarded in the radioactive waste.

SPUN-COLUMN CHROMATOGRAPHY

This method is used to separate DNA, which passes through the gel-filtration matrix, from lower-molecular-weight substances that are retained on the column. Spun-column chromatography is particularly useful when separating labeled DNA from radioactive precursors (e.g., in nick-translation reactions). However, it is also used extensively for other purposes, for example, to change the buffer in which small amounts of DNA are dissolved or to free crude preparations of minipreparations of plasmid or bacteriophage λ DNA from inhibitors that prevent cleavage by restriction enzymes. Several samples of DNA can be handled simultaneously. In this respect, spun-column chromatography is much superior to conventional column chromatography.

1. Plug the bottom of a 1-ml disposable syringe with a small amount of sterile glass wool. This is best accomplished by using the barrel of the syringe to tamp the glass wool in place.

2. Fill the syringe with Sephadex G-50 or Bio-Gel P-60, equilibrated in $1 \times$ TEN buffer (pH 8.0). Start the buffer flowing by tapping the side of the syringe barrel several times. Keep adding more resin until the syringe is completely full.

> **$1 \times$ TEN buffer**
>
> 10 mM Tris·Cl (pH 8.0)
> 1 mM EDTA (pH 8.0)
> 100 mM NaCl

Both Sephadex G-50 and Bio-Gel P-60 work equally well to separate unincorporated dNTPs from nucleic acids or oligonucleotides 16 or more nucleotides in length.

3. Insert the syringe into a 15-ml disposable plastic tube. Centrifuge at 1600g for 4 minutes at room temperature in a swinging-bucket rotor in a bench centrifuge. Do not be alarmed by the appearance of the column. The resin packs down and becomes partially dehydrated during centrifugation. Continue to add more resin and recentrifuge until the volume of the packed column is approximately 0.9 ml and remains unchanged after centrifugation.

4. Add 0.1 ml of $1 \times$ TEN buffer to the columns, and recentrifuge as in step 3.

5. Repeat step 4 twice more.

Spun columns may be stored at this stage if desired. Several spun columns can be prepared simultaneously and stored at 4°C for periods of a month or more before being used. Fill the syringes with $1 \times$ TEN buffer and wrap Parafilm around them to prevent evaporation. Store the columns upright at 4°C. Spun columns stored in this way should be washed once with sterile $1 \times$ TEN buffer as described in step 4 just before they are used.

6. Apply the DNA sample to the column in a total volume of 0.1 ml (use 1 × TEN buffer to make up the volume). Place the spun column in a fresh disposable tube containing a decapped microfuge tube (see Figure E.6).

7. Recentrifuge as in step 3, collecting the effluent from the bottom of the syringe (∼ 100 μl) into the decapped microfuge tube.

8. Remove the syringe, which will contain unincorporated radiolabeled dNTPs or other small components. Using forceps, carefully recover the decapped microfuge tube, which contains the eluted DNA, and transfer its contents to a capped, labeled microfuge tube.

A rough estimate of the proportion of radioactivity that has been incorporated into nucleic acid may be obtained by holding the syringe and the eluted DNA to a hand-held minimonitor.

9. If the syringe is radioactive, carefully discard it in the radioactive waste. Store the eluted DNA at −20°C until needed.

Note

Not all resins are suitable for spun-column centrifugation: DEAE-Sephacel forms an impermeable lump during centrifugation and the larger grades of Sephadex (G-100 and up) cannot be used because the beads are crushed by centrifugation. If a coarser-sieving resin is required, use Sepharose CL-4B.

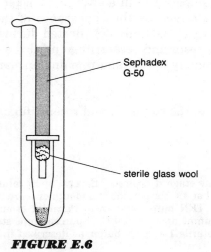

Sephadex
G-50

sterile glass wool

FIGURE E.6

PREPARATION OF DIALYSIS TUBING

1. Cut the tubing into pieces of convenient length (10–20 cm).

2. Boil for 10 minutes in a large volume of 2% (w/v) sodium bicarbonate and 1 mM EDTA (pH 8.0).

3. Rinse the tubing thoroughly in distilled water.

4. Boil for 10 minutes in 1 mM EDTA (pH 8.0).

5. Allow the tubing to cool, and then store it at 4°C. Be sure that the tubing is always submerged. From this point onward, always handle the tubing with gloves.

6. Before use, wash the tubing inside and out with distilled water.

Note

Instead of boiling for 10 minutes in 1 mM EDTA (pH 8.0) (step 4), the tubing can be autoclaved at 20 lb/sq. in. for 10 minutes on liquid cycle in a loosely capped jar filled with water.

Appendix F: Subcloning

Subcloning fragments of DNA from one type of vector to another, for example, from a recombinant bacteriophage λ to a plasmid, or from one type of plasmid to another is one of the most frequently used procedures in molecular cloning. Subcloning is a simple matter when the restriction sites at the termini of the target fragment are identical to, or compatible with, those of the new vector. The target fragment can then be ligated to the vector without enzymatic manipulation of either piece of DNA. In many cases, however, the termini of the target fragment and vector are incompatible. It is then necessary to convert one or both termini of the DNAs into forms that can be ligated easily. There are four ways that such conversion is commonly accomplished:

- *Partial filling of incompatible recessed 3' termini with the Klenow fragment of* E. coli *DNA polymerase I* (Zabarovsky and Allikmets 1986). As discussed in Chapter 5, this frequently generates cohesive termini from recessed 3' termini that are otherwise incompatible.

- *Complete filling of incompatible recessed 3' termini with the Klenow fragment of* E. coli *DNA polymerase I.* This generates blunt-ended DNA molecules that can be ligated to any other blunt-ended DNA.

- *Removal of protruding 3' termini.* This can be accomplished in a number of ways, including treating the DNA with nuclease S1, mung-bean nuclease, or the Klenow fragment of *E. coli* DNA polymerase I. However, the enzyme of choice for removal of protruding 3' termini is bacteriophage T4 DNA polymerase, because of its exceptionally strong 3' → 5' exonuclease activity. In the presence of high concentrations of all four dNTPs, net removal of nucleotides from the 3' terminus ceases when the enzyme reaches the double-stranded region of the DNA molecule.

- *Addition of synthetic linkers to blunt-ended DNA.* Synthetic linkers are an equimolar mixture of self-complementary, chemically synthesized oligomers that can form blunt-ended duplexes containing one or more recognition sites for restriction enzymes. Ligation of linkers to blunt-ended DNA therefore results in addition of one or more restriction sites that can be used in subcloning. The variety of synthetic linkers is now so large that it is almost always possible to tailor the termini of the target DNA and vector into forms that are ideal for the tasks at hand.

FILLING RECESSED 3′ TERMINI

Recessed 3′ termini can be filled by the polymerase activity of the Klenow fragment of *E. coli* DNA polymerase I in the presence of the appropriate dNTPs. Which of the four dNTPs are added to the reaction depends on (1) the sequence of the protuding 5′ termini at the end(s) of the DNA and (2) whether partial or complete filling is required. For example, to fill recessed 3′ termini created by cleavage of DNA by *Eco*RI, only dATP and dTTP need to be present in the reaction:

$$\text{———}G_{OH}\,^{3'} \quad \xrightarrow[\substack{\text{Klenow fragment of} \\ \textit{E. coli}\ \text{DNA polymerase I}}]{\substack{\text{dATP} \\ \text{dTTP}}} \quad \text{———}G\,A\,A\,T\,T_{OH}\,^{3'}$$
$$\text{———}C\,T\,T\,A\,A_{p\,5'} \qquad\qquad\qquad \text{———}C\,T\,T\,A\,A_{p\,5'}$$

On the other hand, all four dNTPs are required to fill recessed termini created by *Hin*dIII:

$$\text{———}A_{OH}\,^{3'} \quad \xrightarrow[\substack{\text{Klenow fragment of} \\ \textit{E. coli}\ \text{DNA polymerase I}}]{\substack{\text{dATP} \\ \text{dTTP} \\ \text{dCTP} \\ \text{dGTP}}} \quad \text{———}A\,A\,G\,C\,T_{OH}\,^{3'}$$
$$\text{———}T\,T\,C\,G\,A_{p\,5'} \qquad\qquad\qquad \text{———}T\,T\,C\,G\,A_{p\,5'}$$

Partial filling of recessed termini created by *Sau*3AI requires the presence of dATP and dGTP. The newly created, shorter protruding terminus is complementary to a partially filled terminus created by cleavage with *Xho*I (Zabarovsky and Allikmets 1986) (see Chapter 5):

$$\text{———}OH\,^{3'} \quad \xrightarrow[\substack{\text{Klenow fragment of} \\ \textit{E. coli}\ \text{DNA polymerase I}}]{\substack{\text{dATP} \\ \text{dGTP}}} \quad \text{———}G\,A_{OH}\,^{3'}$$
$$\text{———}C\,T\,A\,G_{p\,5'} \qquad\qquad\qquad \text{———}C\,T\,A\,G_{p\,5'}$$

1. In a 20-μl reaction, digest 0.2–5 μg of DNA with the appropriate restriction enzyme(s).

2. When digestion is complete, add 1 μl of a solution containing each of the desired dNTPs at a concentration of 1 mM.

3. Add 1 unit of the Klenow fragment of *E. coli* DNA polymerase I for each microgram of DNA in the reaction. Incubate the reaction for 15 minutes at room temperature.

   The Klenow fragment of *E. coli* DNA polymerase I works well in virtually all buffers used for digestion of DNA with restriction enzymes. There is no need to purify the DNA prior to filling recessed 3′ termini created by restriction enzymes.

   DNA fragments that have been purified by gel electrophoresis before filling of recessed 3′ termini should be redissolved in TE (pH 7.6). MgCl$_2$ should then be added to a final concentration of 5 mM before the appropriate dNTPs and the Klenow fragment of *E. coli* DNA polymerase I are added.

4. Inactivate the Klenow fragment of *E. coli* DNA polymerase I and the restriction enzyme(s) present in the reaction by one of the following methods:

- Heat to 75°C for 10 minutes.

  Not all restriction enzymes are completely inactivated by this procedure. Check the manufacturer's specifications that are supplied with the enzyme.

- Add an equal volume of TE (pH 7.6) and extract the solution with phenol:chloroform. Collect the DNA by precipitation with 2 volumes of ethanol.

Notes

i. If desired, the DNA can be separated from unincorporated dNTPs by chromatography on, or centrifugation through, small columns of Sephadex G-50 (see Appendix E). This is not necessary when the filled fragment of DNA is to be used in ligation reactions. Bacteriophage T4 DNA ligase is not inhibited by the presence of dNTPs and works adequately in virtually all buffers used for digestion of DNA with restriction enzymes.

ii. For a full description of the method used to fill recessed 3′ termini of genomic DNA in preparation for cloning, see Chapter 9, page 9.29.

Both the Klenow fragment of *E. coli* DNA polymerase I and bacteriophage T4 DNA polymerase carry a $3' \rightarrow 5'$ exonuclease activity that can be used to remove protruding nucleotides from the 3' termini of DNA. Both enzymes lack a $5' \rightarrow 3'$ exonuclease activity and carry a $5' \rightarrow 3'$ polymerizing activity. Despite its higher cost, the bacteriophage enzyme is generally preferred for removal of protruding 3' termini because its 3' exonuclease activity is more than 200-fold more active than that of the Klenow enzyme.

Bacteriophage T4 DNA polymerase has a pH optimum of 8–9 and displays about 50% of maximal activity in all buffers that are commonly used for digestion of DNA with restriction enzymes. The exonuclease reaction can therefore be carried out by adding the polymerase directly to the digestion mixture together with high concentrations of the four dNTPs. The enzyme removes protruding 3' nucleotides until it reaches the double-stranded region of the DNA molecule, at which point further removal is balanced by incorporation of dNTPs.

1. In a 20-μl reaction, digest 0.2–5 μg of DNA with the appropriate restriction enzyme(s).

2. When digestion is complete, add 1 μl of a solution containing each of the desired dNTPs at a concentration of 2 mM.

3. Add 1–2 units of bacteriophage T4 DNA polymerase for each microgram of DNA in the reaction. Incubate the reaction for 15 minutes at 12°C.

   DNA fragments that have been purified by gel electrophoresis before filling of recessed 3' termini should be redissolved in TE (pH 8.0). $MgCl_2$ should then be added to a final concentration of 5 mM before the appropriate dNTPs and the bacteriophage T4 DNA polymerase are added as described above.

   At 37°C, the turnover number of the 3' exonucleolytic activity of bacteriophage T4 DNA polymerase is some threefold higher than that of its polymerizing activity (Huang and Lehman 1972). At lower temperature, there is a smaller differential between the two activities, and exonucleolytic digestion is therefore carried out at 12°C.

4. Inactivate the bacteriophage T4 DNA polymerase and restriction enzyme(s) present in the reaction by one of the following methods:

   • Heat to 75°C for 10 minutes.

     Not all restriction enzymes are completely inactivated by this procedure. Check the manufacturer's specifications that are supplied with the enzyme.

   • Add an equal volume of TE (pH 7.6) and extract the solution with phenol:chloroform. Collect the DNA by precipitation with 2 volumes of ethanol.

Note

If desired, the DNA can be separated from unincorporated dNTPs by chromatography on, or centrifugation through, small columns of Sephadex

G-50 (see Appendix E). This is not necessary when the filled fragment of DNA is to be used in ligation reactions. Bacteriophage T4 DNA ligase is not significantly inhibited by the presence of dNTPs and works adequately in virtually all buffers used for digestion of DNA with restriction enzymes.

RAPID CLONING IN PLASMID VECTORS

The slowest step in cloning in plasmids is the electrophoretic purification of the desired restriction fragment of foreign DNA and the appropriate segment of plasmid DNA. In the protocol given below (adapted from Struhl 1985 by S. Michaelis, pers. comm.), ligation of plasmid and foreign DNAs is carried out directly in the melted slabs of agarose recovered from the gel used for purification. The method works for both blunt-end ligation and ligation of cohesive termini, although it requires a large amount of ligase and its efficiency is about an order of magnitude lower than the standard procedure.

1. Digest the foreign DNA with the appropriate restriction enzyme(s). The amount of foreign DNA digested should be sufficient to yield approximately 0.2 μg of the target fragment. The digestion should be carried out in a volume of 20 μl or less. In a separate tube, digest 0.5 μg of the vector DNA with the appropriate restriction enzyme(s) in a total reaction volume of 20 μl or less.

   If the vector DNA carries identical cohesive termini, it should be treated with phosphatase as follows: When digestion with the restriction enzyme(s) is complete, add 2.5 μl of 100 mM Tris·Cl (pH 8.3), 10 mM $ZnCl_2$. Add 0.25 unit of calf intestinal alkaline phosphatase and incubate for 30 minutes at 37°C.

2. Separate the desired fragments by electrophoresis on an agarose gel. The gel must be cast with low-melting-temperature agarose and it must be poured and run in 1 × TAE electrophoresis buffer (see Appendix B) containing ethidium bromide (0.5 μg/ml) rather than the conventional 0.5× TBE.

   Caution: Ethidium bromide is a powerful mutagen and is moderately toxic. Gloves should be worn when working with solutions that contain this dye. After use, these solutions should be decontaminated by one of the methods described in Appendix E.

3. Examine the gel by long-wavelength ultraviolet illumination. From the relative fluorescent intensities of the desired bands, estimate the amounts of DNA that they contain (see Appendix E). Using a razor blade, cut out the desired bands in the smallest possible volume of agarose (usually 40–50 μl). Place the excised slices of gel in separate, labeled microfuge tubes.

   Caution: Ultraviolet radiation is dangerous, particularly to the eyes. To minimize exposure, make sure that the ultraviolet light source is adequately shielded and wear protective goggles or a full safety mask that efficiently blocks ultraviolet light.

4. Heat the tubes to 70°C for 10–15 minutes to melt the agarose.

5. Combine aliquots of the melted gel slices in a fresh tube prewarmed to 37°C. The final volume of the combined aliquots should be 10 μl or less, and the molar ratio of foreign DNA:plasmid vector should be approximately 2:1.

In separate tubes, set up two additional ligations as controls, one containing the plasmid vector alone and the other containing only the fragment of foreign DNA.

6. Incubate the three tubes for 5–10 minutes at 37°C, and then add to each tube 10 μl of ice-cold 2× bacteriophage T4 DNA ligase mixture. Mix the contents of the tubes well before the agarose hardens. Incubate the reactions for 12–16 hours at 16°C.

   2× Bacteriophage T4 DNA ligase mixture is prepared as follows:

1 M Tris · Cl (pH 7.6)	1.0 μl
100 mM MgCl$_2$	1.0 μl
200 mM dithiothreitol	1.0 μl
10 mM ATP	1.0 μl
H$_2$O	5.5 μl
bacteriophage T4 DNA ligase	1 Weiss unit

   Mix the components in a tube stored in an ice bath.

   For definition of Weiss unit, see page F.9.

7. Towards the end of the ligation, remove from storage at −70°C three tubes containing 200 μl of frozen competent *E. coli* each (see Chapter 1, page 1.80). As soon as the cells have thawed, place them in an ice bath. Immediately proceed to step 8.

8. Remelt the agarose in the ligation mixtures by heating them to 70°C for 10–15 minutes.

9. Immediately add 5 μl of one of the ligation mixtures to 200 μl of competent *E. coli*. Mix the contents of the tube quickly by gentle shaking. Repeat this procedure with 5 μl taken from each of the remaining ligation mixtures. Store the transformation mixtures on ice for 30 minutes.

10. Proceed with the remainder of the transformation protocol (see Chapter 1, page 1.80, step 13 onward).

ADDITION OF LINKERS TO BLUNT-ENDED DNA

Subcloning with synthetic linkers involves two ligation reactions. In the first reaction, blunt-ended double-stranded linkers are attached to the DNA of interest (which must also be blunt-ended). Because the synthetic linkers are very small (8–12 bp), it is relatively easy to achieve the high concentration of termini required for blunt-end ligation of DNA. In fact, the first ligation reaction is set up so that the concentration of linker termini (4–20 μM) is two to three orders of magnitude greater than the concentration of termini of the fragment of interest. The reaction is therefore "driven" by the linkers, which polymerize onto the blunt-ended termini of the target fragment.

After the synthetic linkers are attached, the ligase is destroyed by heat and the DNA is cleaved by the appropriate restriction enzyme in order to generate cohesive termini. The remnants of the linkers are then removed by gel electrophoresis or chromatography on Sepharose CL-4B and, in a second ligation reaction, the target DNA is joined to another fragment of interest that carries compatible termini.

Synthetic linkers can be purchased in two forms that carry either a phosphate group or a hydroxyl group on their 5′ termini. Only phosphorylated molecules are substrates for bacteriophage T4 DNA ligase, and non-phosphorylated linkers must therefore be treated with bacteriophage T4 polynucleotide kinase and ATP before they can be joined to DNA. Because enzymatic phosphorylation is never completely efficient, linkers prepared in this way do not work as well as those that have been chemically phosphorylated by the manufacturer. Chemically phosphorylated linkers are therefore preferred, even though they are slightly more expensive.

Enzymatic Phosphorylation of Nonphosphorylated Linkers

1. Mix:

10 × linker-kinase buffer	1.0 μl
nonphosphorylated linkers	0.5–2.0 μg
10 mM ATP	1.0 μl
H$_2$O to 10 μl	

> **10 × Linker-kinase buffer**
>
> 0.66 M Tris·Cl (pH 7.6)
> 0.1 M MgCl$_2$
> 100 mM dithiothreitol
> 2 mg/ml bovine serum albumin (Fraction V; Sigma) (optional)
>
> This buffer should be stored in small aliquots at −20°C.

2. Add 2 units of bacteriophage T4 polynucleotide kinase and incubate the reaction for 1 hour at 37°C.

   At the end of the phosphorylation reaction, the linkers can be used without further purification in the ligation reactions described below.

Ligation of Phosphorylated Linkers to Blunt-ended Target Fragments

1. Mix in the order given:

blunt-ended DNA	0.1–0.5 μg
	(in a volume
	of 7 μl or less)
phosphorylated linkers	1–2 μg
	(in a volume
	of 8 μl or less)
H$_2$O to 15 μl	
5 mM ATP	2 μl
10 × blunt-end ligation buffer	2 μl
bacteriophage T4 DNA ligase	
(1–2 Weiss units)	1 μl

Incubate the reaction for 6–16 hours at 16°C.

---

10 × Blunt-end ligation buffer

0.66 M Tris · Cl (pH 7.6)
50 mM MgCl$_2$
50 mM dithiothreitol
1 mg/ml bovine serum albumin (Fraction V; Sigma) (optional)
10 mM hexamminecobalt chloride (optional)
5 mM spermidine HCl (optional)

This buffer should be stored in small aliquots at −20°C.

The spermidine and hexamminecobalt chloride [(NH$_3$)$_6$CoCl$_3$; Aldrich] combination increases the efficiency of ligation of linkers approximately fivefold.

At least three different units are used to measure the activity of bacteriophage T4 DNA ligase. Most manufacturers (apart from New England Biolabs) now calibrate the enzyme in Weiss units (Weiss et al. 1968). One Weiss unit is the amount of enzyme that catalyzes the exchange of 1 nmole of ^{32}P from pyrophosphate into [γ,β-^{32}P]ATP in 20 minutes at 37°C. One Weiss unit corresponds to 0.2 unit determined in the exonuclease resistance assay (Modrich and Lehman 1970) and to 60 cohesive-end units (as defined by New England Biolabs). 0.015 Weiss unit of bacteriophage T4 DNA ligase therefore will ligate 50% of the HindIII fragments of bacteriophage λ (5 μg) in 30 minutes at 16°C. Throughout this manual, bacteriophage T4 DNA ligase is given in Weiss units.

Usually, the ligation reaction goes well, and there is no need to check the products of the reaction before proceeding. If necessary, however, the ligation of linkers can be checked in two ways. During ligation, phosphorylated linkers should ligate to form dimers, trimers, tetramers, and longer oligomers that can be detected as a visible smear after an aliquot of the ligation mixture is analyzed by electrophoresis through a 1.8% agarose gel. Alternatively, a small amount of linkers that have been radiolabeled to high specific activity with [γ-^{32}P]ATP can be included in the ligation reaction. If ligation is successful, the radioactive linkers should form a series of radioactive bands that can be resolved by electrophoresis of the ligation products through a 15% polyacrylamide gel (see Chapter 13). These bands appear as a ladder on autoradiographs of the gel.

---

2. At the end of the incubation, inactivate the ligase by heating the reaction mixture to 65°C for 15 minutes.

3. Cool the reaction mixture on ice, and then add:

H₂O	70 μl
appropriate 10× restriction enzyme buffer	10 μl
appropriate restriction enzyme	20–50 units

Mix, and then incubate the reaction for 4 hours at the optimal temperature for the restriction enzyme.

4. Add an additional 10 units of the restriction enzyme, and continue incubation for another hour.

5. At the end of the incubation, add EDTA (pH 8.0) to a final concentration of 0.01 M. Extract the digested DNA once with an equal volume of phenol: chloroform, and separate the organic and aqueous phases by centrifugation at 12,000*g* for 30 seconds at room temperature in a microfuge. Transfer the aqueous phase to a fresh microfuge tube.

6. Remove the fragments of linkers by one of the following methods.

   • Chromatography through a spun column (2 ml) of Sepharose CL-4B equilibrated in 0.01 M Tris · Cl (pH 7.6), 0.1 M NaCl, 1 mM EDTA (pH 8.0) (see Appendix E): Mix the sample with 15 μl of gel-loading buffer IV (containing bromophenol blue, but not xylene cyanol FF [see Appendix B]), and apply the mixture directly to the column without further treatment. Start collecting fractions (~125 μl) immediately after the DNA is applied to the column. Locate the DNA of interest (high-molecular-weight DNA [> 300 bp] elutes with the bromophenol blue) by gel electrophoresis of an aliquot of each fraction. Concentrate the DNA by precipitation with 2 volumes of ethanol at 0°C.

   • Electrophoresis through an agarose or polyacrylamide gel of the appropriate porosity: Concentrate the DNA from the aqueous phase by precipitation with 2 volumes of ethanol at 0°C. Redissolve the DNA in 10–15 μl of TE (pH 7.6). Add 3–4 μl of the appropriate gel-loading buffer, and load the sample on the gel. After electrophoresis, recover the desired fragment of DNA from the gel as described in Chapter 6.

7. Ligate the fragment of DNA carrying the newly added linkers to a fragment that carries compatible termini.

Note

The effectiveness of the restriction enzyme digestion can be tested as follows:

a. Before adding the enzyme in step 3, remove 2 μl from the reaction and store at 4°C.

b. Withdraw another 2-μl aliquot after adding and mixing the enzyme.

c. Add each of the aliquots to separate tubes containing 100 ng (in a volume of 0.5 μl or less) of a linearized plasmid that contains an internal site for the particular restriction enzyme used.

d. Incubate the small-scale reaction containing the enzyme along with the large-scale reaction. Store the small-scale reaction without the enzyme at 4°C.

e. Analyze both small-scale reactions on a 1% agarose gel. If the reaction worked well, the plasmid DNA should be cleaved into two fragments in the sample containing the restriction enzyme.

Appendix G: List of Suppliers

Listed below are the major commercial suppliers of materials commonly used in molecular cloning. Of course, businesses can come and go and names and addresses can change, but at the time this book was prepared, this list was correct. In the USA, use 800 numbers primarily for out-of-state calls if two numbers are listed.

Aldrich Chemical Company, P.O. Box 2060, Milwaukee, WI 53201, USA. Telephone 800-558-9160. Fax 414-273-4979.

American Bioanalytical, 10 Huron Drive, Natick, MA 01760, USA. Telephone 800-443-0600. Fax 508-655-2754.

American Hoechst Corporation, Route 202–206 North, P.O. Box 2500, Somerville, NJ 08876, USA. Telephone 201-231-2000 or 800-235-2637. Fax 201-231-3225. In Europe: Hoechst AG., Postfach 800320, 6320–Frankfurt am Main 80, Federal Republic of Germany. Telephone (49)-(0)69-3050. In UK: Hoechst UK Ltd., Hoechst House, Salisbury Road, Hounslow, Middlesex TW4 6JH, UK. Telephone (44)-(0)1-570-7712. Fax (44)-(0)1-236-6336.

American Scientific Products. *See* Baxter Healthcare Corporation.

Amersham Corporation, 2636 South Clearbrook Drive, Arlington Heights, IL 60005, USA. Telephone 312-593-6300 or 800-323-9750. Fax 312-593-8236. In UK: Amersham International plc, Research Products Division, White Lion Road, Amersham, Buckinghamshire HP7 9LL, UK. Telephone (44)-(0)2404-4444. Fax (44)-(0)296-85190.

Amicon Corporation, 24 Cherry Hill Drive, Danvers, MA 01923, USA. Telephone 508-777-3622. Fax 508-777-6204.

Applied Biosystems Inc., 850 Lincoln Centre Drive, Foster City, CA 94404, USA. Telephone 415-570-6667. Fax 415-572-2743.

Baker. *See* J.T. Baker Inc.

Baxter Healthcare Corporation, Scientific Products Division, 1430 Waukegan Road, McGaw Park, IL 60085, USA. Telephone 312-689-8410 or 800-633-7370. Fax 312-973-6508.

Beckman Instruments Inc., 45 Belmont Drive, P.O. Box 6764, Somerset, NJ 08875-6764, USA. Telephone 201-560-0076 or 800-742-2345. Fax 201-560-1448. In Europe: Beckman Instruments International SA., 22 Rue Juste-Olivier, CH-1260 Nyon, Switzerland. Telephone (41)-(0)22-631181. Fax (41)-(0)22-621810.

Becton Dickinson, 2375 Garcia Avenue, Mountain View, CA 94043, USA. Telephone 415-968-7744. Fax 415-966-8614. In Europe: P.O. Box 13, Erembodegem 9440, Belgium. Telephone (32)-5378-7830.

Becton Dickinson, Labware, 2 Bridgewater Lane, Lincoln Park, NJ 07035, USA. Telephone 201-628-1144 or 800-235-5953. Fax 201-628-1533.

Becton Dickinson, Primary Care Diagnostics, 1 Becton Drive, Franklin Lakes, NJ 07417, USA. Telephone 201-848-6500. Fax 201-848-6475.

Bellco Biotechnology, P.O. Box B, 340 Erudo Road, Vineland, NJ 08360, USA. Telephone 609-691-1075 or 800-257-7043. Fax 609-691-3247.

Bethesda Research Laboratories Inc. (BRL), P.O. Box 6009, 8717 Grovemont Circle, Gaithersburg, MD 20877, USA. Telephone 301-840-8000. Fax 301-258-8238.

Bio-Rad Laboratories, 1414 Harbor Way, Richmond, CA 94804, USA. Telephone 415-234-4130 or 800-227-5589. Fax 415-232-4257. In UK: Bio-Rad Laboratories Ltd., Caxton Way, Watford Business Park, Watford, Hertfordshire WD1 8RP, UK. Telephone (44)-(0)923-240322. Fax (44)-(0)923-247825.

Boehringer Mannheim Biochemicals, Orders: P.O. Box 50414, Indianapolis, IN 46250, USA. Telephone 317-849-9350 or 800-262-1640. Fax 317-576-2754. Other Information: 9115 Hague Road, Indianapolis, IN 46256, USA. In Europe: Sandhoferstrasse 116, Postfach 310120, D6800 Mannheim, Federal Republic of Germany. Telephone (49)-(0)621-7591. Fax (49)-(0)621-7592890. In UK: BCL-Boehringer, Boehringer-Mannheim House, Bell Lane, Lewes, East Sussex BN7 1LG, UK. Telephone (44)-(0)273-480444. Fax (44)-(0)273-480266.

Brinkmann Instruments Inc., Cantiague Road, Westbury, NY 11590, USA. Telephone 516-334-7500. Fax 516-334-7506. In Europe: Eppendorf-Netheler-Hinz GmbH., P.O. Box 65 06 70, D-2000 Hamburg 65, Federal Republic of Germany. Telephone (49)-(0)40-538-01-0. Fax (49)-(0)40-538-01-556. In UK: BDH Ltd., Apparatus Division, P.O. Box 8, Dagenham, Essex RM8 1RY, UK. Telephone (44)-(0)1-597-8821. Fax (44)-(0)1-597-8300.

BRL. *See* Bethesda Research Laboratories Inc.; GIBCO.

Calbiochem, 10933 North Torrey Pines Road, La Jolla, CA 92037, USA. Telephone 619-450-9600 or 800-854-3417. Fax 619-453-3552. In UK: Cambridge Bioscience, 42 Devonshire Road, Cambridge CB1 2BL, UK. Telephone (44)-(0)223-316855. Fax (44)-(0)223-460396.

Cetus. *See* Perkin Elmer Cetus.

Ciba-Corning Diagnostics, 333 Coney Street, East Walpole, MA 02032, USA. Telephone 508-668-5000 or 800-343-0893. Fax 508-668-4591.

Clay Adams. *See* Becton Dickinson, Primary Care Diagnostics.

Collaborative Research Inc., 2 Oak Park, Bedford, MA 01730, USA. Telephone 617-275-0004. Fax 617-275-0043.

Corning Glass Works, Science Products Division, MP-21-5-8, Corning, NY 14831, USA. Telephone 607-737-1667. Fax 607-737-1636.

Curtin Matheson Scientific Inc., 9999 Veterans Memorial Drive, Houston, TX 77038, USA. Telephone 713-820-9898 or 713-820-1661. Fax 713-878-2221.

Difco Laboratories, P.O. Box 331058, Detroit, MI 48232-7058, USA. Telephone 313-961-0800 or 800-521-0851. Fax 313-591-3530. In UK: Difco Laboratories, P.O. Box 14B, Central Avenue, East Molesey, Surrey KT8 0SE, UK. Telephone (44)-(0)1-979-9951. Fax (44)-(0)1-979-2506.

Dow Chemical Company, Ag-Organics Department, Building 9001-9008, Midland, MI 48641-1706, USA. Telephone 517-636-1000. Fax 517-636-3373.

Dow Corning Corporation, Dow Corning Center, Midland, MI 48686-0994, USA. Telephone 517-496-4000. Fax 517-496-6974.

Drummond Scientific Corporation, 500 Parkway, Broomall, PA 19008, USA. Telephone 215-353-0200. Fax 215-353-6204.

du Pont. *See* E.I. du Pont de Nemours & Company Inc.

Eastman Kodak Special Products Division, 2400 Mt. Read Blvd., Rochester, NY 14650, USA. Telephone 716-588-3124 or 800-242-2424. In UK: Kodak Ltd., Laboratory and Special Products Division, Acornfield Road, Knowsley Industrial Park North, Liverpool L33 7UF, UK. Telephone (44)-(0)51-548-6560. Fax (44)-(0)51-547-2404.

E.I. du Pont de Nemours & Company Inc., Medical Department, Biotechnology Division, P.O. Box 80024, Barley Mill Plaza, Wilmington, DE 19880-0024, USA. Telephone 302-992-3416 or 800-551-2121. Fax 302-992-3474. In UK: Dupont (UK) Ltd., Industrial Products Division, Wedgewood Way, Stevenage, Hertfordshire SG1 4QN, UK. Telephone (44)-(0)438-734000. Fax (44)-(0)438-734154.

Electro-Nucleonics Laboratories Inc. *See* Pharmacia ENI Diagnostics Inc.

Eppendorf-Netheler-Hinz GmbH. *See* Brinkmann Instruments Inc.

Ericomp, 10055 Barnes Canyon Road, Suite G, San Diego, CA 92121, USA. Telephone 619-457-1888 or 800-541-8471.

Falcon. *See*, e.g., Baxter Healthcare Corporation; Becton Dickinson, Labware.

Fisher Scientific, 52 Fadem Road, Springfield, NJ 07081, USA. Telephone 201-379-1400. Fax 201-379-7638.

Flow Laboratories Inc., 7655 Old Springhouse Road, McLean, VA 22102, USA. Telephone 703-893-5925 or 800-368-3569. Fax 703-893-6727. In UK: Flow Laboratories Ltd., Woodcock Hill, Harefield Road, Rickmansworth, Hertfordshire WD3 1PQ, UK. Telephone (44)-(0)923-774666. Fax (44)-(0)923-777005.

Fluka Chemical Corporation, 980 South 2nd Street, Ronkonkoma, NY 11779-7204, USA. Telephone 516-467-0980. Fax 516-467-0663. In Europe: Fluka Chemie AG., Industriestrasse 25, CH-9470 Buchs, Switzerland. Telephone (41)-(0)85-69511. Fax (41)-(0)85-65449. In UK: Fluka Chemicals Ltd., Peakdale Road, Glossop, Derbyshire, UK. Telephone (44)-(0)4574-62518. Fax (44)-(0)4574-4307.

FMC Marine Colloids, Bioproducts Department, 5 Maple Street, Rockland, ME 04841, USA. Telephone 207-594-3360 or 800-341-1574. Fax 207-594-3391.

Fuji Medical Systems, 90 Viaduct Road, Stamford, CT 06907, USA. Telephone 203-353-0300 or 800-431-1850. Fax 203-353-0926.

Gelman Sciences Inc., 600 South Wagner Road, Ann Arbor, MI 48106, USA. Telephone 313-665-0651. Fax 313-761-1208.

GIBCO, 3175 Staley Road, Grand Island, NY 14072, USA. Telephone 716-773-0790 or 800-828-6686. Fax 800-331-2286. In UK: GIBCO-BRL Ltd., Unit 4, Cowley Mill Trading Estate, Longbridge Way, Uxbridge UB8 2YG, UK. Telephone (44)-(0)895-36355. Fax (44)-(0)895-53159.

Gilson Medical Electronics Inc., 3000 West Beltline Highway, P.O. Box 27, Middleton, WI 53562, USA. Telephone 608-836-1551 or 800-445-7667. Fax 608-831-4551. In Europe: Gilson Medical Electronics SA., 72 Rue Gambetta, BP 45, Villiers-le-Bel, 95400 France. Telephone (33)-139-905441.

Hamilton Company, 4970 Energy Way, Reno, NV 89502, USA. Telephone 702-786-7077 or 800-648-5950.

Health Products. *See* Pierce Chemical Co.

Heat Systems Ultrasonics Inc., 1938 New Highway, Farmingdale, NY 11735, USA. Telephone 516-694-9555 or 800-645-9846. Fax 516-694-9412.

Hoechst. *See* American Hoechst Corporation.

Hoefer Scientific Instruments, P.O. Box 77387, 654 Minnesota Street, San Francisco, CA 94107, USA. Telephone 415-282-2307 or 800-227-4750. Fax 415-821-1081.

IBI. *See* International Biotechnologies Inc.

ICN Biochemicals, P.O. Box 28050, Cleveland, OH 44128, USA. Telephone 216-831-3000 or 800-321-6842. Fax 216-831-2569.

ICN Immunobiological Inc., P.O. Box 5023, Costa Mesa, CA 92626, USA. Telephone 800-348-7465. Fax 714-957-3018. In UK: ICN Biomedicals Ltd., Lincoln Road, High Wycombe, Bucks HP12 3XJ, UK. Telephone (44)-(0)494-443826. Fax (44)-(0)494-436048.

ICN Radiochemicals, P.O. Box 19536, Irvine, CA 92713, USA. Telephone 714-545-0113 or 800-854-0530. Fax 714-557-4872.

IEC. *See* International Equipment Company.

International Biotechnologies Inc. (IBI), P.O. Box 9558, 25 Science Park, New Haven, CT 06535, USA. Telephone 203-786-5600 or 800-243-2555. Fax 203-786-5694.

International Equipment Company (IEC), Division of Damon Corporation, 300 Second Avenue, Needham Heights, MA 02194, USA. Telephone 617-449-0800 or 800-225-8856. Fax 617-444-6743. In UK: Damon/IEC UK Ltd., Unit 7, Lawrence Way, Brewers Hill Road, Dunstable, Bedfordshire LU6 1BD, UK. Telephone (44)-(0)582-604669. Fax (44)-(0)582-609257.

ISCO, Instrument Division, P.O. Box 5347, Lincoln, NB 68505-9987, USA. Telephone 402-464-0231 or 800-228-4250. Fax 402-464-4543.

J.T. Baker Inc., 222 Red School Lane, Phillipsburg, NJ 08865, USA. Telephone 201-859-2151 or 800-582-2537. Fax 201-859-9318.

Kirin Breweries. *See* ICN Immunobiological Inc.

Kodak. *See* Eastman Kodak Special Products Division.

Kontes, Spruce Street, P.O. Box 729, Vineland, NJ 08360, USA. Telephone 609-692-8500. Fax 609-692-3242.

Life Sciences Inc., 2900 72nd Street North, St. Petersburg, FL 33710, USA. Telephone 813-345-9371. Fax 813-347-2957.

Linbro Scientific. *See* Flow Laboratories Inc.

LKB. *See* Pharmacia LKB Biotechnology Inc.

Mallinckrodt. *See* Baxter Healthcare Corporation.

Miles Laboratories Inc., 1127 Myrtle Street, Elkart, IN 46515, USA. Telephone 219-264-8111 or 800-348-7414. Fax 219-262-6747. In UK: Miles Laboratories Ltd., Miles Scientific, Stoke Court, Stoke Poges, Slough SL2 4LY, UK. Telephone (44)-(0)2814-5151. Fax (44)-(0)2814-3893.

Millipore Corporation, 80 Ashby Road, Bedford, MA 01730, USA. Telephone 617-275-9200 or 800-225-1380. In UK: Millipore (UK) Ltd., Millipore House, The Boulevard, Ascot Road, Croxley Green, Watford WD1 8YW, UK. Telephone (44)-(0)923-816375. Fax (44)-(0)923-818297.

Nalge. *See,* e.g., Baxter Healthcare Corporation; Fisher Scientific.

NEN. *See* New England Nuclear.

New England Biolabs Inc., 32 Tozer Road, Beverly, MA 01915, USA. Telephone 508-927-5054 or 800-632-5227. Fax 508-921-1350.

New England Nuclear (NEN), 549 Albany Street, Boston, MA 02118, USA. Telephone 617-482-9595 or 800-551-2121. In Europe: du Pont de Nemours (Deutschland) GmbH.,

Biotechnology Systems Division, NEN Research Products, Postfach 40 12 40, 6072 Dreieich 4, Federal Republic of Germany. Telephone (49)-(0)6103-803-155. Fax (49)-(0)6103-897.

NL Chemicals, P.O. Box 700, Hightstown, NJ 08520, USA. Telephone 609-443-2000 or (west of Rockies) 800-257-9478. Fax 609-443-2422.

Nunc Inc., 2000 North Aurora Road, Naperville, IL 60566, USA. Telephone 312-983-5700. Fax 312-416-2519. In UK: GIBCO Ltd., P.O. Box 35, Trident House, Renfrew Road, Paisley, Scotland PA3 4EF. Telephone (44)-(0)41-889-6100. Fax (44)-(0)41-887-1167. In Europe: Nunc A/S., Postbox 280, Kamstrup, DK 4000, Roskilde, Denmark. Telephone (45)-(0)2-359065. Fax (45)-(0)2-350105.

Perkin Elmer Cetus, 761 Main Avenue, Norwalk, CT 06859, USA. Telephone 800-762-4001. Fax 203-761-9645.

Pharmacia ENI Diagnostics Inc., Cell Science Division, 12050 Tech Road, Silver Springs, MD 20904, USA. Telephone 301-622-4218. Fax 301-622-2705.

Pharmacia LKB Biotechnology Inc., 800 Centennial Avenue, Piscataway, NJ 08854, USA. Telephone 201-457-8000 or 800-558-7110. Fax 201-457-0557. In UK: Pharmacia LKB Biotechnology Ltd., Pharmacia House, Midsummer Boulevard, Milton Keynes, Buckinghamshire MK9 3HP, UK. Telephone (44)-(0)908-661101. Fax (44)-(0)908-690091. In Europe: Pharmacia LKB Biotechnology AB., P.O. Box 175, Björkgatan 30, 751 82 Uppsala, Sweden. Telephone (46)-(0)18-163000. Fax (46)-(0)18-143820.

Pierce Chemical Company, P.O. Box 117, Rockford, IL 61105, USA. Telephone 815-968-0747 or 800-874-3723. Fax 815-968-7316.

Promega Biotec, 2800 South Fish Hatchery Road, Madison, WI 53711, USA. Telephone 608-274-4330 or 800-356-9526. Fax 608-273-6967.

Rainin Instrument Company, Mack Road, Woburn, MA 01801, USA. Telephone 617-935-3050. Fax 617-938-8157.

Rohm & Haas Company, Independence Mall West, Philadelphia, PA 19105, USA. Telephone 215-592-3000. Fax 215-592-3377. In UK: Rohm & Haas (UK) Ltd., Lennig

House, 2 Mason's Avenue, Croydon CR9 3NB, UK. Telephone (44)-(0)1-686-8844. Fax (44)-(0)1-681-3207.

Sarstedt Inc., Research Products Division, Box 468, Newton, NC 28658-0468, USA. Telephone 704-465-4000 or 800-257-5101.

Sartorius Corporation, 140 Wilbur Place, Bohemia, NY 11716, USA. Telephone 800-368-7178 or 800-227-2842. Fax 516-563-5065.

Savant Instruments Inc., 110-103 Bi-county Blvd., Farmingdale, NY 11735, USA. Telephone 516-249-4600 or 800-634-8886. Fax 516-249-4639.

Schleicher & Schuell Inc., 10 Optical Avenue, Keene, NH 03431, USA. Telephone 603-352-3810. Fax 603-357-3627.

Scientific Manufacturing Industries. *See* Baxter Healthcare Corporation.

Scientific Products. *See* Baxter Healthcare Corporation.

Serva Biochemicals Inc., 200 Shames Drive, Westbury, NY 11590, USA. Telephone 516-333-1575. Fax 516-333-1582.

Sigma Chemical Company, P.O. Box 14509, St. Louis, MO 63178, USA. Telephone 314-771-5750 or 800-325-5052. Fax 314-771-5757. In UK: Sigma Chemical Company Ltd., Fancy Road, Poole, Dorset BH17 7NH, UK. Telephone (44)-(0)202-733114. Fax (44)-(0)202-715460.

Sorvall. *See* E.I. du Pont de Nemours & Company Inc.

Sterilin Instruments Ltd., Lampton House, Lampton Road, Hounslow, Middlesex TW3 4EE, UK. Telephone (44)-(0)1-572-2468. Fax (44)-(0)1-572-7301.

Stratagene, 11099 North Torrey Pines Road, La Jolla, CA 92037, USA. Telephone 619-535-5400 or 800-424-5444. Fax 619-535-5430. In Europe: Stratagene GmbH., Postfach 105466, D-6900, Heidelberg, Federal Republic of Germany. Telephone (49)-(0)6221-40-06-34. Fax (49)-(0)6221-40-06-39.

Technicon, Route 11 South, Middletown, VA 22645, USA. Telephone 703-869-3200 or 800-431-1970.

Thomas Scientific Company, 99 High Hill Road at I-295, Swedesboro, NJ 08085, USA. Telephone 215-988-0533. Fax 609-467-3087.

3M Company, 3M Center, St. Paul, MN 55144, USA. Telephone 612-733-5454.

Ultraviolet Products. *See* UVP Inc.

United States Biochemical Corporation (USB), P.O. Box 22400, Cleveland, OH 44122, USA. Telephone 216-765-5000 or 800-321-9322. Fax 216-464-5075.

USB. *See* United States Biochemical Corporation.

UVP Inc. (Ultraviolet Products), P.O. Box 1501, 5100 Walnut Grove Avenue, San Gabriel, CA 91776, USA. Telephone 818-285-3123 or 800-452-6788. Fax 818-285-2940.

Value Plastics Inc., 626 West 66th Street, Loveland, CO 80537, USA. Telephone 303-669-4351. Fax 303-223-0953.

Vangard International, 1111A Green Grove Road, Neptune, NJ 07753, USA. Telephone 201-922-4900. Fax 201-922-0557.

Van Waters and Rogers, 1600 Norton Building, 801 2nd Avenue, Seattle, WA 98104-1564, USA. Telephone 206-447-5911. Fax 206-340-1988.

VWR Scientific, P.O. Box 7900, San Francisco, CA 94120, USA. Telephone 415-468-7150. Fax 415-330-4185.

Waters, Division of Millipore, 34 Maple Street, Milford, MA 01757, USA. Telephone 508-478-2000 or 800-252-4752. Fax 508-872-1990 (for electronically transmitted orders) or 508-478-2000 (extension 3016).

Whatman Laboratory Products Inc., 9 Bridewell Place, Clifton, NJ 07014, USA. Telephone 201-773-5800. Fax 201-472-6949. In UK: Whatman Ltd., Springfield Mill, Maidstone, Kent ME14 2LE, UK. Telephone (44)-(0)622-692022. Fax (44)-(0)622-691425.

Wheaton Scientific, Division of Wheaton Industries, 1000 N. Tenth Street, Millville, NJ 08332, USA. Telephone 609-825-1400 or 800-225-1437. Fax 609-825-1131.

Worthington Biochemical, Halls Mill Road, Freehold, NJ 07728, USA. Telephone 201-462-3838 or 800-445-9603. Fax 201-308-4453. In UK: Cambridge Bioscience, 42 Devonshire Road, Cambridge CB1 2BL, UK. Telephone (44)-(0)223-316855. Fax (44)-(0)223-460396.

References

Appleyard, R.K. 1954. Segregation of new lysogenic types during growth of a doubly lysogenic strain derived from *Escherichia coli* K12. *Genetics* **39:** 440.

Arber, W., L. Enquist, B. Hohn, N.E. Murray, and K. Murray. 1983. Experimental methods for use with lambda. In *Lambda II* (ed. R.W. Hendrix et al.), p. 433. Cold Spring Harbor Laboratory, Cold Spring Harbor, New York.

Bajorath, J., S. Raghunathan, W. Hinrichs, and W. Saenger. 1989. Long-range structural changes in proteinase K triggered by calcium ion removal. *Nature* **337:** 481.

Bensaude, O. 1988. Ethidium bromide and safety—Readers suggest alternative solutions. Letter to editor. *Trends Genet.* **4:** 89.

Bertani, G. and J.J. Weigle. 1953. Host controlled variation in bacterial viruses. *J. Bacteriol.* **65:** 113.

Bethesda Research Laboratories. 1986. BRL pUC host: *E. coli* DH5α™ competent cells. *Bethesda Res. Lab. Focus* **8(2):** 9.

Bolivar, F. and K. Backman. 1979. Plasmids of *Escherichia coli* as cloning vectors. *Methods Enzymol.* **68:** 245.

Bolivar, F., R.L. Rodriguez, P.J. Greene, M.C. Betlach, H.L. Heyneker, H.W. Boyer, J.H. Crosa, and S. Falkow. 1977. Construction and characterization of new cloning vehicles. II. A multipurpose cloning system. *Gene* **2:** 95.

Bonner, W.M. and R.A. Laskey. 1974. A film detection method for tritium-labelled proteins and nucleic acids in polyacrylamide gels. *Eur. J. Biochem.* **46:** 83.

Borck, K., J.D. Beggs, W.J. Brammar, A.S. Hopkins, and N.E. Murray. 1976. The construction in vitro of transducing derivatives of phage lambda. *Mol. Gen. Genet.* **146:** 199.

Boyer, H.W. and D. Roulland-Dussoix. 1969. A complementation analysis of the restriction and modification of DNA in *Escherichia coli*. *J. Mol. Biol.* **41:** 459.

Brawerman, G., J. Mendecki, and S.Y. Lee. 1972. A procedure for the isolation of mammalian messenger ribonucleic acid. *Biochemistry* **11:** 637.

Brent, R. and M. Ptashne. 1981. Mechanism of action of the *lexA* gene product. *Proc. Natl. Acad. Sci.* **78:** 4204.

Bullock, W.O., J.M. Fernandez, and J.M. Short. 1987. XL1-Blue: A high efficiency plasmid transforming recA *Escherichia coli* strain with beta-galactosidase selection. *BioTechniques* **5:** 376.

Campbell, A. 1965. The steric effect in lysogenization by bacteriophage lambda. I. Lysogenization of a partially diploid strain of *Escherichia coli* K12. *Virology* **27:** 329.

Campbell, J.L., C.C. Richardson, and F.W. Studier. 1978. Genetic recombination and complementation between bacteriophage T7 and cloned fragments of T7 DNA. *Proc. Natl. Acad. Sci.* **75:** 2276.

Chamberlain, J.P. 1979. Fluorographic detection of radioactivity in polyacrylamide gels with the water-soluble fluor, sodium salicylate. *Anal. Biochem.* **98:** 132.

Daniels, D.L., J.L. Schroeder, W. Szybalski, F. Sanger, A.R. Coulson, G.F. Hong, D.F. Hill, G.B. Petersen, and F.R. Blattner. 1983. Appendix II: Complete annotated lambda sequence. In *Lambda II* (ed. R.W. Hendrix et al.), p. 519. Cold Spring Harbor Laboratory, Cold Spring Harbor, New York.

Dayhoff, M.O., ed. 1972. *Atlas of protein sequence and structure*, vol. 5. National Biomedical Research Foundation, Silver Spring, Maryland.

Denhardt, D.T. 1966. A membrane-filter technique for the detection of complementary DNA. *Biochem. Biophys. Res. Commun.* **23:** 641.

Dente, L., G. Cesareni, and R. Cortese. 1983. pEMBL: A new family of single stranded plasmids. *Nucleic Acids Res.* **11:** 1645.

Dickerson, R.E. 1983. The DNA helix and how it is read. *Sci. Am.* **249(6):** 94.

Dickerson, R.E., H.R. Drew, B.N. Conner, M.L. Kopka, and P.E. Pjura. 1983. Helix geometry and hydration in A-DNA, B-DNA, and Z-DNA. *Cold*

Spring Harbor Symp. Quant. Biol. **47:** 13.

Felton, J. 1983. M13 host strain JM103 contains two restriction systems. *BioTechniques* **1:** 42.

Frischauf, A.-M., H. Lehrach, A. Poustka, and N. Murray. 1983. Lambda replacement vectors carrying polylinker sequences. *J. Mol. Biol.* **170:** 827.

Gibson, T.J. 1984. "Studies on the Epstein-Barr virus genome." Ph.D. thesis, Cambridge University, England.

Green, A.A. 1933. The preparation of acetate and phosphate buffer solutions of known pH and ionic strength. *J. Am. Chem. Soc.* **55:** 2331.

Guarente, L. and M. Ptashne. 1981. Fusion of *Escherichia coli lacZ* to the cytochrome *c* gene of *Saccharomyces cerevisiae*. *Proc. Natl. Acad. Sci.* **78:** 2199.

Hanahan, D. 1983. Studies on transformation of *Escherichia coli* with plasmids. *J. Mol. Biol.* **166:** 557.

———. 1985. Techniques for transformation of *E. coli*. In *DNA cloning: A practical approach* (ed. D.M. Glover), vol. 1, p. 109. IRL Press, Oxford.

Hohn, B. 1979. *In vitro* packaging of λ and cosmid DNA. *Methods Enzymol.* **68:** 299.

Hohn, B. and K. Murray. 1977. Packaging recombinant DNA molecules into bacteriophage particles *in vitro*. *Proc. Natl. Acad. Sci.* **74:** 3259.

Huang, W.M. and I.R. Lehman. 1972. On the exonuclease activity of phage T4 deoxyribonucleic acid polymerase. *J. Biol. Chem.* **247:** 3139.

ISCO. 1982. *ISCOTABLES: A handbook of data for biological and physical scientists*, 8th edition. ISCO, Inc., Lincoln, Nebraska.

Jendrisak, J., R.A. Young, and J.D. Engel. 1987. Cloning cDNA into λgt10 and λgt11. *Methods Enzymol.* **152:** 359.

Johnson, D.A., J.W. Gautsch, J.R. Sportsman, and J.H. Elder. 1984. Improved technique utilizing nonfat dry milk for analysis of proteins and nucleic acids transferred to nitrocellulose. *Gene Anal. Tech.* **1:** 3.

Karn, J., S. Brenner, L. Barnett, and G. Cesareni. 1980. Novel bacteriophage λ cloning vector. *Proc. Natl. Acad. Sci.* **77:** 5172.

Kirby, K.S. 1956. A new method for the isolation of ribonucleic acids from mammalian tissues. *Biochem. J.* **64:** 405.

Kunkel, T.A., J.D. Roberts, and R.A. Zakour. 1987. Rapid and efficient site-specific mutagenesis without phenotypic selection. *Methods Enzymol.* **154:** 367.

Laskey, R.A. 1980. The use of intensifying screens or organic scintillators for visualizing radioactive molecules resolved by gel electrophoresis. *Methods Enzymol.* **65:** 363.

Laskey, R.A. and A.D. Mills. 1975. Quantitative film detection of ^{3}H and ^{14}C in polyacrylamide gels by fluorography. *Eur. J. Biochem.* **56:** 335.

———. 1977. Enhanced autoradiographic detection of ^{32}P and ^{125}I using intensifying screens and hypersensitized film. *FEBS Lett.* **82:** 314.

Leder, P., D. Tiemeier, and L. Enquist. 1977. EK2 derivatives of bacteriophage lambda useful in the cloning of DNA from higher organisms: The λgtWES system. *Science* **196:** 175.

Levinson, A., D. Silver, and B. Seed. 1984. Minimal size plasmids containing an M13 origin for production of single-strand transducing particles. *J. Mol. Appl. Genet.* **2:** 507.

Low, B. 1968. Formation of merodiploids in matings with a class of Rec⁻ recipient strains of *Escherichia coli* K12. *Proc. Natl. Acad. Sci.* **60:** 160.

Lunn, G. and E.B. Sansone. 1987. Ethidium bromide: Destruction and decontamination of solutions. *Anal. Biochem.* **162:** 453.

Maxwell, I.H., F. Maxwell, and W.E. Hahn. 1977. Removal of RNase activity from DNase by affinity chromatography on agarose-coupled aminophenylphosphoryl-uridine-2'(3')-phosphate. *Nucleic Acids Res.* **4:** 241.

Meissner, P.S., W.P. Sisk, and M.L. Berman. 1987. Bacteriophage λ cloning system for the construction of directional cDNA libraries. *Proc. Natl. Acad. Sci.* **84:** 4171.

Meselson, M. and R. Yuan. 1968. DNA restriction enzyme from *E. coli*. *Nature* **217:** 1110.

Messing, J. 1979. A multipurpose cloning system based on single-stranded DNA bacteriophage M13. *Recomb. DNA Tech. Bull.* **2(2):** 43.

Messing, J., R. Crea, and P.H. Seeburg. 1981. A system for shotgun DNA sequencing. *Nucleic Acids Res.* **9:** 309.

Messing, J., B. Gronenborn, B. Müller-Hill, and P.H. Hofschneider. 1977. Filamentous coliphage M13 as a cloning vehicle: Insertion of a HindII fragment of the lac regulatory region in M13 replicative form in vitro. Proc. Natl. Acad. Sci. 74: 3642.

Miller, J.H. 1972. Experiments in molecular genetics. Cold Spring Harbor Laboratory, Cold Spring Harbor, New York.

Modrich, P. and I.R. Lehman. 1970. Enzymatic joining of polynucleotides. IX. A simple and rapid assay of polynucleotide joining (ligase) activity by measurement of circle formation from linear deoxyadenylate-deoxythymidylate copolymer. J. Biol. Chem. 245: 3626.

Murray, N.E., W.J. Brammar, and K. Murray. 1977. Lambdoid phages that simplify the recovery of in vitro recombinants. Mol. Gen. Genet. 150: 53.

Nader, W.F., T.D. Edlind, A. Huettermann, and H.W. Sauer. 1985. Cloning of Physarum actin sequences in an exonuclease-deficient bacterial host. Proc. Natl. Acad. Sci. 82: 2698.

Nagai, K. and H.C. Thøgersen. 1984. Generation of β-globin by sequence-specific proteolysis of a hybrid protein produced in Escherichia coli. Nature 309: 810.

Oka, T., S. Sakamoto, K. Miyoshi, T. Fuwa, K. Yoda, M. Yamasaki, G. Tamura, and T. Miyake. 1985. Synthesis and secretion of human epidermal growth factor by Escherichia coli. Proc. Natl. Acad. Sci. 82: 7212.

Okayama, H. and P. Berg. 1982. High-efficiency cloning of full-length cDNA. Mol. Cell. Biol. 2: 161.

Patterson, T.A. and M. Dean. 1987. Preparation of high titer lambda phage lysates. Nucleic Acids Res. 15: 6298.

Peacock, S.L., C.M. McIver, and J.J. Monahan. 1981. Transformation of E. coli using homopolymer-linked plasmid chimeras. Biochim. Biophys. Acta 655: 243.

Quillardet, P. and M. Hofnung. 1988. Ethidium bromide and safety—Readers suggest alternative solutions. Letter to editor. Trends Genet. 4: 89.

Raleigh, E.A. and G. Wilson. 1986. Escherichia coli K-12 restricts DNA containing 5-methylcytosine. Proc. Natl. Acad. Sci. 83: 9070.

Remaut, E., P. Stanssens, and W. Fiers. 1981. Plasmid vectors for high-efficiency expression controlled by the p_L promoter of coliphage lambda. Gene 15: 81.

Rosenberg, S.M. 1985. EcoK restriction during in vitro packaging of coliphage lambda DNA. Gene 39: 313.

Rüther, U. and B. Müller-Hill. 1983. Easy identification of cDNA clones. EMBO J. 2: 1791.

Shapiro, D.J. 1981. Quantitative ethanol precipitation of nanogram quantities of DNA and RNA. Anal. Biochem. 110: 229.

Shimatake, H. and M. Rosenberg. 1981. Purified λ regulatory protein cII positively activates promoters for lysogenic development. Nature 292: 128.

Singh, L. and K.W. Jones. 1984. The use of heparin as a simple cost-effective means of controlling background in nucleic acid hybridization procedures. Nucleic Acids Res. 12: 5627.

Smith, H.O. 1980. Recovery of DNA from gels. Methods Enzymol. 65: 371.

Smith, O.P. 1988. Small-scale preparation of RNase-free DNase. BioTechniques 6: 846.

Struhl, K. 1985. A rapid method for creating recombinant DNA molecules. BioTechniques 3: 452.

Studier, F.W. and B.A. Moffatt. 1986. Use of bacteriophage T7 RNA polymerase to direct selective high-level expression of cloned genes. J. Mol. Biol. 189: 113.

Sutcliffe, J.G. 1978. Nucleotide sequence of the ampicillin resistance gene of Escherichia coli plasmid pBR322. Proc. Natl. Acad. Sci. 75: 3737.

———. 1979. Complete nucleotide sequence of the Escherichia coli plasmid pBR322. Cold Spring Harbor Symp. Quant. Biol. 43: 77.

Svaren, J., S. Inagami, E. Lovegren, and R. Chalkley. 1987. DNA denatures upon drying after ethanol precipitation. Nucleic Acids Res. 15: 8739.

Swanstrom, R. and P.R. Shank. 1978. X-ray intensifying screens greatly enhance the detection by autoradiography of the radioactive isotopes ^{32}P and [125]I. Anal. Biochem. 86: 184.

Tartof, K.D. and C.A. Hobbs. 1987. Improved media for growing plasmid and cosmid clones. Bethesda Res. Lab. Focus 9: 12.

Taylor, W.R. 1986. The classification of amino acid conservation. J. Theor. Biol. 119: 205.

Vieira, J. and J. Messing. 1987. Production of single-stranded plasmid DNA. *Methods Enzymol.* **153:** 3.

Weiss, B., A. Jacquemin-Sablon, T.R. Live, G.C. Fareed, and C.C. Richardson. 1968. Enzymatic breakage and joining of deoxyribonucleic acid. VI. Further purification and properties of polynucleotide ligase from *Escherichia coli* infected with bacteriophage T4. *J. Biol. Chem.* **243:** 4543.

Wilkie, N.M. and R. Cortini. 1976. Sequence arrangement in herpes simplex virus type 1 DNA: Identification of terminal fragments in restriction endonuclease digests and evidence for inversions in redundant and unique sequences. *J. Virol.* **20:** 211.

Williams, B.G. and F.R. Blattner. 1979. Construction and characterization of the hybrid bacteriophage lambda Charon vectors for DNA cloning. *J. Virol.* **29:** 555.

Wood, W.B. 1966. Host specificity of DNA produced by *Escherichia coli:* Bacterial mutations affecting the restriction and modification of DNA. *J. Mol. Biol.* **16:** 118.

Wyman, A.R. and K.F. Wertman. 1987. Host strains that alleviate underrepresentation of specific sequences: Overview. *Methods Enzymol.* **152:** 173.

Yanisch-Perron, C., J. Vieira, and J. Messing. 1985. Improved M13 phage cloning vectors and host strains: Nucleotide sequences of the M13mp18 and pUC19 vectors. *Gene* **33:** 103.

Young, R.A. and R.W. Davis. 1983a. Efficient isolation of genes by using antibody probes. *Proc. Natl. Acad. Sci.* **80:** 1194.

———. 1983b. Yeast RNA polymerase II genes: Isolation with antibody probes. *Science* **222:** 778.

Zabarovsky, E.R. and R.L. Allikmets. 1986. An improved technique for the efficient construction of gene libraries by partial filling-in of cohesive ends. *Gene* **42:** 119.

Zagursky, R.J. and M.L. Berman. 1984. Cloning vectors that yield high levels of single-stranded DNA for rapid DNA sequencing. *Gene* **27:** 183.

Zoller, M.J. and M. Smith. 1987. Oligonucleotide-directed mutagenesis: A simple method using two oligonucleotide primers and a single-stranded DNA template. *Methods Enzymol.* **154:** 329.

Index

A

Acetic acid
  common commercial strengths, B.2
  pH values of stock solutions, B.3
Acetonitrile, precautions for laboratory
    safety, 11.29
Acetyl coenzyme A. *See* Chloramphenicol
    acetyltransferase
Acrylamide. *See also* Polyacrylamide gel
    electrophoresis
  precautions for laboratory safety, 6.39,
    B.9
  properties, 6.36, 18.49
  stock solutions, 6.39, 13.47, B.9
    /urea for sequencing, 13.48
Actinomycin D
  inhibits second-strand cDNA synthesis,
    5.54, 8.56
  precautions for laboratory safety, 5.54,
    B.9
  stock solution, B.9
ADA. *See* Adenosine deaminase
Adenine and related compounds, molecu-
    lar structure, C.4–C.5
Adenosine, structure, C.4
Adenosine deaminase (ADA)
  amplification system (pMT3SV$_2$), 16.29
  selectable marker in mammalian cells,
    16.15
Adenosine 5′-diphosphate (5′-ADP),
    structure, C.5
Adenosine 5′-monophosphate (5′-AMP),
    structure, C.4
Adenosine 5′-triphosphate (5′-ATP)
  stock solution, B.10
  structure, C.5
S-Adenosyl methionine, stock solution,
    8.56
5′-ADP. *See* Adenosine 5′-diphosphate
Affinity purification
  of antibodies to be used for immunologi-
    cal screening, 12.13, 12.27–12.28
  of immunospecific antisera using anti-
    gen immobilized on nitrocellulose
    filters, 18.17–18.18
AG50 resin, use in removal of ethidium
    bromide from plasmid DNA, 1.47
Agar, recipe for media containing, A.4
Agarose
  labeling DNA in presence of low-
    melting-temperature, 10.16–
    10.17
  properties, 6.3

  rapid cloning in plasmid vectors using
    low-melting-temperature, F.6–F.7
  restriction enzyme digestion in agarose
    blocks for pulsed-field gel elec-
    trophoresis, 6.57
  top agarose, recipe, A.4
Agarose gel. *See also* Pulsed-field gel
    electrophoresis
  alkaline
    analysis of first- and second-strand
      cDNA synthesis, 8.5
    analysis of single-stranded DNA,
      6.20–6.21
    bromocresol green as tracking dye in
      electrophoresis, 6.12, 6.21
    gel-loading buffer, 6.12, B.24
    preparation, 6.20
  electrophoresis apparatus, 6.8
  electrophoresis buffers, 6.6–6.7
  electrophoresis of restriction fragments
    of mammalian DNA, 9.32–9.33
  factors that influence DNA migration,
    6.3–6.7, 6.13
  -loading buffers, 6.12, B.24
  minigels, 6.14
  photography of gels, 6.19
  preparation, 6.9–6.12
  recovery of DNA, 6.22–6.23
    electroelution into dialysis bags,
      6.28–6.29
    electrophoresis onto DEAE-cellulose
      membrane, 6.24–6.27
    purification of recovered DNA,
      6.32–6.35
    use of low-melting-temperature aga-
      rose, 6.30–6.31
  size fractionation of denatured RNA for
    northern hybridization, 7.40–7.42,
    7.43–7.45
  size fractionation of mRNA in presence
    of methylmercuric hydroxide,
    7.30–7.34
  staining of DNA with ethidium bro-
    mide, 6.15
  staining of RNA with ethidium bro-
    mide, 7.31, 7.42, 7.44–7.45, 7.51
  transfer of nucleic acids to solid
    supports
    capillary, 7.46–7.50, 9.34–9.35
    electrophoretic, 9.35–9.36
    vacuum, 9.36–9.37
Alkaline chromatography on Sepharose
    CL-4B, to isolate small DNA
    probes, 10.25

Alkaline electrophoresis buffer, preparation, 6.7, B.23
Alkaline gel-loading buffer, B.24
Alkaline lysis procedure
 for extraction of bacteriophage M13 DNA, 4.31
 for extraction of plasmid DNA from bacteria
  large-scale preparation, 1.38–1.39
  small-scale preparation, 1.25–1.28
  buffer preparation, B.22
Alkaline phosphatase, enzyme-coupled conjugates for immunodetection, 12.20, 18.74
Alkaline phosphatase promoter (phoA), and signal sequence expression vectors, 17.31–17.33
Alkaline phosphatases. See BAP; CIP
α-Complementation of defective β-galactosidase
 histochemical identification of recombinant clones, 1.85–1.86, 4.7–4.8
 and lacZΔM15 on F′ in bacteriophage M13 host strains, 4.12
 and plaque screening of bacteriophage M13, 4.7–4.8, 4.22–4.23, 4.37–4.38
 vectors carrying lacZ′, 1.8–1.9, 1.13, 1.14, 1.17, 1.18, 1.20
α-Thiophosphate dNTPs, mutagenesis by misincorporation of nucleotides, 15.108
Amber suppressors
 in bacteriophage λ vectors and host strains, 2.16, 2.55, 2.57–2.59
 in bacteriophage M13 host strains, 4.13–4.15
 of nonsense mutations used in molecular cloning, D.1
 suppression for direct selection of recombinant clones, 1.6, 1.19, 9.8
Amino acids
 classification, D.6
 and codons (the genetic code), D.1
 properties, D.2–D.5
Aminoglycoside phosphotransferase (APH) in dominant selection systems, 16.10–16.14. See also Antibiotics; Selection
Aminopterin, selection, 3.18, 16.9. See also HAT medium; Selection
Ammonium acetate, stock solution, B.10
Ammonium hydroxide
 common commercial strengths, B.2
 pH values of stock solutions, B.3

Ammonium persulfate, catalysis of acrylamide polymerization, 6.36, 6.40–6.41, 13.48
 stock solution, B.10
5′-AMP. See Adenosine 5′-monophosphate
Ampicillin
 mechanism of action, 1.6
 stock solution, A.6
 use as selective agent, 1.5
  for maintaining expression plasmids, 17.16
Amplification, 1.3
 of bacteriophage λ plaques on nitrocellulose filters, 2.112
 of DNA by the polymerase chain reaction. See Polymerase chain reaction
 of gene copy number in mammalian cells
  using adenosine deaminase, 16.29
  using dihydrofolate reductase, 16.28–16.29
 of genomic DNA library in bacteriophage λ, 9.30
 of plasmid DNA using chloramphenicol, 1.3, 1.21, 1.23, 1.33. See also Chloramphenicol
 and storage of cosmid libraries
  liquid culture method, 3.50–3.51
  replica filter method, 3.46–3.49
  selection of method, 3.44–3.45
  transduction method, 3.52–3.53
AmpliTaq. See Taq DNA polymerase
Antibiotics
 ampicillin
  mechanism of action, 1.6
  stock solution, A.6
  use as selective agent, 1.5
   for maintaining expression plasmids, 17.16
 carbenicillin, 1.81, A.6
 chloramphenicol
  amplification of plasmids, 1.3, 1.21, 1.23, 1.33
   colony hybridization, 1.92, 1.95, 1.97
  mechanism of action, 1.6
  as selective agent, 1.5
  stock solution, 1.33, A.6
 geneticin (G418), use as selective agent, 16.10
 kanamycin
  mechanism of action 1.6
  resistance gene in pHSG274, cosmid

Autoradiography, E.21–E.23
  for detection of DNA
    alkaline agarose gels, 6.21
    polyacrylamide gels, 6.45, 13.56–13.57
  sensitivity of different methods, E.25
  setting up, E.26–E.29
Avian reverse transcriptase. *See* Reverse
  transcriptase

B

Bacteria. *See E. coli*
Bacterial alkaline phosphatase (BAP). *See*
  BAP
Bacteriophage-encoded DNA-dependent
  RNA polymerases. *See also*
    Chromosome walking; Probes,
    RNA; T7 bacteriophage; Tran-
    scription, in vitro
  activities and uses, 5.58–5.59
  expression system using, 5.58, 17.14–
    17.16
  promoter-containing vectors
    bacteriophage λ, 2.35–2.39, 2.52–2.54
    cosmid, 3.7, 3.20–3.23
    plasmid, 1.9, 1.15–1.18, 1.20, 2.52–
      2.54, 16.21, 17.14–17.16
  and synthesis of radiolabeled RNA
    probes, 10.27–10.37
Bacteriophage λ. *See* λ bacteriophage
Bacteriophage M13. *See* M13 bacterio-
    phage. *See also* Filamentous bac-
    teriophages; M13KO7; Phage-
    mids
Bacteriophage, single-stranded, filamen-
    tous. *See* Filamentous bacterio-
    phages; M13 bacteriophage
Bacteriophage SP6. *See* SP6 bacteriophage
Bacteriophage T3. *See* T3 bacteriophage
Bacteriophage T4. *See* T4 bacteriophage
Bacteriophage T7. *See* T7 bacteriophage
Bacto-agar, recipes, A.4
BAL 31 nuclease, activities and uses,
    5.73–5.77
  buffer, B.28
  engineering restriction sites, 17.20–
    17.24
  generation of bidirectional sets of dele-
    tion mutants, 15.14–15.15,
    15.20–15.26
  generation of nested sets of deletion mu-
    tants for DNA sequence analysis,
    13.34–13.35
BAP (bacterial alkaline phosphatase)

activity and uses, 5.72
inactivation after reaction, 5.72
suppression of self-ligation of plasmid
  cloning vectors, 1.56. *See also* CIP
BB4, *E. coli* strain
  genotype and uses, 2.59, A.9
  and λZAP, 8.45, 8.75
BCIP (5-bromo-4-chloro-3-indolyl phos-
    phate), and NBT for detection of
    alkaline phosphatase-coupled im-
    mune complexes, 12.20, 18.74
  stock solution, B.10
Benton and Davis method for screening
    bacteriophage plaques by hybrid-
    ization, 2.108–2.113
BES (*N,N*-bis[2-hydroxyethyl]-2-amino-
    ethanesulfonic acid)-buffered
    saline, stock solution, for transfec-
    tion of DNA into mammalian
    cells, 16.39, B.10
β-Aspartyl hydroxamate, aspartic acid an-
    alog, 16.15
β-Galactosidase
  and α-complementation
    in bacteria that contain recombinant
      plasmids, 1.85–1.86
    vectors carrying *lacZ′*, 1.8–1.9, 1.13,
      1.14, 1.17, 1.18, 1.20
    by bacteriophage M13 derivatives,
      4.7–4.8, 4.22–4.23, 4.37–4.38
  *lacZ*ΔM15 mutant, 4.12
  monitoring expression of, 17.34–17.35
  as reporter gene in mammalian cells,
    16.58
    assays, 16.66–16.67
β-Lactamase, resistance to ampicillin, 1.6,
    1.81
β-Mercaptoethanol, stock solution, 8.56,
    B.12
BHB2688, *E. coli* strain
  checking genotype, 2.96
  genotype and uses, 2.59, A.9
  preparation of packaging extracts from,
    2.102–2.103
BHB2690, *E. coli* strain
  checking genotype, 2.96
  genotype and uses, 2.59, A.9
  preparation of packaging extracts from,
    2.100–2.101
*bio, bio*256, genetic markers in bacterio-
    phage λ, 2.16
Bio-Gel A-50m chromatography, for re-
    moval of linkers, 8.25
Bio-Gel A-150m chromatography, for re-

moval of RNA from DNA, 1.52

Bio-Gel P-60 chromatography, for purification of radiolabeled oligonucleotides, 11.37–11.38

Biotin-avidin conjugates for immunodetection, 12.14

Bisacrylamide (N,N'-methylenebis-acrylamide), chemical properties, 6.36. *See also* Polyacrylamide gel electrophoresis
stock solution for acrylamide, B.9

N,N-Bis(2-hydroxyethyl)-2-aminoethane-sulfonic acid-buffered saline. *See* BES-buffered saline

BL21(DE3), *E. coli* strain, genotype and uses, A.9

Blocking agents
in prehybridization solutions, 9.48–9.50, B.15
for western blotting, 18.69, 18.73

Blocks in sequencing gels, 13.74

BLOTTO (Bovine Lacto Transfer Technique Optimizer)
preparation, 9.49, B.15
uses as blocking agent in hybridization, 1.102, 9.48–9.50, B.15

Bluescript M13+, M13−, 1.20. *See also* pBluescript and phagemid systems
in bacteriophage λ, 2.16, 2.52–2.54
map, 1.20, 2.54

Blunt ends, creating. *See* Klenow fragment (large fragment) of *E. coli* DNA polymerase I; Nuclease; T4 bacteriophage, DNA polymerase

BMT monkey cells that express T antigen, 16.20. *See also* COS monkey cells that express T antigen

BNN93 (C600), *E. coli* strain, genotype and uses, 2.58, A.9

BNN102 (C600*hflA*), *E. coli* strain, genotype and uses, 2.58, A.9

Boiling lysis procedure for extraction of DNA from bacteria
large-scale preparation, 1.34–1.35
small-scale preparation, 1.29–1.30

Bolton-Hunter reagent, 18.24

Bovine Lacto Transfer Technique Optimizer. *See* BLOTTO

Bovine papillomavirus (BPV)
extrachromosomal replication, 16.6
vectors, 16.23–16.26

BPV-1, bovine papillomavirus vector, 16.23
map, 16.24

5-Bromo-4-chloro-3-indolyl-β-D-galactoside. *See* X-gal

5-Bromo-4-chloro-3-indolyl phosphate. *See* BCIP

Bromocresol green, as tracking dye in alkaline agarose gel electrophoresis, 6.12, 6.21

5-Bromodeoxyuridine as thymidine analog, 16.9

Bromophenol blue
migration in polyacrylamide gels, 6.37
in denaturing gels, 7.70
use in agarose gel electrophoresis, 6.12

Buffer-gradient polyacrylamide gels, preparation, 13.47–13.53

Buffers
commonly used, B.20
phosphate, B.21
electrophoresis, B.23
enzyme, B.26–B.28
gel-loading, B.24–B.25
lysis, for minipreps of plasmid DNA, B.22

Butanol
concentrating nucleic acids by extraction with, E.16
extraction of ethidium bromide, 1.46

C

C-1a, *E. coli* strain, genotype and uses, 2.59, A.9

C127 murine cells, as host for bovine papillomavirus vectors, 16.25

c2RB, cosmid vector for propagation of eukaryotic DNA, 9.13
description, 3.13
directional cloning using double *cos* site, 3.10–3.15
map, 3.12
preparation of arms, 3.42–3.43

C600 (BNN93), *E. coli* strain, genotype and uses, 2.58, A.9

C600*hflA* (BNN102), *E. coli* strain, genotype and uses, 2.58, A.9

CAD protein and uridine biosynthesis, 16.14–16.15

Calcium chloride
stock solutions, B.10
transformation of *E. coli*, 1.82–1.84. *See also* Transformation

Calcium phosphate-DNA transfection of mammalian cells, 16.30, 16.32
adherent cells, method, 16.33–16.36

5'-CDP. *See* Cytidine 5'-diphosphate

Cell-free translation
  of hybridization-selected RNA, method,
      18.79–18.80
  of mRNA, 8.4

cer locus, equal partitioning of plasmids
      during cell division, 1.12

CES200, *E. coli* strain, genotype and uses,
      2.58, A.9

CES201, *E. coli* strain, genotype and uses,
      2.58, A.9

Cesium chloride (CsCl)
  cushion for RNA isolation, 7.19–7.22
  equilibrium centrifugation for purifica-
      tion of bacteriophage λ DNA, 2.79
  solutions for step gradients prepared in
      SM, 2.74–2.76

Cesium chloride–ethidium bromide
      gradients
  collection of superhelical DNA, 1.45
  discontinuous gradient, 1.42–1.45
  equilibrium centrifugation, 1.23, 1.42–
      1.45
  removal of ethidium bromide, 1.46–1.48

Cetylpyridinium bromide, precipitation of
      radiolabeled oligonucleotides,
      11.35–11.36

Charomid vectors
  description, 3.5–3.6, 3.25–3.26
  map, 3.24
  in restriction mapping of foreign DNA,
      3.25–3.26

Charon series of bacteriophage λ vectors,
      maps and descriptions. *See also* λ
      bacteriophage
  4A, 2.18–2.19
  21A, 2.18, 2.20–2.21
  32, 2.18, 2.22–2.23
  33, 2.18, 2.24–2.25
  34, 2.18, 2.26–2.27
  35, 2.18, 2.27, 2.28–2.29
  40, 2.18, 2.30–2.31

CHEF (contour-clamped homogeneous
      field electrophoresis). *See* Pulsed-
      field gel electrophoresis

Chemical degradation of DNA method of
      DNA sequencing (Maxam and Gil-
      bert). *See* Maxam and Gilbert
      chemical degradation of DNA
      method for DNA sequence analysis

chi sites for recombination in bacterio-
      phage λ, 2.12, 2.13, 2.16

Chloramine T, radioiodination of immu-
      noglobulin G using, 12.29, 18.24–

18.25

Chloramphenicol
  amplification of plasmids, 1.3, 1.21,
      1.23, 1.33
    colony hybridization, 1.92, 1.95, 1.97
  mechanism of action, 1.6
  as selective agent, 1.5
  stock solution, 1.33, A.6

Chloramphenicol acetyltransferase (CAT)
  assays, 16.59
    diffusion into scintillation fluid,
        16.64–16.65
    extraction with organic solvents,
        16.63
    thin-layer chromatography,
        16.60–16.63
  as reporter gene, 16.56
  vectors, 16.57

4-Chloro-1-naphthol, detection of
      antibody-horseradish peroxidase
      complexes, stock solution, 12.19

Chloroquine treatment following transfec-
      tion of mammalian cells, 16.32,
      16.35

Chromatography
  affinity, removing anti-*E. coli* anti-
      bodies by, 12.27–12.28
  Bio-Gel A-50m, for removal of linkers,
      8.25
  Bio-Gel A-150m, for removal of RNA
      from DNA, 1.52
  Bio-Gel P-60, for purification of radiola-
      beled oligonucleotides, 11.37–
      11.38
  column
    general method, E.36
    precautions for laboratory safety with
        radioactivity, E.36
  gel-filtration, general method, E.34–
      E.36
  hydroxyapatite
    preparation of subtracted probes, 8.49,
        10.43
    separation of single- and double-
        stranded DNA, E.30–E.34
  reversed-phase on silica gel, for purifica-
      tion of oligonucleotides, 11.29–
      11.30
  Sepharose CL-4B, alkaline, to isolate
      small probes, 10.25. *See also* Seph-
      arose CL-4B chromatography
  Sep-Pak C_{18} column, to isolate oligonu-
      cleotides, 11.39
  spun-column, E.37–E.38

Chromatography *(continued)*
  thin-layer, on silica gels, 16.57, 16.60–16.61
Chromosome crawling using the polymerase chain reaction, 14.12–14.13
Chromosome walking
  in cosmids, 3.2, 3.9, 3.23, 9.3
  in pWE15 and pWE16, 3.21
  using RNA probes generated by in vitro transcription of cosmids, 3.23
CIP (calf intestinal alkaline phosphatase)
  activity and uses, 5.72
  dephosphorylation buffer, 1.60, 3.28, B.27
  inactivation by proteinase K, 3.28, 3.36, 3.38, 5.72
  preparation of bacteriophage λ arms, 2.90–2.91
  preparation of radiolabeled DNA and RNA probes, 10.64–10.65
  suppression of self-ligation of cloning vectors, 1.60–1.61
  treatment of cosmid vector DNA
    digestion with one restriction enzyme and dephosphorylation in single *cos* site vector, 3.28
    digestion with two restriction enzymes and dephosphorylation in double *cos* site vector, 3.42–3.43
    digestion with two restriction enzymes and dephosphorylation in single *cos* site vector, 3.36–3.37
  treatment of eukaryotic DNA prior to ligation, 3.38–3.39
  versus size fractionation to eliminate self-ligation, 3.35, 3.38
Citric acid, pH values of stock solutions, B.3
CJ236 *(dut⁻ ung⁻ F′)*, *E. coli* strain, genotype and use in mutagenesis, 15.76, A.9
Cloning of DNA
  in bacteriophage λ, 2.82–2.107. *See also* cDNA synthesis and cloning; λ bacteriophage
  in cosmid vectors, 3.4–3.6, 3.27–3.55. *See also* Cosmid
  in plasmid vectors, 1.53–1.73. *See also* Plasmid
5′-CMP. *See* Cytidine 5′-phosphate
CMT monkey cells that express T antigen, 16.20. *See also* COS monkey cells that express T antigen
Coamplification systems, 16.28–16.29

Codons and amino acids (the genetic code), D.1
ColE1, origin of replication
  in cosmids, 3.5
  in plasmids, 1.2, 1.4
Colicins, plasmid production of, 1.2
Collodion bags, for isolation of high-molecular-weight DNA, 9.14
Colony hybridization, 1.90–1.104. *See also* Hybridization
Competent *E. coli* for DNA cloning. *See also* Transformation
  prepared by calcium chloride method, 1.82–1.84
  prepared by hexamminecobalt chloride/DMSO method, 1.76–1.81
  for transfection of bacteriophage M13 vectors, 4.36–4.38
  transformation with plasmid DNA, 1.74–1.84
Complementation of defective β-galactosidase. *See* α-Complementation of defective β-galactosidase
Compressions in sequencing gels, 13.10, 13.74–13.75
Condensing agents, use in DNA ligation, 1.70–1.71
Conjugation and mobilization of plasmids, 1.5
Contour-clamped homogeneous field electrophoresis (CHEF). *See* Pulsed-field gel electrophoresis
Coomassie brilliant blue dye, staining SDS-polyacrylamide gels, 18.55
Copy number of plasmids in bacterial hosts, 1.4
cos202, cosmid vector for transfection of mammalian cells, 3.19
cos203, cosmid vector for transfection of mammalian cells, 3.19
Cosmid
  cloning
    capacity, 3.2–3.3, 3.5–3.6
    and differential growth of clones, 3.3
    directional, 3.11
    partial digestion of eukaryotic DNA with *Mbo*I or *Sau*3AI, 3.32, 9.24–9.28
    treatment of eukaryotic DNA with alkaline phosphatase, 3.38–3.39
  libraries
    advantages for genomic cloning, 3.2, 3.27, 9.4

DE-81 filters, adsorption of radiolabeled nucleotides to, E.19
DEAE-cellulose membrane, use in recovery of DNA from agarose gels, 6.24–6.27
DEAE-dextran, transfection of mammalian cells, 16.30, 16.41, 16.42–16.46
DEAE-Sephacel, use in purification of DNA, 6.32–6.33
7-Deaza-2′-deoxyguanosine-5′-triphosphate. See 7-Deaza-dGTP
7-Deaza-dGTP (7-deaza-2′-deoxyguanosine-5′-triphosphate), use in resolving artifacts in DNA sequence analysis, 13.10. See also Compressions in sequencing gels
Δ(lac-proAB), deletion of lac operon, genetic marker in E. coli host strain, 4.12
Denaturing gradient polyacrylamide gels, 6.49, 13.47–13.53
Denhardt's reagent
  preparation, 9.49, B.15
  uses as blocking agent in hybridization, 9.48–9.50, B.15
2′-Deoxyadenosine 5′-triphosphate (dATP), structure, C.5
2′-Deoxycytidine 5′-triphosphate (dCTP), structure, C.7
2′-Deoxyformycin, inhibits adenosine deaminase, 16.15
2′-Deoxyguanosine 5′-triphosphate (dGTP), structure, C.9
2′-Deoxyinosine-5′-triphosphate. See dITP in analysis of sequencing artifacts
Deoxyribonuclease I. See DNAase I
Deoxyribonucleoside triphosphates (dNTPs)
  α-thiophosphate, 15.108
  optimal concentration for Klenow fragment of E. coli DNA polymerase I, 10.20
  stock solutions, 8.56, B.10–B.11
  structures, C.5, C.7, C.9, C.10
2′-Deoxythymidine, structure, C.10
2′-Deoxythymidine 5′-phosphate (TMP), structure, C.10
2′-Deoxythymidine 5′-triphosphate (dTTP), structure, C.10
DEPC (diethyl pyrocarbonate)
  precautions for laboratory safety, 7.3
  use as RNAase inhibitor, 7.3–7.4

Dephosphorylation of DNA. See BAP; CIP
Dextran sulfate, hybridization of nucleic acids, 9.50
dGTP. See 2′-Deoxyguanosine 5′-triphosphate
DH1, E. coli strain, genotype and uses, 3.53, A.9
DH5, E. coli strain, genotype and uses, A.10
DH5a, E. coli strain, genotype and uses, A.10
dhfr. See Dihydrofolate reductase
Dialysis tubing, preparation, E.39
Diaminobenzidine as substrate for horseradish peroxidase conjugates for immunodetection, 12.14, 12.19, 18.74
Diazobenzyloxymethyl paper. See DBM paper
Diazophenylthio paper. See DPT paper
Dichlorodimethylsilane
  precautions for laboratory safety, E.1
  siliconizing glassware, plasticware, and glass wool, E.1–E.2
Dideoxy-mediated chain-termination (Sanger) method for DNA sequence analysis, 13.3–13.6. See also Oligonucleotides, synthetic; Polymerase chain reaction
  denaturing polyacrylamide gels, 6.49, 13.45–13.46
    autoradiography, 13.56–13.57
    loading and running, 13.54–13.55
    preparation of buffer-gradient gels, 13.47–13.53
    reading sequence from gels, 13.58
    troubleshooting, 13.74–13.77
  DNA polymerases, 13.7–13.9
    Klenow fragment of E. coli DNA polymerase I, 13.59–13.64
    Sequenases, 13.65–13.69
    troubleshooting, 13.73–13.74, 13.76–13.77
  microtiter plates, 13.42–13.43
  primers, 13.6, 13.42
  reaction mixtures, 13.43–13.44
    analogs for solving artifacts in gels, 13.10, 13.74–13.75
    structures of nucleotides, C.5, C.7, C.9, C.10, C.13
    for use with Klenow fragment of E. coli DNA polymerase I, 13.44, 13.60

Guanine and related compounds, molecular structures, C.8–C.9
Guanosine, structure, C.8
Guanosine 5′-diphosphate (5′-GDP), structure, C.9
Guanosine 5′-monophosphate (5′-GMP), structure, C.8
Guanosine 5′-triphosphate (5′-GTP), structure, C.9
Guessmers
  designing, 11.11–11.15
  as hybridization probes, 11.52–11.53
  labeling, 11.15–11.16

H

^{3}H, isotopic data, B.8
HAT (hypoxanthine, aminopterin, and thymidine) medium, use as selective agent, 16.9, 16.14
HB101, *E. coli* strain
  genotype and uses, A.10
  propagation of bacteriophage λ, 2.59
  purification of plasmid DNA from, 1.22
Hemin, stock solution, for in vitro translation, 18.78
Heparin
  preparation, 9.49, B.15
  uses as blocking agent in hybridization, 9.48–9.50, B.15
HEPES (*N*-2-hydroxyethylpiperazine-*N*′-2-ethanesulfonic acid), pK_a and buffering range, B.1
HEPES-buffered saline, stock, B.11
Hexamminecobalt chloride
  use in ligation of blunt-ended DNA, 1.70–1.71
  use in preparation of competent bacteria, 1.76–1.81
hfl (high frequency of lysogenization), genetic marker in *E. coli* host strain, 2.11–2.12, 2.57
HGPRT. *See* Hypoxanthine-guanine phosphoribosyl transferase
Histochemical identification of recombinant clones, 1.85–1.86. *See also* α-Complementation of defective β-galactosidase; λORF8; pMR100
HMS174, *E. coli* strain, genotype and uses, A.10
Homologous recombination
  of cosmids and plasmid sequences, 3.6–3.7

in plasmid vectors, 1.19
Homopolymeric tailing in cDNA synthesis and cloning, 8.11–8.12, 8.21–8.23, 8.34–8.35
Horizontal slab gel electrophoresis. *See* Agarose gel
Horseradish peroxidase (HRP) conjugates for immunodetection, 12.14, 12.19, 18.74
HRP. *See* Horseradish peroxidase conjugates for immunodetection
hsdM, genetic marker in *E. coli* host strain, 2.57
hsdR, genetic marker in *E. coli* host strain, 2.57, 4.13
hsdS, genetic marker in *E. coli* host strain, 2.57
Human growth hormone, as reporter gene in mammalian cells, 16.58
Hybridization
  of bacteriophage λ plaques (Benton and Davis method), 2.108–2.113
  of bacteriophage M13 single-stranded DNA, 4.40–4.41
  blocking agents to suppress background, 9.48–9.50, B.15
  differential, of cDNA libraries, 8.49, 10.38
  DNA. *See* Oligonucleotides, synthetic; Probes; Southern hybridization
  dot and slot, of RNA, 7.37, 7.53–7.57
  effect of length and degeneracy of oligonucleotide on specificity, 11.7–11.8
  effects of mismatches, 11.47
  filter, of RNA, 7.38
  formamide solution, 1.101
  Grunstein and Hogness method, 1.90–1.104
  in situ
    of bacterial colonies, 1.90–1.104
    of bacteriophage λ plaques, 2.108–2.117
    of bacteriophage M13 plaques, 4.41
  northern, of RNA, 7.37, 7.39–7.52
  prehybridization solution, 1.101–1.102, 9.52–9.53, 11.56
  of radiolabeled probes to immobilized nucleic acids, 9.47–9.57. *See also* Southern hybridization
  rate, 9.50
  RNA, 7.37–7.52
  screening of bacterial colonies for re-

combinant plasmids, 1.90–1.104.
See also Plasmid
of cDNA libraries, 8.46–8.49
screening with degenerate oligonucleotide probes, conditions, 11.12–
11.13, 11.19, 11.45–11.57. *See also*
Oligonucleotides, synthetic
slot, of RNA, 7.37, 7.53–7.57
solution for nylon membranes, 9.54
solution, of RNA, 7.37
Southern, for analysis of genomic DNA.
See Southern hybridization
temperature, 9.50
T_m, 9.50–9.51
calculation, 11.46, 11.50
empirical determination, 11.55–11.57
washing conditions, 9.50
Hybrid selection of cDNA clones, 8.33,
8.50–8.51
Hydrazine, for Maxam and Gilbert DNA
sequence analysis. *See* Maxam and
Gilbert chemical degradation of
DNA method for DNA sequence
analysis
precautions for laboratory safety, 13.85
stock solution, 13.84
Hydrochloric acid (HCl)
common commercial strengths, B.2
pH values of stock solutions, B.3
precautions for laboratory safety, 17.41
Hydroxyapatite chromatography
preparation of subtracted probes, 8.49,
10.43
separation of single- and double-
stranded DNA, E.30–E.34
Hygromycin B phosphotransferase (*hyg*)
in cos202, cos203, cosmid vectors for
transfection of mammalian cells,
3.19
in mammalian expression vector pHyg,
16.13
as selectable marker in mammalian
cells, 16.14
Hypoxanthine
as selective agent, 16.9. *See also* HAT
medium
structure, C.12
Hypoxanthine-guanine phosphoribosyl
transferase (HGPRT)
in pCV series, cosmid vectors for transfection of mammalian cells, 3.19
as selectable marker in mammalian
cells, 16.14

I

^{125}I, ^{131}I
isotopic data, B.8
precautions for laboratory safety, 18.24
*imm*21, 2.16
*imm*80, 2.16
*imm*434, 2.16
Immunoadsorption
of anti-*E. coli* antibodies from sera to be
used for immunological screening,
12.13, 12.26
of cross-reacting antibodies from antisera, 18.15
Immunological assays
immunoprecipitation. *See* Immunoprecipitation
solid-phase radioimmunoassay, 18.19–
18.20
using two antibodies, 18.21–18.23
western blotting. *See* Western blotting
Immunological screening
of bacterial colonies, 12.21–12.28
lysis buffer, 12.24
of bacteriophage λ cDNA libraries, 2.14,
8.49–8.50
method, 12.16–12.20
vectors, 2.42–2.54, 8.36–8.45
choosing the antibody, 12.11–12.12
methods used to detect antibodies bound
to proteins expressed in *E. coli*,
12.14
purification of antisera, 12.13
removing anti-*E. coli* antibodies
by affinity chromatography,
12.27–12.28
with *E. coli* lysates, 12.26
by pseudoscreening, 12.25
uses, 12.2–12.3
Immunoprecipitation
formation of antigen-antibody complexes, 18.42–18.46
lysis of bacteria, 18.40–18.41
lysis of mammalian cells, 18.30–18.34
buffers, 18.32
lysis of yeast
enzymatic, 18.36–18.37
mechanical, 18.35
rapid (RIPA buffer), 18.38
NET-gel buffer, 18.44
preclearing the cell lysate, 18.43
radiolabeling
mammalian cells, 18.26–18.28

Immunoprecipitation, radiolabeling
(continued)
yeast and bacteria, 18.29
and SDS-polyacrylamide gel electrophoresis of proteins, 18.47–18.59
Inclusion bodies, protein purification from, 17.37
cell lysis, 17.38
purification and washing, 17.39–17.40
refolding of proteins, 17.41
solubilization, 17.41
Incompatibility groups, plasmid, 1.3–1.4
Inosine
as a neutral base at positions of degeneracy in degenerate probes, 11.17–11.19, 11.54
structure, C.12
Insertional inactivation of antibiotic resistance genes in plasmid, 1.87–1.89
In situ hybrization. *See* Hybridization, in situ
*int*29 in bacteriophage λ, 2.16
Intensifying screen, use in detecting different isotopes, E.25
Interference-resistant helper bacteriophages, 4.18–4.19
Intergenic regions of filamentous bacteriophages, 4.3, 4.7–4.8
in phagemids, 4.17–4.19
Intra-allelic complementation. *See* α-Complementation of defective β-galactosidase
Iodination, radiolabeling of immunoglobulin G using chloramine T, 12.29, 18.24–18.25
IPTG (isopropylthio-β-D-galactoside)
in bacteriophage M13 cloning, 4.33, 4.37–4.38
induction of β-galactosidase, 1.8–1.9, 1.86
induction of *lac* and *tac* promoters, 17.12–17.13
stock solution, 4.22, 8.57, B.11
IR1, helper virus recommended for use with pEMBL phagemid, 4.18
Isopropanol precipitation of nucleic acids, E.10–E.15
Isopropylthio-β-D-galactoside. *See* IPTG
Isoschizomers
and methylation, 5.25
of restriction enzymes, 5.4–5.8, 5.14
Isotopic data, B.8

J

JM101, *E. coli* strain
genotype and uses, A.10
use in propagation of bacteriophage M13 vectors, 4.14, 4.15, A.13
JM103, *E. coli* strain, 4.15, A.13
JM105, *E. coli* strain
genotype and uses, A.10
use in propagation of bacteriophage M13 vectors, 4.14, A.13
JM106, *E. coli* strain, 4.15, A.13
JM107, *E. coli* strain
genotype and uses, A.10
use in propagation of bacteriophage M13 vectors, 4.14, 4.15, A.13
JM108, *E. coli* strain, 4.15, A.13
JM109, *E. coli* strain
genotype and uses, A.10
use in propagation of bacteriophage M13 vectors, 4.14, 4.15, 4.50, A.13
JM110, *E. coli* strain
genotype and uses, A.10
use in propagation of bacteriophage M13 vectors, 4.14, A.13

K

K802, *E. coli* strain, genotype and uses, 2.58, A.10
Kanamycin
mechanism of action 1.6
resistance gene in pHSG274, cosmid vector for transfection of mammalian cells, 3.19
stock solution, A.6
use as selective agent, 1.5, 16.10
Keyhole limpet hemocyanin (KLH), coupling of peptides to, 18.8–18.9
KGB buffer, for restriction enzyme digestion, 5.28–5.31, B.26
Kinase, bacteriophage T4 polynucleotide. *See* Polynucleotide kinase, bacteriophage T4
KK2186, *E. coli* strain
genotype and uses, A.10
as host strain for phagemids, 4.18
use in propagation of bacteriophage M13 vectors, 4.15, 4.18, A.13
Klenow buffer, B.26
Klenow fragment (large fragment) of *E. coli* DNA polymerase I, reactions and uses, 5.34, 5.40–5.43

DNA sequence analysis by dideoxy-mediated chain-termination method, 13.7–13.8, 13.59–13.64, 13.73–13.74

labeling of synthetic oligonucleotides, 11.40–11.44

labeling 3′ termini of double-stranded DNA, 10.51–10.53

primer repair, 17.18–17.19

primer:template ratios and nucleotide concentrations, effect on yield of radioactive probes, 10.20–10.21

repair of DNA to create blunt end, 8.27, 17.20–17.23

synthesis of single-stranded DNA probes, 10.22–10.24

synthesis of uniformly labeled DNA probes using random oligonucleotide primers, 10.13–10.17

KLH. *See* Keyhole limpet hemocyanin

Kunkel method of oligonucleotide-mediated, site-directed mutagenesis, 15.74–15.79

L

*lac*5, 2.16

*lacI*q

-encoding plasmid pMC9, 2.57

genetic marker in *E. coli* host strain, 2.57, 4.12, 17.13

marker in λORF8, 2.51

lac operon

effect of *lacI*q mutation, 4.12

Δ(*lac-proAB*), deletion of lac operon, 4.12

Δ(*lac-proAB*), deletion of *lac* operon, genetic marker in *E. coli* host strain, 4.12

Lactamase. *See* β-Lactamase

lacZ

fusion proteins, expression in *E. coli*, 17.4–17.9, 17.29–17.30

gene in α-complementation in bacteriophage M13 vectors, 4.7–4.8

gene product. *See* β-Galactosidase

lacZ′ in bacteriophage λ, 2.16, 2.51, 2.57

*lacZ*ΔM15, genetic marker in *E. coli* host strain, 2.57, 4.12

λ2001, bacteriophage λ vector map and description, 2.12, 2.34–2.35. *See also* λ bacteriophage

λ bacteriophage

adsorption, 2.3

for amplification and storage of cosmid libraries, 3.44, 3.52–3.53, 9.30

arms

amber mutations, 9.8

ligation to fragments of foreign DNA, 2.94

purification, 2.85–2.89

assembly pathway, 2.7

att sites, 2.8

deletions of *b* region that damage, 2.16, 2.95

and cDNA cloning, 8.36–8.45

Cro protein, 2.5

D ("decoration") protein, 2.7, 2.95

diluent (buffer), 2.118, A.8

DNA

digestion with restriction enzymes, 2.83–2.85, 2.92–2.93

preparation of markers for pulsed-field gel electrophoresis, 6.58–6.59

preparation of vector, 2.82–2.90

purification and extraction, 2.60–2.81

using cesium chloride step gradients, 2.74–2.77

using glycerol step gradients, 2.78

yield, 2.81

replication, 2.5

E protein, 2.7, 2.95

exonuclease, activity and uses, 5.86

expression libraries, 12.10

advantages, 12.4–12.7

immunological screening, 2.42–2.54, 8.36–8.45, 12.16–12.20

screening with synthetic oligonucleotides, 12.30–12.38

fusion proteins, preparation of lysates containing, 12.39–12.40

genes, 2.6

genetic map, 2.6

genetic markers

*c*I, *c*II, and *c*III genes, 2.8

checking for *ts* mutation, 2.96

*c*Its857 or *imm*434 *c*Its, 2.16, 2.95

chi sites for recombination, 2.12, 2.13, 2.16

cos site of prophage, 2.98

exo bet (*red*), 2.16

gam gene, 2.5, 2.12, 2.13, 2.16

in vectors for construction of genomic libraries, 9.8

*int*am, 2.16

wild-type, map, 2.17

Xis protein, 2.8

λDASH, bacteriophage λ vector map and description, 2.12, 2.35, 2.36–2.37. *See also* λ bacteriophage

λFIX, bacteriophage λ vector map and description, 2.35, 2.37, 2.38–2.39. *See also* λ bacteriophage

λgt10, bacteriophage λ vector, 2.11, 2.41. *See also* λ bacteriophage

cloning cDNA in, 8.36–8.37, 8.38, 8.78

map, 2.40

λgt11, bacteriophage λ vector, 2.12, 2.43. *See also* λ bacteriophage

cloning cDNA in, 8.37–8.38, 8.79

expression libraries, 12.10

lysogens, preparation of lysates containing fusion proteins, 12.39–12.40

map, 2.42

λgt18–23, bacteriophage λ vectors, 2.14, 2.43. *See also* λ bacteriophage

expression libraries, 12.10

λgt18 and λgt19, 2.45

map, 2.44

use in cDNA cloning, 8.41–8.42

λgt20 and λgt21, 2.47

map, 2.46

use in cDNA cloning, 8.42

λgt22 and λgt23, 2.49

map, 2.48

use in cDNA cloning, 8.42

λORF8, bacteriophage λ vector, 2.14, 2.51. *See also* λ bacteriophage

expression libraries, 12.10

map, 2.50

use in cDNA cloning, 8.39–8.41

λZAP and λZAPII, bacteriophage λ vector, 2.53. *See also* λ bacteriophage

Bluescript plasmid in, 2.54

cDNA cloning in, 8.44–8.45, 8.75, 8.79

expression libraries, 12.10

map, 2.52

LB medium, recipe, A.1

LE392, *E. coli* strain, genotype and uses, 2.58, A.11

Leupeptins, protease inhibitors, properties, 18.31

LG90, *E. coli* strain, genotype and uses, 17.4, A.11

Libraries. *See* cDNA synthesis and cloning; Expression, libraries; Genomic DNA libraries; Immunological screening; λ bacteriophage

Ligases, 5.61

DNA. *See also* Ligation of DNA

bacteriophage T4

buffer, 1.68, B.27

properties and uses, 5.60, 5.61, 5.62–5.63

Weiss units, 5.62

E. coli, 5.64–5.65

RNA, bacteriophage T4, 5.61, 5.66–5.67

Ligation of DNA. *See also* Cosmid; λ bacteriophage; Ligases, DNA; M13 bacteriophage, cloning; Plasmid

biochemical reaction, 1.63–1.67

blunt-ended, 1.70–1.71, 5.11

inhibition by high concentrations of ATP, 3.13, 3.42

buffers, 1.62, B.27

cleaved by restriction enzymes, 5.10–5.13

of cohesive termini

compatible, 1.68–1.69, 5.10

incompatible, 5.11–5.13

effects of DNA length and concentration, 1.63–1.67

in low-melting-temperature agarose, 1.72–1.73, F.6–F.7

of partially filled termini, 5.13. *See also* Genomic DNA libraries

reaction, 1.63–1.67

selection of appropriate concentration of plasmid and foreign DNA, 1.65–1.67

strategies, 1.53–1.59

test reactions

for bacteriophage λ arms to foreign DNA, 2.94

for construction of cosmid DNA libraries, 3.29

and transformation of plasmid DNA, 1.62

Linear DNA (form III), electrophoretic properties, 6.5

Linker(s)

addition to blunt-ended DNA in subcloning, F.8–F.10. *See also* Oligonucleotides, synthetic

cDNA synthesis and cloning

sequential addition of different linkers, 8.27–8.28

use of synthetic DNA linkers and adapters, 8.23–8.24, 8.26

ligation to, 8.68–8.69

RNA
blotting, 7.37, 7.39–7.52. *See also* RNA, hybridization
denaturation with glyoxal/DMSO, 7.40–7.42
determining concentration, 7.8
:DNA hybrids, stability, 9.51
electrophoresis. *See also* RNA, mapping
after denaturation with glyoxal/DMSO, 7.40–7.42
in gels containing formaldehyde, 7.43–7.45
relative mobility through polyacrylamide/urea, 7.76
extraction buffer, 7.6, 7.12
glyoxal/DMSO
denaturation, 7.40–7.42
removal of glyoxal for hybridization, 7.48
homogenization buffer, for tissue, 7.16
hybridization, 9.47–9.55
dot and slot, 7.37, 7.53–7.57
filter, 7.38
northern, 7.37, 7.39–7.52
solution, 7.37
inhibitors of RNAase activity. *See* RNAase inhibitors
isolation. *See also* RNA, messenger
cesium chloride cushion, 7.19
of cytoplasmic RNA from mammalian cells, 7.12–7.15
from eggs and embryos, 7.16–7.17
oligo(dT)-cellulose, 7.26–7.29
poly(U)-Sephadex, 7.29
from tissue, 7.22
of total RNA from mammalian cells, 7.6–7.11
using denaturants, 7.18
guanidine HCl and organic solvents, 7.23–7.25
guanidinium thiocyanate, 7.19–7.22
ligase, bacteriophage T4, 5.61, 5.66–5.67
mapping
with nuclease S1, 7.58–7.61
using double-stranded DNA probes, 7.62–7.65
using single-stranded DNA probes, 7.66–7.70
by primer extension, 7.79–7.83
with RNAase and radiolabeled RNA probes, 7.71–7.78
messenger (mRNA)

abundance, 8.6
capping of synthetic, for translation in vitro, 18.81
and cDNA synthesis. *See* cDNA synthesis and cloning
enrichment for low abundance, 8.6–8.8
poly(A)$^+$, selection of, 7.26–7.29
polyadenylation and mRNA stability, 18.81
size fractionation in presence of methylmercuric hydroxide, 7.30
through agarose gels, 7.31–7.34
through sucrose gradients, 7.35–7.36
synthesis of synthetic, 18.82–18.84
translation, in vitro, of synthetic, 18.81–18.85
northern hybridization (RNA blotting), 7.37, 7.39–7.52
poly(A)$^+$ mRNA, selection of, 7.26–7.29
polymerases. *See* RNA polymerases
primer extension, 7.79–7.83
probes. *See* Probes
quantitation, E.5
removal of oligodeoxyribonucleotides, 7.9
removal from plasmid DNA preparations
centrifugation through 1 M sodium chloride, 1.51
chromatography through Bio-Gel A-150m or Sepharose CL-4B, 1.52
precipitation with lithium chloride from minipreps, 13.72
ribosomal (rRNA), 7.2
spectrophotometric determination of the amount, E.5
staining
with ethidium bromide before transfer to solid support, 7.31, 7.42, 7.44–7.45, 7.51
with methylene blue after transfer to solid support, 7.51
synthesis, in vitro, 10.32–10.37. *See also* Bacteriophage-encoded DNA-dependent RNA polymerases; Transcription, in vitro
transfer, capillary
to nitrocellulose filters, 7.46–7.48
to nylon filters, 7.49–7.50
transfer (tRNA), as carrier for precipitating DNA, 6.26
translation. *See* Translation

RNA I and RNA II, control of plasmid
    replication, 1.3–1.5
RNAase (ribonuclease)
  DNAase-free
    preparation, B.17
    removal of RNA from plasmid DNA,
      1.51–1.52
  inhibitors, 7.3–7.5
  mapping of RNA, 7.71–7.78
    digestion mixtures, 7.75
RNAase A, activity and uses, 5.81
RNAase H
  processing for DNA synthesis, 1.2–1.3
  and replacement synthesis of second-
    strand cDNA, 8.15, 8.64–8.65
RNAase inhibitors
  diethyl pyrocarbonate, 7.3–7.4
  guanidine HCl, 7.5
  guanidinium thiocyanate, 7.5
  Macaloid, 7.5
  placental protein inhibitor, 7.4, 8.12–
    8.13, 8.58
  vanadyl-ribonucleoside complexes,
    7.4–7.5
RNAase T, activity and uses, 5.82
RNA-dependent DNA polymerase. *See* Re-
    verse transcriptase
RNA ligase, bacteriophage T4, 5.61,
    5.66–5.67
RNA polymerases, DNA-dependent, reac-
    tions and uses, 5.58–5.59. *See also*
    Bacteriophage-encoded DNA-
    dependent RNA polymerases
  expression system using, 5.58, 17.11–
    17.16
  and synthesis of radiolabeled RNA
    probes, 10.27–10.37. *See also*
    Probes, RNA; Transcription, in
    vitro
RNA probes. *See* Probes, RNA
Rop protein, control of plasmid replica-
    tion, 1.3–1.5, 1.13
rpsL, genetic marker in *E. coli* host
    strain, 2.57
RR1, *E. coli* strain, genotype and uses,
    A.11
rRNA, 7.2

S

71/18, *E. coli* strain
  genotype and uses, A.9
  as host strain for phagemids, 4.18

  use in propagation of bacteriophage
    M13 vectors, 4.15, 4.18
[35]S, isotopic data, B.8
  radiolabeling DNA for sequencing,
    13.9–13.10, 13.61, 13.67–13.68
  radiolabeling target protein for immu-
    noprecipitation, 18.26–18.29
S1 nuclease, activity and uses, 5.78–5.79
  buffer, 4.43, B.28
  determining size of inserts cloned into
    bacteriophage M13 vectors, proce-
    dure, 4.43
  digestion of hairpin loop, in cDNA syn-
    thesis, 8.15
  digestion of nicked/gapped circular
    DNA, for mutagenesis, 15.40–
    15.41
  RNA mapping, 7.58–7.70
  buffer, 7.64
Salicylate, precautions for laboratory
    safety, E.24
Salicylate fluorography, E.24
Salmon sperm DNA, denatured,
    fragmented
  preparation, 9.49, B.15
  uses as blocking agent in hybridization,
    9.48–9.50, B.15
Sample-loading buffers for agarose and
    acrylamide gels. *See* Loading
    buffers
Sanger method for DNA sequence analy-
    sis. *See* Dideoxy-mediated chain-
    termination (Sanger) method for
    DNA sequence analysis
Sarkosyl. *See* Sodium lauryl sarcosinate
sbcA, genetic marker in *E. coli* host
    strain, 2.57
sbcB, genetic marker in *E. coli* host
    strain, 2.57
Screening. *See also* Immunological
    screening
  bacterial colonies by hybridization,
    1.90–1.104
  bacteriophage M13 recombinants, 4.29,
    4.39–4.43, 4.49. *See also* α-Com-
    plementation of defective β-
    galactosidase
  of cDNA libraries, 8.46–8.52
  recombination, 1.19, 9.8
SDS (sodium dodecyl sulfate), stock solu-
    tion, B.13
SD sequence. *See* Shine-Dalgarno
    sequence

SDS gel-loading buffer, B.25
SDS-polyacrylamide gel electrophoresis of
    proteins, 18.47–18.48
  drying gels, 18.58–18.59
  gel-loading buffer, 18.53
  immunoblotting. *See* Western blotting
  preparation, 18.49–18.54
  range of separation versus acrylamide
    concentration, 18.48
  reagents, 18.49
  staining gels
    Coomassie brilliant blue, 18.55
    silver salts, 18.56–18.57
  transfer to solid supports. *See* Western
    blotting
  Tris-glycine electrophoresis buffer,
    18.53
Selection. *See also* Amber suppressors;
    Antibiotics; Plasmid
  adenosine deaminase, 16.15
    amplification system (pMT3SV$_2$),
      16.29
  aminoglycoside phosphotransferase,
    16.10–16.14
  asparagine synthetase, 16.15
  in bacteria, 1.5–1.6
  CAD and uridine biosynthesis,
    16.14–16.15
  in cosmid vectors for transfection of
    mammalian cells, 3.19
  dihydrofolate reductase, 16.9–16.10,
    16.20–16.22
    amplification system, 16.28–16.29
  geneticin (G418), 16.10
  hygromycin B phosphotransferase,
    16.13–16.14
  kanamycin, 3.19, 16.10
  neomycin, 3.18, 16.10, 16.12
  L-phosphonacetyl-L-aspartate and CAD,
    16.14–16.15
  thymidine kinase gene, 16.9, 16.11. *See
    also* HAT medium
  xanthine-guanine phosphoribosyl trans-
    ferase, 16.14
Sephadex G50, stock solution, 8.58
Sepharose CL-4B chromatography
  alkaline chromatography, for isolation
    of small DNA probes, 10.25
  preparation of medium, 8.59
  for removal of linkers, 8.25
  for removal of RNA from plasmid DNA,
    1.52
  for size selection of cDNA, 8.70–8.72

Sep-Pak C$_{18}$ column chromatography, for
    purification of radiolabeled oligo-
    nucleotides, 11.39
Sequenase enzymes. *See also* T7 bacterio-
    phage, DNA polymerase
  use in DNA sequence analysis, 13.8–
    13.9, 13.65–13.70, 13.77
Sequencing gel-loading buffer, B.25
Sex pili encoded by F factor and infection
    of bacterial strains with filamen-
    tous virus particles, 4.3, 4.12
Shine-Dalgarno (SD) sequence, ribosome-
    binding site in *E. coli*, 17.12–
    17.21
Sib selection of cDNA clones, 8.50
Siliconizing glassware, plasticware, and
    glass wool, E.1–E.2
Silver staining of polypeptides in SDS-
    polyacrylamide gels, 18.56–18.57
Single-stranded DNA. *See also* Filamen-
    tous bacteriophages; M13 bacte-
    riophage; Phagemids
  analysis on alkaline agarose gels,
    6.20–6.21
  -binding protein (SSB), 5.87, 5.88
  filamentous bacteriophage produce
    phagemids, 4.20
  generation from plasmids, 1.9, 1.14
  large-scale preparation, 4.32
  probes for mapping RNA with nuclease
    S1, 7.66–7.70
  production of phagemid, 4.48–4.50
  small-scale preparation, 4.29–4.30
Single-stranded, filamentous bacterio-
    phage. *See* Filamentous bacterio-
    phages; M13 bacteriophage
Site-directed mutagenesis of cloned DNA.
    See Mutagenesis, site-directed
SMC-A, SMC-B buffer, for bacteriophage
    λ packaging extracts, 2.107
SMR10, *E. coli* strain
  genotype and uses, 2.59, A.11
  preparation of packaging extracts from,
    2.105–2.106
SOB medium, recipe, A.2
SOC medium, recipe, A.2
Sodium acetate
  stock solution, B.13
  use in precipitating nucleic acids,
    E.10–E.15
Sodium azide
  precautions for laboratory safety, 1.48
  use in BLOTTO, 1.102

Sodium bicarbonate, pH values of stock solutions, B.3
Sodium bisulfite mutagenesis of DNA, 15.106
Sodium butyrate increases efficiency of mammalian cell transfection, 16.32, 16.35
Sodium carbonate, pH values of stock solutions, B.3
Sodium chloride
   gradients, for purification of bacteriophage λ vector arms, 2.88–2.89
   removal of RNA from plasmid DNA preparations by centrifugation through 1 M sodium chloride, 1.51
   stock solution, B.13
   use in precipitating nucleic acids, E.10–E.15
Sodium dodecyl sulfate. *See* SDS
Sodium hydroxide (NaOH)
   common commercial strengths, B.2
   pH values of stock solutions, B.3
Sodium lauryl sarcosinate, in solutions for RNA isolation, 7.20
Sodium lauryl sulfate. *See* SDS
Sodium phosphate buffers, preparation, B.21
Sodium salicylate fluorography, E.24
Solid-phase radioimmunoassay (RIA), 18.19–18.20
   using two antibodies, 18.21–18.23
Sonication buffer for packaging extracts, 2.100
Southern hybridization, for analysis of genomic DNA, 9.31–9.57. *See also* Genomic DNA; Hybridization; Probes; Pulsed-field gel electrophoresis
   agarose gel electrophoresis for DNA separation, 9.32–9.33
   blocking agents to suppress background, 9.48–9.50, B.15
   prehybridization solutions, 9.52–9.53
   of radiolabeled oligonucleotides to genomic DNA, 9.56–9.57
   transfer of DNA to solid supports
      capillary, 9.34–9.35, 9.38–9.46
      electrophoretic, 9.35–9.36
      vacuum, 9.36–9.37
   using dextran sulfate, 9.50
   using formamide, 9.52–9.55
   washing conditions, 9.50
SP6 bacteriophage, DNA-dependent RNA polymerase, activities and uses, 5.58–5.59. *See also* Bacteriophage-encoded DNA-dependent RNA polymerases
Spectinomycin, antibiotic to amplify plasmids, 1.3
Spectrophotometric determination of the amount of DNA/RNA, E.6
Spermidine
   effect on ligation of blunt-ended DNA fragments, 1.70
   versus putrescine in buffers for preparing packaging extracts in cosmids, 3.33
Spin dialysis using Centricon microconcentrators, 14.22–14.24
Splicing signals in eukaryotic expression modules, 16.7–16.8
SSB. *See* Single-stranded DNA-binding protein
SSC, stock solution, B.13
SSPE, stock solution, B.13
Stab cultures, for bacterial storage, recipe, A.5
Stains
   Coomassie brilliant blue dye for proteins, 18.55
   ethidium bromide for nucleic acids. *See* Ethidium bromide
   India ink for immobilized proteins, 18.68
   methylene blue for immobilized RNA, 7.51
   Ponceau S for immobilized proteins, 18.67
   silver staining of polypeptides in SDS-polyacrylamide gels, 18.56–18.57
STE (sodium chloride, Tris, EDTA) buffer for extracting plasmid DNA from bacteria, 1.34
   preparation of buffer, B.20
STET (sodium chloride, Tris, EDTA, Triton) buffer for boiling lysis extraction of plasmid DNA from bacteria, 1.29, 1.34
   preparation of buffer, B.20
Storage media, for bacteria, recipes, A.5
Storage of cosmid libraries. *See* Amplification, and storage of cosmid libraries
Strand separating gels, 6.49
Streptomycin
   mutation in *rpsL* for resistance, 2.57
   stock solution, A.6
Subcloning, F.1

TBE (Tris-borate/EDTA electrophoresis buffer), 6.7, B.23
TB medium. *See* Terrific Broth
TBS (Tris-buffered saline), stock solution, B.14
TCA (trichloroacetic acid)
  precipitation of nucleic acids, E.18
  stock solution, B.13
TE (Tris-EDTA), preparation of buffer, B.20
TEACl. *See* Tetraethylammonium chloride
TEMED (*N,N,N',N'*-tetramethylethylenediamine), use in polyacrylamide gel electrophoresis, 6.36, 18.49
TEN buffer for ion-exchange method of removing ethidium bromide from DNA, 1.47, B.20
ter function of bacteriophage λ gene *A* protein, 3.5
Terminal transferase (terminal deoxynucleotidyl transferase), reactions and uses, 5.56–5.57
  addition of homopolymeric tails for second-strand cDNA synthesis, 8.17
Terrific Broth (TB medium), recipe, A.2
Tetracycline
  mechanism of action, 1.6
  stock solution, A.6
  use as selective agent, 1.5
Tetraethylammonium chloride (TEACl), for hybridization of pools of oligonucleotides, 11.48–11.51
Tetramethylammonium chloride (TMACl), for hybridization of pools of oligonucleotides, 11.48–11.51
N,N,N',N'-Tetramethylethylenediamine (TEMED), use in polyacrylamide gel electrophoresis, 6.36
TFB buffer for preparation of competent *E. coli*, 1.78
TG1, *E. coli* strain
  competent cells, preparation, 4.36
  genotype and uses, A.12
  use in propagation of bacteriophage M13 vectors, 4.14
TG2, *E. coli* strain
  competent cells, preparation, 4.36
  genotype and uses, A.12
  use in propagation of bacteriophage M13 vectors, 4.14

Thin-layer chromatography on silica gels, 16.57, 16.60–16.61
Thiophosphate dNTPs. *See* α-Thiophosphate dNTPs
Thymidine kinase gene (*tk*), as selectable marker in mammalian cells, 16.9, 16.11. *See also* HAT medium
Thymine and related compounds, structures, C.10
tk. See Thymidine kinase gene
TLCK (tosyllysine chloromethyl ketone), protease inhibitor, properties, 18.31
TMACl. *See* Tetramethylammonium chloride
T_m (melting temperature) of hybrid between probe and target sequence
  empirical determination for oligonucleotide hybridized to target sequence, 11.55–11.57
  estimating, 9.50–9.51, 11.46–11.49
  for mutagenesis, 15.55
  irreversible, 11.48–11.49
Tn*10*, transposon encoding resistance for tetracycline as genetic marker in *E. coli* host strain, 2.57
TNT (Tris, NaCl, Tween 20), preparation of buffer, B.20
Toluenesulfonyl fluoride. *See* PMSF
tonA, genetic marker in *E. coli* host strain, 2.57
Topoisomerase I, calf thymus
  activity, 5.89
  treatment of mutated DNA for size fractionation, 15.36, 15.50
Tosyllysine chloromethyl ketone. *See* TLCK
Tosylphenylalanine chloromethyl ketone. *See* TPCK
TPCK (tosylphenylalanine chloromethyl ketone), protease inhibitor, properties, 18.31
TPE (Tris-phosphate/EDTA electrophoresis buffer), 6.7, B.23
Tracking dyes
  migration in polyacrylamide gels, 6.37
    in denaturing polyacrylamide gels, 7.70
  use in agarose gel electrophoresis, 6.12
    in alkaline agarose gel electrophoresis, 6.12, 6.21
traD36, mutation that suppresses conjugal transfer of F factors, 4.13

Uridine, structure, C.11
Uridine biosynthesis and CAD protein, as selectable marker in mammalian cells, 16.14
Uridine 5′-phosphate (5′-UMP), structure, C.11
Uridine 5′-triphosphate (5′-UTP), structure, C.11
5′-UTP. *See* Uridine 5′-triphosphate

V

Vacuum transfer of nucleic acids from agarose gels to solid supports. *See* Southern hybridization
Vanadyl-ribonucleoside complexes, 7.3–7.4
Vectors
    bacterial expression. *See E. coli*, expression vectors
    bacteriophage λ. *See* λ bacteriophage, vectors
    bacteriophage M13mp series. *See* M13 bacteriophage, mp series of vectors
    cosmid. *See* Cosmid, vectors
    filamentous bacteriophages. *See* M13 bacteriophage
    mammalian expression. *See* Mammalian cells, expression vectors
    phagemids. *See* Phagemids
    plasmid. *See* Plasmid, vectors

W

Weiss unit of bacteriophage T4 DNA ligase, 5.62
Western blotting, 18.60
    binding primary antibody to target protein, 18.70–18.71
    blocking solution, 18.69, 18.73
    chromogenic substrates for enzyme-coupled antibodies, 18.74–18.75
    preparation of samples, 18.61
    secondary reagents
        enzyme-coupled, 18.73–18.75
        radiolabeled, 18.72–18.73
    staining
        with India ink, 18.68
        with Ponceau S, 18.67
    transfer of proteins from SDS-polyacrylamide gels to solid supports, 18.64–18.66

transfer buffer, 18.65
Whatman 541 filter paper, and hybridization screening, 1.90
WL113, deletion in bacteriophage λ, 2.16

X

Xanthine, structure, C.12
Xanthine-guanine phosphoribosyl transferase (XGPRT), as selectable marker in mammalian cells, 16.14
X-gal (5-bromo-4-chloro-3-indolyl-β-D-galactoside)
    in bacteriophage M13 cloning, 4.8, 4.22–4.23, 4.33, 4.37–4.38
    chromogenic substrate used in testing α-complementation, 1.85–1.86
    stock solution, 1.86, B.14
XGPRT. *See* Xanthine-guanine phosphoribosyl transferase
XL1-Blue, *E. coli* strain
    genotype and uses, 2.59, A.12
    as host for phagemids, 4.18
    and λZAP, 8.45, 8.75
    use in propagation of bacteriophage M13 vectors, 4.14
XS101, *E. coli* strain
    genotype and uses, A.12
    as host for phagemids, 4.18
    for propagation of bacteriophage M13 vectors, 4.15
XS127, *E. coli* strain
    genotype and uses, A.12
    as host for phagemids, 4.18
    for propagation of bacteriophage M13 vectors, 4.15
Xylene cyanol FF
    migration in polyacrylamide gels, 6.37
    in denaturing gels, 7.70
    use in agarose gel electrophoresis, 6.12
9-β-D-Xylofuranosyl adenine (Xyl-A), selection for adenosine deaminase, 16.15

Y

Y1089, *E. coli* strain, genotype and uses, A.12
Y1090*hsdR*, *E. coli* strain, genotype and uses, 2.58, A.12
YAC. *See* Yeast artificial chromosome

Yeast, lysis for immunoprecipitation
enzymatic, 6.55, 18.36–18.37
mechanical, 18.35
rapid (RIPA buffer), 18.38
Yeast artificial chromosome (YAC), genomic DNA libraries, 9.5–9.6
YK537, *E. coli* strain, genotype and uses, A.12

YT medium, recipe, A.3

Z

Zymolase, for enzymatic lysis of yeast cells, 18.36–18.37